EAA series - Textbook

Editors

EAA series is successor of the EAA Lecture Notes and supported by the European Actuarial Academy (EAA GmbH), founded on the 29 August, 2005 in Cologne (Germany) by the Actuarial Associations of Austria, Germany, the Netherlands and Switzerland. EAA offers actuarial education including examination, permanent education for certified actuaries and consulting on actuarial education.

actuarial-academy.com

For further volumes:
www.springer.com/series/7879

Michael Koller

Life Insurance Risk Management Essentials

Springer

Michael Koller
Mathematics Department
ETH Zurich
Rämistr. 101
Zurich 8092
Switzerland
mikoller@ethz.ch

ISSN 1869-6929 e-ISSN 1869-6937
ISBN 978-3-642-20720-4 e-ISBN 978-3-642-20721-1
DOI 10.1007/978-3-642-20721-1
Springer Heidelberg Dordrecht London New York

Library of Congress Control Number: 2011928505

Cover design: deblik, Berlin

Printed on acid-free paper

Springer is part of Springer Science+Business Media (www.springer.com)

For Luisa, Giulia and Anna

Introduction

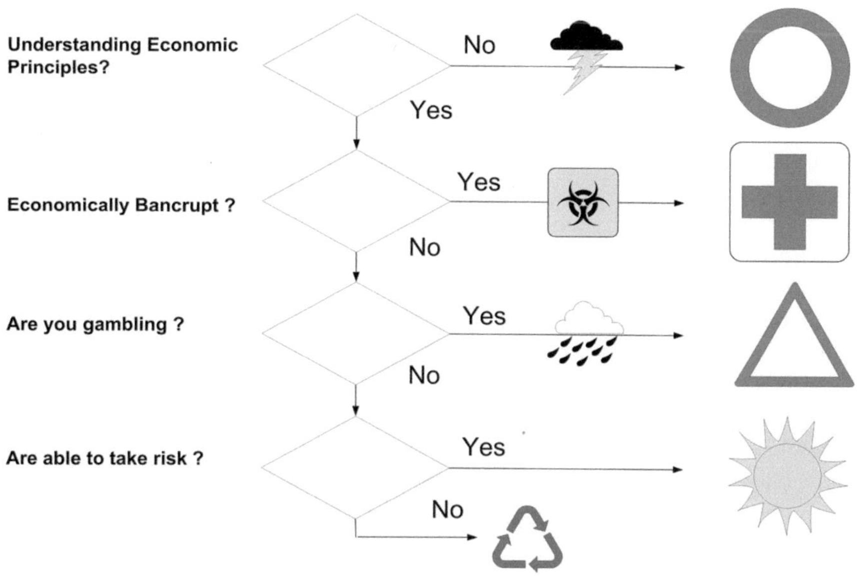

Fig. 0.1 Risk Management in a Nutshell

When starting to write a book, one always needs to ask oneself why one wants to do that and who is the target audience. Furthermore one needs also to know what is required from his audience.

To start with the first question. I have performed quite a wide range of jobs in different international insurance groups and found that some patterns always repeat and that there is a need for education in relation to risk management. So at every place there were pieces missing and the overall risk management structure had to be improved. Over time I had to repeat parts of this educational process again and again. So I decided that it would be worthwhile to gather and compile all these different facts and leanings in order that other people can profit from it. I thought that it would be in particular valuable to have a book which covers a wide range of different topics in relation to risk management in such a way and tangible enough to be readily applicable.

Hence the book does not focus on abstract concepts for the sake of mathematical beauty, but rather with the aim to concretely solve problems and in order to be able to set up a top performing risk management organisation. I have put considerable focus in order to make the book as practical and applicable as possible and I have tried to show a lot of concrete reporting templates including methodology and examples, in order to better understand the context. My intention was, that one could read this book in order to afterwards be able to solve and implement some 80 % of the typical questions which occur when setting up a concrete risk management organisation. Two things need to be stressed: on one hand there are always a lot of different ways which lead to Rome, hence there might in some instances exist better solutions than those presented in this document. On the other hand I believe that this book is also suited for the advanced reader due to its aim to be quite extensive. The overall focus of this book is mainly on financial risk management, insurance risk management, economic steering of an insurance company, insurance processes and products. I would also like to mention, that the methods presented in this book are not only applicable to life insurance companies, but also to pension funds, applying them mutatis mutandis.

I wrote this book because I believed this would be fun and help me to better understand risk management. This book is intended for both those who want to learn risk management starting at a beginner's level, but also for readers who want to widen their horizon. I have also tried to include more practical questions, such as what can go wrong with particular products in order to help to avoid them.

The book is aimed to be self contained and I expect the reader to understand the basics of analysis and some probability theory, such as [JP04]. Since the application of the theory and methods is in the centre, we use different types of mathematical approaches without proof. The required mathematics has been placed in different appendices. Some of them require advanced mathematics such as measure theory, functional analysis and stochastic integration. Here we refer to the following text books for the underlying theory: measure theory – [DS57], functional analysis – [DS57], [Con91], [Ped89], stochastic integration – [IW81], [Pro90]. Finally I would like to give some references in respect of arbitrage free pricing theory – [Pli97], [KM03], [Duf92] and interest rate and equity models – [BS73], [CIR85], [Hul97], [BM01]. For the relevant actuarial life literature we refer to [Ger95] and [Kol10].

Finally I would like to thank the many people who helped me to make this project happen. In particular I would like to thank my wife Luisa and my children who always support me.

Michael Koller, 2010

Contents

List of Figures

Chapter 1
What is Risk Management

1.1 Introduction

1.1.1 Raison D'être of Risk Management

In order to understand the need for risk management it is necessary to look at the different building blocks of a holistic risk management. Figure 1.1 tries to decompose the risk management into its generic components. The overall aim is to manage the risks a company is facing. All employees in the company are expected to some larger or smaller extent to manage risks in order to limit a potentially adverse out-

M. Koller, *Life Insurance Risk Management Essentials*, EAA Series,
DOI 10.1007/978-3-642-20721-1_1, © Springer-Verlag Berlin Heidelberg 2011

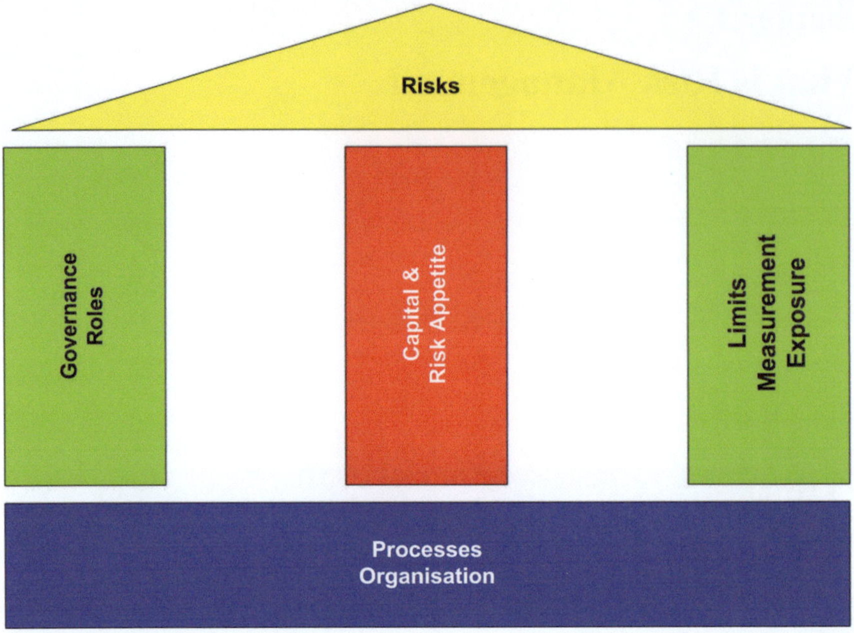

Fig. 1.1 Overview

come and to generate profit and stability for the stakeholders of the company. The corresponding risk culture is of paramount importance. An open communication and a clear and unbiased view in respect of the risks are essential in order to become a professional risk-taker.

In order to control these risks it is first necessary to analyse and categorise the risks into its components. This decomposition is described in section 1.2.1.

After understanding the risks and their impact on the stakeholders, it is essential to understand what are the foundations and the pillars which allow us to operate in such a way that we manage risks in an optimal manner.

The foundations of each company are its organisation and its processes. Therefore it is necessary to define the relationship between these foundations and the risks. The second foundation of each organisation are its processes. Here the escalation processes (section 15.1.3) are particularly important, since they define how to behave in the case a risk "gets out of control" and "needs to be fixed". The generic risk management process, in chapter 5, helps to better analyse the risks in a consistent way and hence ensures better communication, understanding and analysis within the insurance company.

The three pillars which ensure that risks are taken in a conscious and value enhancing way are:

- Governance and roles,

- Capital and risk appetite, and
- Measurement, limits and exposure.

Each of the pillars has a particular purpose:

Governance and Roles: In order to ensure a "fit and proper" management it is essential to have adequate governance in place. A common understanding of the various parts within the organisation is of paramount importance. The corresponding definitions, that cover part of the roles and responsibilities are documented in section 15.1.1. Together with the generic governance principles (section 15.1.2) they form the corner stones of the company's governance structures. The governance structures are further detailed in section 15.10.

Capital and Risk Appetite: Risk can be defined as a potential adverse outcome, and it can normally be measured in monetary terms. The capital resources available to the company serve as a buffer in order to limit the need for fresh capital to prevent bankruptcy. Hence it is of utmost importance to know the available resources (which can serve as buffer) and to ensure that the risk appetite is commensurate with the company's strategic aims (e.g. rating, capital level, etc.) and the limits imposed by its stakeholders (Board of Directors, regulators, etc.). This relationship is documented in chapters 2, 4 and 10.

Measurement, Limits and Exposure: The last pillar defines how to measure risk. This is particularly important in order to have reliable information for knowing the actual risk profile. To ensure that the company operates within its risk appetite, some of the risks are limited by a *limit system*. An example could be, that the company does not want to invest more than 10% of its assets in shares. How this is done, is documented in chapters 6 and 7.

Having all the before mentioned parts in place, means that the insurance company is a professional risk taker, which aims to outperform the market and its peers. This can *only* be achieved if everybody is responsible for risk management. The risk management function acts as a enabler and consolidator.

1.1.2 The Role of Risk Management

The role of risk management can be summarised as follows:

- To ensure risk appetite is clearly articulated for all risk categories.
- To ensure the businesses operates within the established risk appetite through monitoring and controls.
- To ensure the level of capital held in the balance sheets is compatible with the risks taken.

- To ensure efficient capital structures operate within the business.
- To ensure compliance with risk policies.
- To ensure an efficient process is in place to identify emerging issues and risks.
- To help mitigate the risks which are outside the risk appetite.
- To define methods and processes to measure the available and required risk capital.

1.1.3 Three Lines of Defence in Risk Management

Finally, it needs to be stressed that risk management is not only carried out by the risk management function, but by the whole organisation.

The organisation can be split into the so called three lines of defence.

First line of defence: The *line management* as first line of defence is of paramount importance in risk management, because this function is essentially responsible for ensuring that all processes in the business adhere to the *risk management policies* and that the company operates within the *limits* as agreed upon by the Board of Directors and the executive.

Second line of defence: The *risk management function* lead by the Chief Risk Officer is the second line of defence. It has the duty to provide a reliable challenge to the first line of defence and it measures the necessary risk capitals and independently monitors the adherence to limits and appetite. In case of limit breaches it initiates together with the first line of defence mitigating actions. The *risk management function* is also responsible for the various risk committees and risk reporting.

Third line of defence: *Internal audit* is the third line of defence. Its main task in respect to risk management is to provide independent assurance to the Board of Directors and the senior executive that the risk management processes are adequately working within the first and second lines of defence.

1.2 Principles

The aim of this section is to define the generally applicable operating principles which are used within the company to ensure adequate and efficient risk management. These principles define on a high level the main risk categories and risk management principles and it is expected that the whole organisation adheres to them.

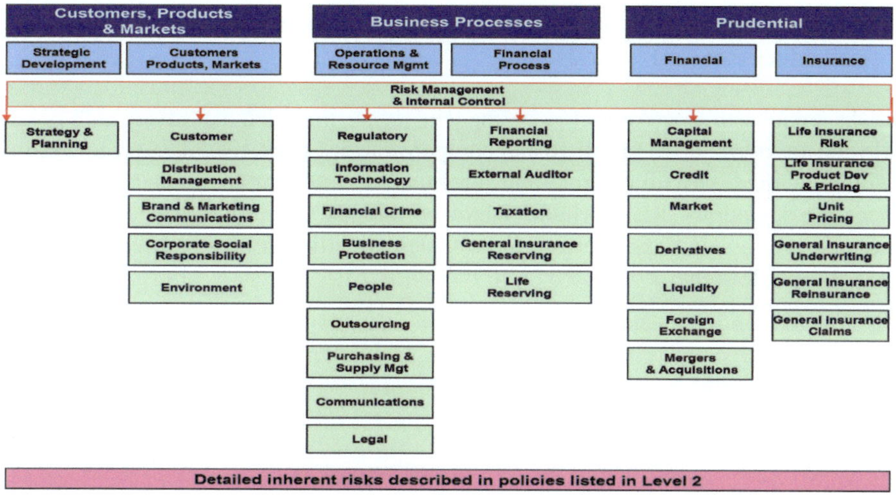

Fig. 1.2 Risk Map

1.2.1 Risk Categories

In order to have a systematic approach to measure, limit and to mitigate the risks the insurer is facing, a so called risk landscape has to be created. The main aim is to have a structured and uniform approach towards risks. Such a landscape normally takes the form of a tree where risks become more and more granular. The depth of the branches corresponds to the level of the model.

Each risk can be characterised by its impact (*severity*) and its probability (*frequency*). Furthermore we speak of *inherent risk* if we look at it before any dedicated controls or mitigating actions are put in place. We speak of a *residual risk* if we measure it taking into account the existence and effectiveness of controls.

Distinguishing *inherent risks* and *residual risks* is necessary in order to know whether a certain control is efficient or sufficient in order to limit a risk to an adequate level. Obviously the full elimination of a risk by using a lot of mitigating actions might not be optimal in the sense that the corresponding costs for the mitigating actions could outweight the potential loss. Hence it is essential to have a commensurate risk appetite which takes this into consideration.

Risk is defined as the potential danger that an actual result will deviate (adversely) from the expected result. Risk is measured according to probabilities and the extent of negative deviations. Risk is defined as:

The magnitude of a risk expressed in terms of impact and probability before any dedicated controls or mitigating actions are put in place or assessed on the basis that the dedicated controls and mitigating actions in place fail.

The impact and probability of an *inherent risk* taking into account the existence and effectiveness of controls.

Risks are measured and assessed in financial terms, provided that this is both possible and appropriate. To weigh up and compare various risks, risk management ratios will be defined, providing consistent information on the probabilities and extent of negative deviations. Risks which cannot be directly quantified (especially operational and strategic risks) are also to be systematically recorded and represented in an appropriate form.

In order to identify, measure and limit certain risks, a systematic approach is needed. In a first step risks are categorised according to a risk map (figure 1.2). Examples of specific risks within the individual categories are:

Market risks	ALM or gap risk, interest rate risk, equity risk, currency risk, real estate risk, commodity risk, etc.
Liquidity risks	Market liquidity risk, funding risk.
Credit risks	Counter-party risk, country risk, concentration risk, risk of rating changes, etc.
Insurance risks	Death, disability, longevity, illness, etc.
Operational risks	Distribution risk, financial crime, legal risk, reputation risk, business protection risk, HR risk, loss of expertise, etc.
Strategic risks	Risk of pursuing the wrong strategy or of being unable to implement the strategy (e.g., market access).

1.3 Risk Management Process

Figure 1.3 defines the generic risk management and controlling process:

Strategy and plans: This is the first step of this generic and cyclic process where, based on risk and reward, a strategy is determined in order to optimise return to shareholders on a risk adjusted basis. Implicit to this task is the high level risk measurement and capital consumption of a certain strategy. This part of the process is owned by the *risk owners*. Risk management information should be used to provide insight, inform the operational planning process and influence resource allocation including capital. Businesses must ensure that changes to their risk profile including control effectiveness are explicitly considered within strategy setting, business planning, objective setting and performance monitoring.

Risk appetite: Based on the plans and the high level risk and capital allocation, the risk appetite is defined and risk limits are set. This process is governed by the risk committee and the owner of this process step is the *risk owner*. Risk

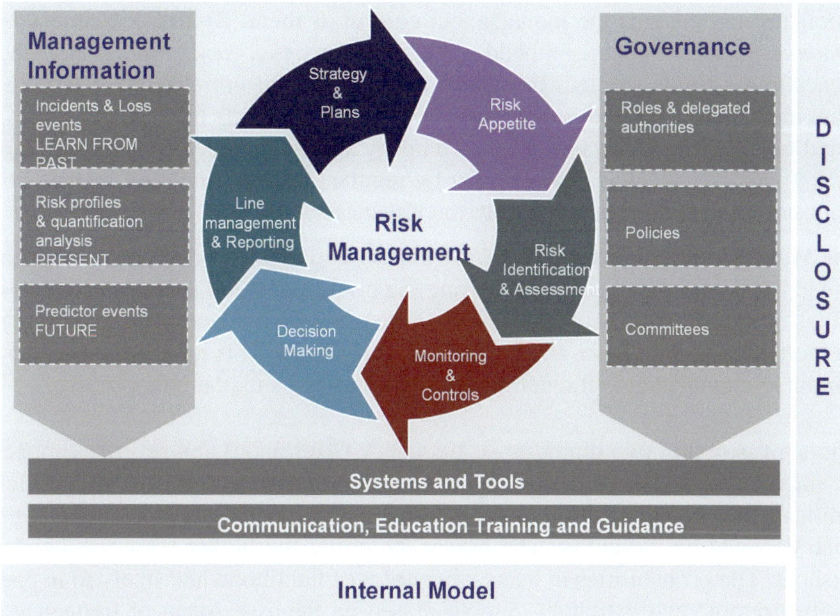

Fig. 1.3 Risk Management Process

appetite statements and tolerances should be clearly defined and refreshed on a regular basis and as an integral part of the planning process. Risk appetite should be defined for a business as usual situation within an established business and also needs to be sufficiently flexible to deal with a variety of situations (e.g. rapid market expansion, managing significant change) and should support rather than constrain sensible risk taking to deliver business objectives.

Risk Identification and Assessment: As next step there is a detailed risk analysis comprising risk identification and assessment. This step, owned by the different *risk experts* and the *risk function* ensures that all risks are properly captured within the systems and processes of the company. Furthermore it ensures that material risks can be quantified adequately with high quality. All material inherent risks must be identified, assessed and recorded. Controls to mitigate each material inherent risk must be documented and assessed for their adequacy and effectiveness in risk mitigation in order to produce a residual risk assessment which is within appetite. The risk model must be used as the basis for considering all types of risk.

Monitoring and Controls: The agreed risk limits are entered into the models monitoring the risk and controls in order to ensure a timely detection of limit and control breaks. This part of the process is owned by the *risk function*.

Decision Making: During the year risks are taken according to the policies defined by the company. *Line management* is responsible for adherence to the risk policies and ensures the management control of them. By doing business *line management* ensures the embedding of the *risk management policies* and adherence to the limits granted. In case of limit breaks, the corresponding processes are initialised. During this task *line management* optimises the risk return profile and hereby generates value for the company and its shareholders. *Risk management* supports the *line management* by regular risk reporting and reports limit consumption and limit breaches to the *risk owners*.

Line Management and Reporting: This last step of the process ensures a proper feed-back over the cycle, by assessing the performance on a risk adjusted base. Risk adjusted returns and limit breaches are prepared by the *risk function* and are reported to the *line management* and the *risk owners*. This information serves as input to management remuneration and the strategy and planning process.

Risk is measured in two dimensions: frequency (likelihood) and severity (impact). The impact can be one of the following in decreasing order: catastrophic, critical, significant and important. Each level of impact corresponds to a monetary amount, which depends on the entity. The bigger the entity the higher the corresponding threshold. The probabilities in decreasing order of likelihood are: likely to happen, possible, remote and extremely remote. Based on the assessment of frequency vs severity the overall risk can be expressed in a more holistic way. The figure 1.4 shows a such an overview:

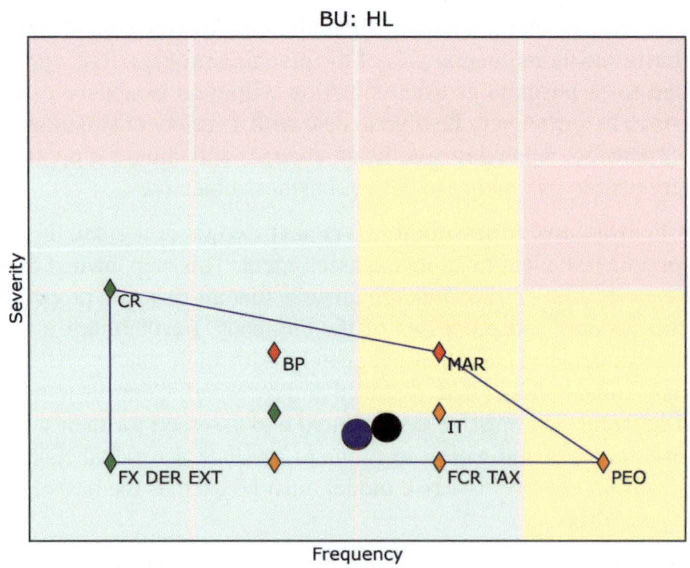

Fig. 1.4 Frequency vs. Severity

In figure 1.4, the frequency of the emergence of a risk is plotted and each of the four categories has a specified probability. Similarly the potential impact is classified into one of four categories. Depending on frequency and severity each of the 16 squares is allocated to one of the three colour codings: green, amber and red. In a next step all policies are mapped against this grid and plotted. The label "PEO" refers for example to people risk, which has in this example a high probability of materialisation with a rather low impact. This policy has been evaluated as amber and hence the marker is amber. In the same sense one can see the market policy ("MAR"), the credit policy ("CR") etc. In order to have an overall picture the circles represent the barycentre of the assessments, in black for the past reporting period, and in blue for the current. From this it can be seen that the over all risk moved slightly south–east, hence has reduced a little.

1.4 Risk Management Policies and Risk Landscape

In order to define minimal standards on how different risks have to be treated and define minimal governance standard, insurance companies codify the corresponding rules and responsibilities in terms of risk management policies which cover risks, which belong together. The following list provides a quite complete list of risk management policies. Obviously there are different ways on how one can arrange these policies.

Brand & Marketing Communications: This risk management guideline describes the risk which is intrinsic in the brand and marketing communication process. Here the main aim is to safeguard the company's reputation and to ensure that the communication is aligned with the core values of the company.

Business Protection: Business protection aims to protect the orderly running of the business and is therefore concerned with things such as physical security, IT security, data recovery, business continuity in case of a damage of a property or in case of a pandemic etc. Hence the aim of the policy is to define the limits of acceptable risk with respect to these topics.

Capital Management: The aim of this guideline is to define the processes and the risk appetite the company has in respect to capital management, hence in respect of levels and quality of capital. Here also the process of raising capital, paying dividends and the risk appetite of becoming insolvent is anchored. One could for example state: "There is no risk appetite that the statutory capital level falls below 120%."

Communications: Communications cover both internal and external communication. Here it is defined how information is treated and who is allowed to communicate internally and externally. The corresponding risks are unhappy employees because of bad internal communication, or externally: reputational issues and communication leaks.

Credit: This is the financial credit risk, where credit migration, credit spread and default risk is addressed. Furthermore guidance is given in respect to concentration limits and the processes used in order to ensure the company operates within a given risk appetite. Hence some of the requirements limit financial risks and others aim to address operational (risk) issues.

Customer: One of the big reputational and regulatory risks of each insurance company is the relationship vis-a-vis the customers. Here it is important to define what "treating customers fairly" means and how the corresponding risk appetite is defined. In consequence governance rules are established in order that the company operates within these boundaries.

Derivatives: Since derivatives imply a much higher (operational and financial) risk than "normal" assets, it is important to define the corresponding governance processes in a stringent and efficient way. Hence this guideline addresses, besides the pure financial risk, also the important operational procedures and hence aims to limit also the operational risk.

Distribution Management: The distribution management policy aims to limit the risks which are induced by the insurer's distribution network. Here risk appetite and processes are set in respect to the quality of people acting as distributors, turnover of distribution managers, remuneration schemes etc.

Environment: Here the company states its risk appetite with respect to environmental issues, such as energy consumption etc.

External Auditor: As a consequence of Enron and Worldcom, the attitude vis-a-vis accounting has become much more stringent and most companies have no appetite to make accounting errors and a lot of them have also implemented quality standards such as SoX 404. This guideline defines the relationship towards the auditors and states which behaviours are not acceptable and which services may not be taken from the own external auditor.

Financial Crime: The financial crime policy states the required behaviour in respect to financial crime, such as fraud, money laundry etc. Most companies do not have the slightest appetite for financial crime and hence these guidelines are normally very prescriptive and restrictive.

Financial Reporting: The financial reporting guideline needs to viewed as a companion of the external auditor guideline with the aim to reduce errors and omissions with respect to financial reporting down to an acceptable (low) level.

Foreign Exchange: The FX guideline is also one of the financial risk guidelines and it has the same aim as all of these guidelines, namely that the company operates within a well defined risk appetite. In consequence the limit setting, monitoring and reporting processes are of utmost importance.

GI Claims: This guideline governs the GI claim processes and defines which measures have to be taken, to prevent fraud and to treat customers fairly. Obviously the claim settlement process for GI claims is of utmost importance, because there is a narrow margin between being too onerous and being too strict. As a consequence we speak here about operational risk, which has a direct financial impact.

GI Reinsurance: Since a lot of GI lines of business are heavily re-insured (say some 25% of the total GI premiums), it is important to have a clear guidance which level of risk is still acceptable and which risks need to reinsured. Besides the insurance risk (such as windstorm, earthquake, ...), it is important to recognise there is also credit risk involved, since reinsurers also might default. Hence a balanced reinsurance portfolio is important in order to avoid severe problems in case of a reinsurer default. In the reinsurance risk guideline the risk appetite is not only relevant quo lines of business but also quo counterparties.

GI Reserving: Looking at the balance sheet of a GI insurer it becomes obvious that a large part of the balance sheet consists of claim reserves. Hence it is important to have a clear risk appetite in order to ensure on one hand adequate reserves, which are on the other hand, not too onerous. Furthermore the GI reserving process involves, besides actuarial techniques, also considerable judgement. Hence in the light of financial reporting risk it is important to have rigid and robust processes in place.

GI Underwriting: The GI Underwriting guideline can be considered as a companion guideline to the GI claims guideline covering the underwriting process. A stringent process is needed in order to ensure an adequate portfolio quality. Let's assume for the moment that a company would attract all "bad" risks. In this case the company would obviously suffer because of an inadequate pricing. Hence also the GI pricing is anchored in this guideline.

Information Technology: Information technique per se is a vast topic and the corresponding intrinsic risks are big. This guideline steers the risk appetite in respect to IT risks, such as infrastructure, IT projects etc.

Legal: The legal risk policy speaks about the company's attitude in respect of legal issues, litigation etc. Here it is important to allocate the responsibilities and duties accordingly. This is in particular relevant when entering into a litigation or s settlement of a claim. The legal risk policy does not only cover the risks the corporate faces, but also risks which are consequences of disputed life and in particular GI claims.

Life Insurance Product Development & Pricing: As we will see in chapter 12 the product development and product pricing process for life insurance policies is a difficult one. As a consequence of the typically big volumes and long contract terms (20 years and more) and the fact that issues become costly quite easily. It is of paramount importance to have a clearly defined risk appetite in respect of product development and pricing, and corresponding robust governance processes. It is also important to recognise that besides the pure financial risks there are also significant operational risks which can materialise in ill-designed products.

Life Insurance Risk: This guideline covers the risk appetite of the pure technical insurance risks which are, for example, mortality, disability, surrender etc. In order to operate within a well defined risk appetite these technical risks are to be limited with a limit system.

Life Reserving: This is the companion guideline of the "GI Reserving".

Liquidity: Liquidity risk guideline governs the process to monitor liquidity and to ensure that the company has always enough liquidity to fulfil its obligations. This guideline is also one of the financial risk guidelines.

Market: From all the financial risk guidelines, this is the most important, covering the market risk of all financial assets (such as equities, bonds, hedge funds etc.) and the corresponding ALM risk if also taking the liabilities into consideration. In consequence governance, limit systems, escalation processes, risk mitigation and risk measurement play an important role in this guideline. Only if these building blocks are robust and accurate is it possible to operate in a well defined environment, taking risks in a conscious manner.

Mergers & Acquisitions: This guideline sets the risk appetite and standards for M&A processes. It is known that these processes are difficult and can lead to substantial problems if done in an inappropriate manner. Hence it is important to have a stringent guideline describing processes, governance arrangements and risk appetite.

Outsourcing: The Outsourcing guideline defines the risk appetite and the protocols to follow in case of outsourcing arrangements. Obviously it is the aim of such a guideline to limit the corresponding operational and counter-party risks.

People: For all financial institutions there are two main resources needed: capital and people (human capital). It is very important to clearly articulate the risk appetite in respect to people to ensure the attractiveness to key performers and to ensure an adequate turnover to get new talent on board.

Purchasing & Supply Management: See "Outsourcing".

Regulatory: This guideline can be compared with the "External Auditor" guideline since it defines the risk appetite in respect to the different regulators of the company.

Risk Management & Internal Control: This guideline defines how risk management works in the corporate environment and covers many issues and questions of this book.

Strategy & Planning: Looking at the main processes of an insurance company, the strategy and planning process is particular, since it defines what the company will do in the following year. It is also known that ill-behaved strategies are one of the root causes for corporate failures. Hence it is important to also control this process and strategic planning in a environment with a well defined risk appetite.

Taxation: This is the companion guideline of "Regulatory" vis-a-vis the tax authorities.

Whereas the risk management policy view is efficient for managing the company another view is needed in order to decompose the risks in their generic risk factors. Assume, for example, credit risk. This risk factor influences more than one risk covered in one of the risk management polices, such as "Credit Risk", "Reinsurance", "Customer", "Outsourcing" etc. Whereas the link is clear for "Credit Risk", the relationship is not always as straight forward. The following table summarises these relationships:

Policy	**Relationship**
Credit	Via the credit default and credit mitigation risk of bonds and mortgages.
Reinsurance	Via the counter-party credit risk of the reinsurance treaties and insurance linked securities.
Customer	The reputational issue if the customer suffers in case of a credit default independently on whether the insurance company bears the risk or not.
Outsourcing	Via the operational risk, which is induced by the default of an outsourcing partner.

As a consequence it is necessary to decompose the risk-universe into its drives. This map is called *risk landscape*. Also here it is possible to have a coarser or finer view on the risk landscape. Figure 1.5 shows a quite high level risk landscape.

These risk factors form, from a mathematical point of view, the base for the risk capital calculations and represent a multi-dimensional random variable or stochastic process. All random fluctuation within a economic capital model are derived from these risk factors. The financial instrument sub-model of the Swiss solvency test uses for example about 80 different risk factors, which are modelled as a multi-dimensional normally distributed random variable ($X \sim \mathcal{N}(\mu, \Sigma)$).

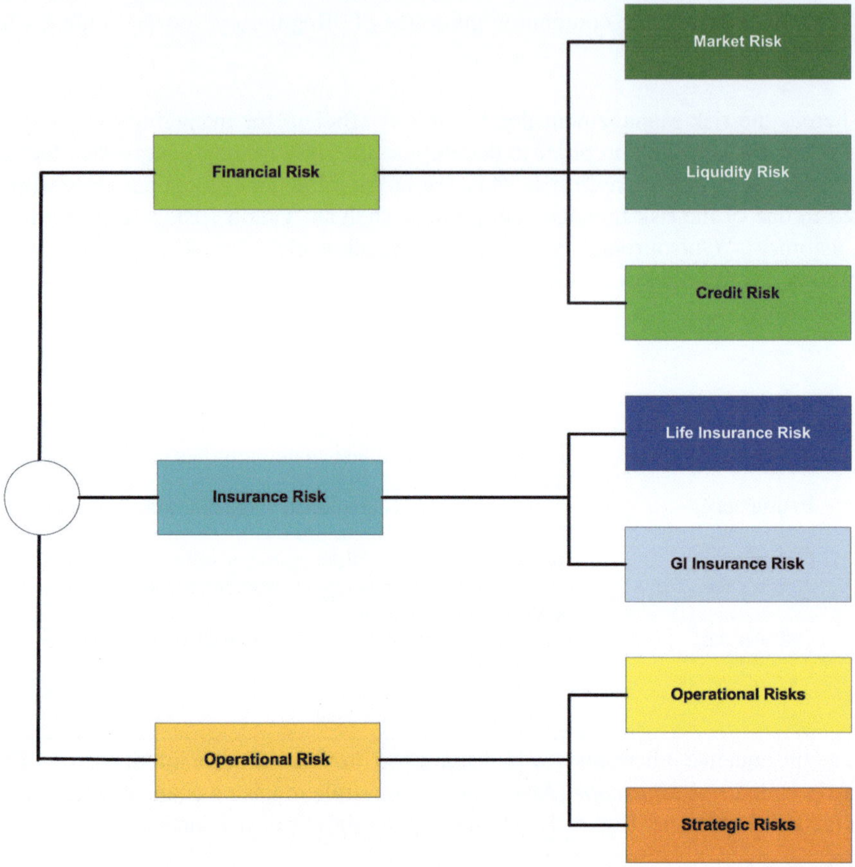

Fig. 1.5 Risk Landscape

Chapter 2
The Role of the Balance Sheets and of Capital

2.1 The Balance Sheet of an Insurance Company

In order to assess the financial strength of an insurance company one often looks at its balance sheet. It is however not quite as easy as it sounds to look at the balance sheet of an insurance company, since the corresponding concepts are quite complex and because there exist different accounting standards. A further complexity is the fact that not all of the assets and liabilities are traded and their value is not directly observable.

M. Koller, *Life Insurance Risk Management Essentials*, EAA Series,
DOI 10.1007/978-3-642-20721-1_2, © Springer-Verlag Berlin Heidelberg 2011

2.2 Role of Valuation

In order to do physics it is essential to measure the different quantities accurately. Independently of the actual length of 1 meter the different laws and formulae are valid and correct. Therefore the meter serves mainly as an objective yardstick for comparison. In the world of economics the common measure is the face amount of the money, for example 1 EUR. Here the situation is however more difficult, in the sense that it is a priori not clear how to value complex financial instruments such as options, illiquid stocks, insurance policies etc. In order to be able to publish reliable financial statements and to undertake sensible risk management it is imperative to base on reliable valuation principles and methods. Without these methods neither financial accounting nor risk management make sense. The aim of this chapter is to give an introduction into economic valuation methods.

2.2.1 Valuation Methods

For a given financial instrument or liability, a valuation can in principle be done based on market or book values. In case of book values the implicit aim is to prudently valuate the assets based on the purchase price. In case of stocks the corresponding principle results in the so called “lower cost or market” valuation, which means that the stock is in the books at the purchase price as long as the market price is not lower. Assume that a stock has been bought at EUR 100000 and has doubled its market price. In this case the book value would still be EUR 100000 and its market value EUR 200000. Correspondingly there is a revaluation reserve of EUR 100000 which is not accounted for in this type of balance sheet. In order to show these hidden values more transparently the so called market value accounting principles were introduced.

The market value of financial instruments can usually be determined looking at deep and liquid markets where these instruments are traded. In case of most stocks this is the case. There are however instruments, which are not regularly traded and here it is necessary to base the valuation on models. Typical instruments where models are required are for example:

Instrument	**Method for valuation**
Illiquid Stocks	Usually last paid price
Synthetic Zero Coupon Bonds	By recursion based on bonds with coupons
Properties	By discounted cash flow method or expert judgement
Options and other derivatives	By mathematical methods such as Black-Scholes-Formula
Insurance Liabilities	Based on synthetic replicating portfolios

The above table clearly shows the need for mathematical methods to approximate market values where such are not directly observable. One of the most useful theories is the arbitrage free pricing theory which will be explained in the next section.

2.2.2 Principle of No Arbitrage

The aim of this section is to give a high level overview about the so called arbitrage free pricing theory. If the reader wants to get a more mathematical representation of the corresponding topic, we refer to appendices D and E. For a general formal approach to abstract valuation we refer to appendix C.

In economics, arbitrage is the practise of taking advantage of a state of imbalance between two or more markets: a combination of matching deals are struck that capitalise upon the imbalance, the profit being the difference between the market prices. When used by academics, an arbitrage is a transaction that involves no negative value at any probabilistic or temporal state and a positive value in at least one state. A person who engages in arbitrage is called an arbitrageur. The term is mainly applied to trading in financial instruments, such as bonds, stocks, derivatives and currencies.

If the market prices do not allow for profitable arbitrage, the prices are said to constitute an arbitrage equilibrium or arbitrage free market. An arbitrage equilibrium is a precondition for a general economic equilibrium. The following example shows an arbitrage opportunity:

Suppose that the exchange rates (after taking out the exchange fees) in London are £5 = \$10 = ¥1000 and the exchange rates in Tokyo are ¥1000 = £6 = \$10. Converting \$10 to £6 in Tokyo and converting that £6 into \$12 in London, for a profit of \$2, would be arbitrage. In reality, this "triangle arbitrage" is so simple that it almost never occurs.

The most important elements of the Arbitrage Free Pricing Theory are:

- Pricing systems,
- Arbitrage and
- Self-financing strategies.

We denote by $S_k(t)$ the price of the asset k at time t, where $S_0(t)$ denotes usually the investment in cash. A portfolio at time t is a vector $\phi_k(t)$ indicating the number of units of the corresponding asset hold at time t. The value of this portfolio at time t equals

$$V(t) =< S(t), \phi(t) >= \sum_k S_k(t) \times \phi_k(t).$$

A self-financing trading strategy is a sequence of portfolios, which fulfils besides some additional mathematical requirements the following equation: $V(t^-) = V(t)$, which can be interpreted as the absence of injecting or withdrawing money during the changes of the portfolio. The trading strategy is called admissible if its value never falls below 0. The idea of arbitrage free pricing is to replicate a financial instrument such as a stock option by a corresponding self-financing trading strategy, which has exactly the same payout pattern as the financial instrument for (almost) all possible states of the financial market. As the strategy was self-financing the value of the instrument at time 0, needs to equal the value of the portfolio of the strategy at inception.

The arbitrage free pricing theory can today be considered as one of the cornerstones for pricing derivatives of financial instruments such as stock options, swaptions etc. From a mathematical point of view this theory is intrinsically linked to martingales - the prototype of a fair game. It can be shown that the absence of arbitrage implies the existence of a so called equivalent martingale measure Q. The price of the derivative is then the expected value of the discounted value of the instrument, not with respect to the original measure P, but with respect to the equivalent martingale measure Q. This can be interpreted that the value process under this new measure follows a fair game.

By using all the theoretical tools available for martingales it is possible to show a lot of nice features of these processes. The so called Itô-calculus allows the analytical and numerical treatment of such instruments.

In relation to the valuation it becomes obvious that options and other derivatives which have no deep and liquid market are priced and valuated based on these concepts. Furthermore they also play a significant role in the risk management of derivatives, because Itô-calculus allows the quantification of the changes in the price depending on the parameters resulting in the so called greeks. They represent the partial derivatives of the price and can be used to approximate the change in value by using a Taylor-approximation.

Another aspect of these tools is the possibility to simulate the price of financial instruments by Monte-Carlo-methods. Arbitrage free pricing theory and the need for equivalent martingale measures for pricing indicate the need to use the equivalent martingale measure for simulations - or equivalently to use so called deflators with respect to the original measure P. Deflators can be considered as a link between the two measures and are closely related to the concept of a Radon-Nikodym density $\frac{dQ}{dP}$. For further details we refer to appendix C.

A section about the Arbitrage Free Pricing Theory is certainly incomplete without mentioning the Black-Scholes Formula. The Black-Scholes model is a model of the evolving price of financial instruments, in particular stocks. The Black-Scholes formula is a mathematical formula for the theoretical value of European put and call stock options derived from the assumptions of the model. The formula was derived by Fischer Black and Myron Scholes and published in 1973. They built on earlier research by Edward Thorpe, Paul Samuelson, and Robert C. Merton. The fundamental insight of Black and Scholes is that the option is implicitly priced if the stock is traded. Merton and Scholes received the 1997 Nobel Prize in Economics for this and related work; Black was ineligible, having died in 1995.

2.2.3 Reconciliation of Balance Sheets

One of the main challenges with respect to economic balance sheets is the missing experience in doing so. Companies are much more used to producing their financial reports based on book value based principles where often virtual assets and liabilities and other "difficult animals" occur, such as:

- Deferred acquisition cost assets,
- Activated software assets,
- Deferred taxes,
- Equalisation reserves,
- Additional technical reserves for all types of insurance cover, etc.

In order to produce reliable economic balance sheets it is therefore advisable to start with an audited balance sheet of the company and to reconcile each position from book values to market values. In case of assets the reconciliation between book values and market values usually equals the revaluation reserves. But for some positions a reconciliation is difficult and therefore even more necessary. Just to mention one of the most difficult positions. What is the market value of a 100% consolidated subsidiary?

After having done the reconciliation it is far easier to explain a economic balance sheet to an audience understanding the traditional accounts. The reconciliation furthermore gives deep insights, where the company suffers small margins or has a lot of fat.

2.3 Bonds

In a typical insurance company most of its assets are bonds or bond like investments. A bond is a financial asset, where the investor pays at the time of buying a fixed amount. In the following years until the maturity of the bond, the investor gets a regular interest payment - a coupon. This payment is based on the nominal value of the bond. At the bond's maturity the investor gets the nominal value of the bond. Depending on the relationship of the relative interest rates the investor can buy the bond below the nominal value, at nominal value or above nominal value. Correspondingly the purchase is called below par, at par or above par.

For bonds there are many different possibilities and variations. Firstly there are so called perpetual bonds, which never mature. In actuarial terms they are perpetual annuities. Furthermore there are so called zero coupon bonds, where the interest rate of the coupon is 0%. Finally it is worth mentioning that there are also callable bonds, where the issuer can call the bond back before its maturity. So one could buy for example a bond which matures in 60 years from now, but which can be called after 10 years every 5 years. The idea behind a callable bond is to provide for a type of capital substitute for the issuer. So the issuer is not forced to refinance the bond in hard times. Normally the buyer of such bonds has to be compensated for this effect in case the bond is not called. A step up facility is such a method. In the above example one could for example get 5% for the first 10 years. After that, one could expect an uplift of 150 bp (eg 1.50%).

In today's environment, bonds are not issued anymore in paper form, but mostly only exist in a virtual form. In the past a bond consisted of a large piece of paper with attached small sections, the so called coupons. When the interest payment was due, these coupons were cut away and brought back to the bank. In exchange the bank payed the interest.

After understanding what a bond is we need to address the risks of a bond. There are two different types of risks, which are intrinsic to a bond, namely the interest rate risk and the credit risk.

As seen above the interest rate is the amount of money which the investor gets for borrowing his money. This price depends on the economical environment and can be higher or lower, depending on the moment in time. The so called *yield curves* describe the interest rate one could get at a certain point of time for borrowing the money for a certain term. In consequence yield curves depend per currency on two parameters, namely the time and the term of the bond. The following table shows the interest rates as at Jan 1st., 2008. It becomes obvious that the interest rates for example in CHF are lower than those in USD. Figure 2.1 shows the corresponding yield curves.

01.01.2008	US	EURO	UK	JAPAN	AUS $
3 Month	3.2745%	3.9564%	4.8201%	0.5069%	6.5797%
6 Month	3.4112%	4.0640%	4.8170%	0.5103%	6.5915%
1 Year	3.3624%	4.0488%	4.5518%	0.6486%	6.6866%
2 Year	3.0825%	4.0549%	4.3958%	0.7062%	6.9174%
3 Year	3.0936%	4.0778%	4.4141%	0.7818%	6.8486%
4 Year	3.3828%	4.1624%	4.4507%	0.9138%	6.7332%
5 Year	3.4584%	4.1592%	4.5251%	1.0195%	6.5823%
7 Year	3.8290%	4.2489%	4.5763%	1.1847%	6.3889%
8 Year	3.9680%	4.2982%	4.6010%	1.2444%	6.3642%
9 Year	4.0681%	4.3348%	4.6026%	1.3916%	6.3350%
10 Year	4.2713%	4.3811%	4.6145%	1.5144%	6.2876%
15 Year	4.5299%	4.5805%	4.5926%	1.8450%	6.2431%
20 Year	4.5214%	4.6512%	4.5123%	2.1060%	–
25 Year	4.4676%	4.6689%	4.4309%	2.2586%	–
30 Year	4.4601%	4.6279%	4.3449%	2.3804%	–

Please note that for some currencies (eg AUS) there exist no long dated bonds (eg for Australia there is no interest rate for 20 yrs) and hence the yield curve is not complete. This not an issue in respect of the bonds, but there is an issue in valuing long term liabilities such as life insurance contracts with a term longer than the longest dated bond.

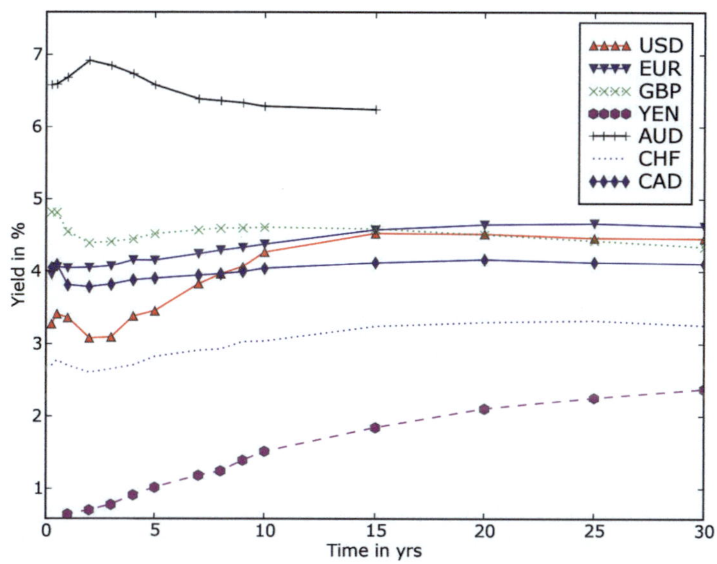

Fig. 2.1 Yield curves as at 1.1.2008

In a next step it is interesting to have a look how the yield curves move over the time. In order to do this we want to have a look at the yield curves just about one year before:

29.12.2006	US	EURO	UK	JAPAN	AUS $
3 Month	4.9999%	3.6216%	5.2126%	0.2996%	6.2142%
6 Month	5.0386%	3.7508%	5.2517%	0.4099%	6.2337%
1 Year	4.9692%	3.8736%	5.2828%	0.5956%	6.2101%
2 Year	4.8293%	3.8806%	5.2632%	0.7925%	6.1771%
3 Year	4.7430%	3.8841%	5.1827%	0.9505%	6.1159%
4 Year	4.6693%	3.9097%	5.1239%	1.1175%	6.0616%
5 Year	4.6673%	3.9180%	5.0511%	1.2481%	6.0178%
7 Year	4.7125%	3.9421%	4.9372%	1.4639%	5.9519%
8 Year	4.7361%	3.9497%	4.8857%	1.5418%	5.9225%
9 Year	4.8751%	3.9453%	4.8241%	1.6153%	5.9001%
10 Year	4.7832%	3.9591%	4.7926%	1.6916%	5.8816%
15 Year	4.9564%	4.0413%	4.6623%	1.9393%	5.8335%
20 Year	4.9309%	4.0914%	4.4779%	2.0721%	–
25 Year	4.8704%	4.1142%	4.3386%	2.2340%	–
30 Year	4.8099%	4.0749%	4.2355%	2.3115%	–

In order to be able to compare the two sets of yield curves the table below provides a comparison between them.

Δ	Time	US	EURO	UK	JAPAN	AUS $
3 Month	2007	4.9999%	3.6216%	5.2126%	0.2996%	6.2142%
3 Month	2008	3.2745%	3.9564%	4.8201%	0.5069%	6.5797%
3 Month	Δ	-1.7254 %	0.3348 %	-0.3925 %	0.2073 %	0.3655%
5 Year	2007	3.4584%	4.1592%	4.5251%	1.0195%	6.5823%
5 Year	2008	4.6673%	3.9180%	5.0511%	1.2481%	6.0178%
5 Year	Δ	1.2089%	-0.2412 %	0.5260%	0.2286 %	-0.5645%
15 Year	2007	4.5299%	4.5805%	4.5926%	1.8450%	6.2431%
15 Year	2008	4.9564%	4.0413%	4.6623%	1.9393%	5.8335%
15 Year	Δ	0.4265%	-0.5392%	0.0697%	0.0943%	-0.4096%
30 Year	2007	4.4601%	4.6279%	4.3449%	2.3804%	–
30 Year	2008	4.8099%	4.0749%	4.2355%	2.3115%	–
30 Year	Δ	0.3498%	-0.5530%	-0.1094%	-0.0689%	–

One can see for example that the interest rate has increased considerably over this year for the 5–year bond in USD.

But how does one determine the yield curves? This question is closely linked to the valuation of the bonds and hence we will speak in a first step about bond valuation. In order to do that we will for the moment denote by π_t the price of a bond or cash flow stream at time t and we denote by $y_t(n)$ the yield of a n-year zero coupon bond ($\mathcal{Z}_{(n)} = (\delta_{nk})_{k\in\mathbb{N}_0}$) at time t. An ordinary bond with a cash flow pattern $\mathcal{B} = (CF_k)_{k\in\mathbb{N}_0}$ has in this context the following value at time t:

$$\begin{aligned}\pi_t(\mathcal{B}) &= \sum_{k=0}^{\infty} CF_k \times \pi_t(\mathcal{Z}_{(k)}) \\ &= \sum_{k=0}^{\infty} CF_k \times (1 + y_t(k))^{-k}.\end{aligned}$$

Now it is possible to determine the yield curve based on a set of ordinary bonds by means of a recursion. Another important concept are forward rates $f_t(n, m)$. The interpretation of $f_t(n, m)$ is the yearly interest rate which we would get from time n to time m ($n \le m$). We denote by:

$$f_t(n, m) = \left(\frac{\pi_t(\mathcal{Z}_{(n)})}{\pi_t(\mathcal{Z}_{(m)})}\right)^{\frac{1}{m-n}} - 1,$$

and remark that the following equation holds, by using a non-arbitrage argument (Barbel-strategy):

$$(1 + y_t(n))^n = \prod_{k=0}^{n-1} (1 + f_t(k, k+1)).$$

After defining the yield curves, it is now possible to introduce different concepts for the valuation of bonds. There is on the one hand the statutory amortised cost valuation method. On the other hand there is the market value valuation for bonds. In order to understand the amortised cost method, we need to look at the relationship between the coupon of a bond and the current interest rate level. As seen above, a bond can be viewed as a vector of cash flows, and it looks normally as follows:

$$\mathcal{B} = (CF_k)_{k\in\mathbb{N}_0}\text{, with}$$

$$CF_k = \begin{cases} 0 & \text{if} \quad k = 0, \\ i & \text{if} \quad 0 < k < n, \\ 1+i & \text{if} \quad k = n, \\ 0 & \text{if} \quad k > n. \end{cases}$$

In the above example we consider a bond with a nominal interest rate i and a term of n years. We say that we purchase the bond

below par if $\pi_t(\mathcal{B}) < 1$,
at par if $\pi_t(\mathcal{B}) = 1$,
above par if $\pi_t(\mathcal{B}) > 1$.

If we apply an amortised cost method to value a bond, we fix the price $\pi_t(\mathcal{B}) < 1$ at purchase date and amortise it until maturity of the bond at time n to the value 1, the nominal value of the bond. If we buy a bond above (below) par the corresponding amortisation leads to a yearly gain (loss), which is recognised in the profit and loss account. This implies in particular that the value of the bond does not change during its lifetime due to interest rate movements.

The other valuation method is based on the market value of the bond. Here the accounting value for the bond equals $\pi_t(\mathcal{B})$ and hence depends on the relative level of interest rates. It is worth noting that the value of a bond such as $\mathcal{B}$, decreases if interest rates increase and it increases in case of a reduction in interest rates.

In a next step we want to have a look at the following bond:

- Nominal value: EUR 100000,
- Term: 5 years,
- 4.0% interest rate,
- Bought at par and we assume a flat yield curve,

and we want to look at the value of its parts:

t	Coupon	Principal	Total CF	PV 3%	PV 4%	PV 5%
0			0	–	–	–
1	4000		4000	3883.49	3846.15	3809.52
2	4000		4000	3770.38	3698.22	3628.11
3	4000		4000	3660.56	3555.98	3455.35
4	4000		4000	3553.94	3419.21	3290.80
5	4000	100000	104000	89711.31	85480.41	81486.72
Total	20000	100000	120000	**104579.70**	**100000**	**95670.52**
Difference				*4579.70*	*0*	*-4329.47*

It becomes obvious that the value of the bond changes some 4.5% per 1% shift in interest rate. In order to quantify this sensibility one normally uses the so called *Macaulay duration* $d(\mathcal{B})$. It is defined by

$$\begin{aligned} d(\mathcal{B}) &= \frac{\sum_{k=0}^{\infty} k \times CF_k \times \pi_t(\mathcal{Z}_{(k)})}{\sum_{k=0}^{\infty} CF_k \times \pi_t(\mathcal{Z}_{(k)})} \\ &= \frac{\sum_{k=0}^{\infty} k \times CF_k \times (1 + y_t(k))^{-k}}{\sum_{k=0}^{\infty} CF_k \times (1 + y_t(k))^{-k}}. \end{aligned}$$

In the concrete example above we have $d(\mathcal{B}) = 4.62$, which is remarkably close to the interest rate sensitivity above. This is not an accident, but rather the reason, why the concept of duration exists. It aims to estimate the interest rate sensitivity in case of a parallel shift of the yield curve. But how can we do this? In a first step we define a modified valuation based on a flat interest curve with an interest rate of x:

$$\tilde{\pi}_t(\mathcal{B})(x) = \sum_{k=0}^{\infty} CF_k \times (1+x)^{-k}.$$

Now using Taylor approximation (eg $\tilde{\pi}_t(\mathcal{B})(x+\Delta x) \approx \tilde{\pi}_t(\mathcal{B})(x)+\Delta x \times \frac{d}{dx}\tilde{\pi}_t(\mathcal{B})(x)$) we get in a first step:

$$\begin{aligned}\frac{d}{dx}\tilde{\pi}_t(\mathcal{B})(x) &= \sum_{k=0}^{\infty} CF_k \times (-k) \times (1+x)^{-k-1} \\ &= \frac{1}{1+x} \times \sum_{k=0}^{\infty} CF_k \times (-k) \times (1+x)^{-k}.\end{aligned}$$

If we now want a relative value for the change we correspondingly get

$$\frac{\frac{d}{dx}\tilde{\pi}_t(\mathcal{B})(x)}{\tilde{\pi}_t(\mathcal{B})(x)} = \frac{1}{\tilde{\pi}_t(\mathcal{B})(x) \times (1+x)} \times \sum_{k=0}^{\infty} CF_k \times (-k) \times (1+x)^{-k}.$$

This last expression is called *modified duration* $d^{mod}(\mathcal{B})$ and we have the

$$d^{mod}(\mathcal{B}) = \frac{d(\mathcal{B})}{1+x}.$$

Since the interest rate is normally small, the approximation above is quite accurate. As mentioned the Taylor approximation (or modified duration) is still more precise. In our case the modified duration amounts to 4.40 which equals the average of 4.57 and 4.23 with an error of less than 0.01.

Until now we have looked at counter-party risk free bonds. This means, that we have assumed that the payments are always paid and that the issuer of the bond can not default. In reality bonds can default and as a consequence there is the possibility that not all coupons and/or the principal are paid. Assume for a moment that we have the following survival probabilities:

time	0	1	2	3	4	5
p(x)	1.000	0.980	0.950	0.910	0.860	0.810

The above table assumes for example that in average 2% of the companies issuing a certain type of bond default in the first year. Obviously such a bond has less value than the one from an issuer which can not default. This latter bond is called risk free bond and it is normally assumed that its issuer is a government. There are two questions, which we want to answer in the sequel, namely how do we valuate such (defaultable) bonds and how can we assign a quality to such a bond, since the value of the bond obviousely depends on its default probabilities.

In a first step we want to have a look at the valuation of such a bond and we want to revisit the example from above. In this case we have:

t	A Total CF nominal	B Survival Prob	C Total CF risk adjusted	D PV nom. 4.00%	E PV risk adj. 4.00%	F PV nom. **8.37%**
0	0	1.00	0	–	–	–
1	4000	0.98	3920	3846.15	3769.23	3690.83
2	4000	0.95	3800	3698.22	3513.31	3405.57
3	4000	0.91	3640	3555.98	3235.94	3142.35
4	4000	0.86	3440	3419.21	2940.52	2899.48
5	104000	0.81	84240	85480.41	69239.13	69559.90
Total	120000		99040	100000	**82698.15**	**82698.15**

Column A in the above table denotes the expected cash flows for the bond in case we assume that it does not default. The corresponding survival probabilities are stated in column B and in consequence we get the expected cash flow payments including the probability to default in column C. Based on this calculation we get the value of the non-defaultable bond in column D and of the defaultable one in column E. Besides this direct calculation, one can also ask how much bigger interest rate is necessary, in order to get the same present value if we base our calculation on the nominal cash flows in column A. This results in an interest rate for discounting of 8.37%, which represents a risk premium of 437 bp. This is the way in which the market values defaultable bonds, by adding a risk premium on top of the risk free yield curve. Hence we get a yield curve for defaultable bonds including a *credit spread*. Figure 2.2 shows the development of the credit spread over time. It needs to be stressed that the increase of the credit spreads in this figure is only partially a consequence of higher default probabilities. The other effect is the fact that during the end of the year 2008 there was a liquidity and capital crunch. The absence of a liquid market can also lead to higher credit spreads, as observed in the crisis of 2008/09.

In a next step we want to understand how different bonds are assessed in terms of credit rating. In order to do this, bonds are evaluated by rating agencies, which put them in (homogeneous) classes which should have the same survival behaviour. To this end S&P classifies bonds between AAA, AA, A, BBB, BB, ...C and D. Bonds with a higher credit quality than BBB are called investment grade and bonds classified as D are in default.

In order to model the credit rating process one normally uses *Markov chains* $(X_t)_{t \in \mathbb{R}+}$ on a finite state space S (see also appendix B). In case of the S&P rating one could define S as

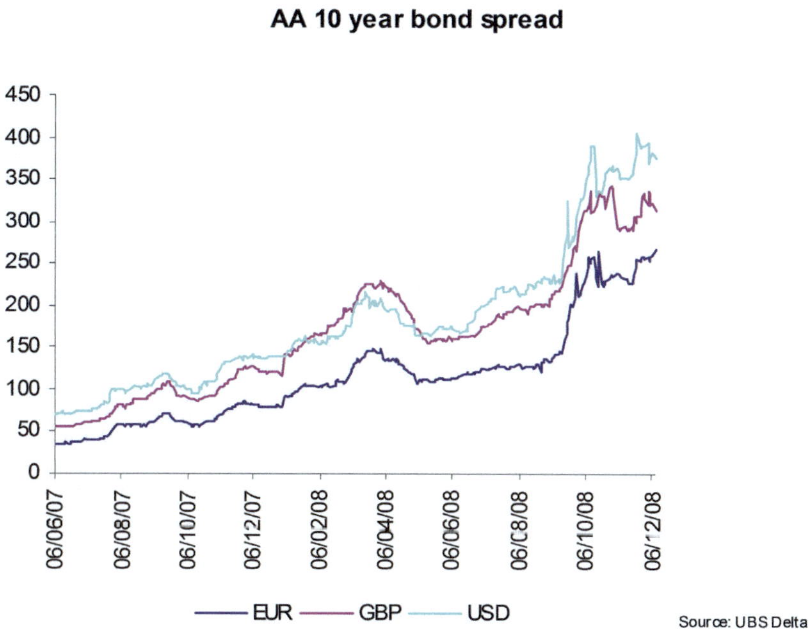

Fig. 2.2 Credit spreads over time

$$S = \{AAA, AA, A, BBB, BB, B, C, D, NR\},$$

where NR stands for not rated. For a Markov chain, one defines the transition probabilities $p_{ij}(s,t)$ as

$$\begin{aligned} p_{ij}(s,t) &= P[X_t = j \mid X_s = i], \text{ and} \\ P(s,t) &= (p_{ij}(s,t))_{(i,j)\in S\times S} \,. \end{aligned}$$

From the *Chapman-Kolmogorov equation* it is known that, we have the following relationship for $s < t < u$:

$$P(s,u) = P(s,t)\,P(t,u).$$

For credit risk it is normally assumed that $P(s, s+1)$ is constant ("time-homogeneous Markov chain"), and hence we can further simplify:

$$P(0,t) = P(0,1)^t.$$

After this formula we can now look at a concrete example of such a transition matrix ($P(0,1)$), remarking that the states "NR" and "D" have been merged:

$j \rightsquigarrow$ i	AAA	AA	A	BBB	BB	B	C	D
AAA	88.12%	11.88%	0.00%	0.00%	0.00%	0.00%	0.00%	0.00%
AA	0.00%	92.45%	3.30%	0.24%	0.00%	0.00%	0.00%	4.01%
A	0.00%	1.54%	91.32%	2.92%	0.08%	0.00%	0.00%	4.14%
BBB	0.00%	0.00%	3.21%	88.77%	2.01%	0.13%	0.00%	5.88%
BB	0.10%	0.00%	0.10%	4.66%	80.55%	5.17%	0.20%	9.22%
B	0.00%	0.00%	0.00%	0.33%	7.22%	72.54%	3.50%	16.41%
C	0.00%	0.00%	0.72%	0.00%	3.6%	18.71%	50.36%	26.62%
D	0.00%	0.00%	0.00%	0.00%	0.00%	0.00%	0.00%	100.00%

From this table one can for example read that within one year 11.88% of all AAA bonds are downgraded to AA. In a next step it is now possible to calculate for example $P(0,10)$:

$j \rightsquigarrow$ i	AAA	AA	A	BBB	BB	B	C	D
AAA	28.23%	48.06%	8.17%	1.22%	0.07%	0.01%	0.00%	14.23%
AA	0.00%	46.81%	15.88%	3.07%	0.22%	0.03%	0.00%	33.98%
A	0.00%	7.34%	43.42%	11.91%	1.15%	0.2%	0.01%	35.96%
BBB	0.03%	1.06%	12.98%	33.54%	5.13%	1.19%	0.1%	45.97%
BB	0.24%	0.33%	2.66%	11.8%	15.17%	5.86%	0.6%	63.34%
B	0.06%	0.07%	0.72%	3.76%	8.37%	7.29%	0.91%	78.81%
C	0.04%	0.13%	1.11%	2.36%	5.53%	5.19%	0.73%	84.91%
D	0.00%	0.00%	0.00%	0.00%	0.00%	0.00%	0.00%	100.00%

From the above transition matrix from year 0 to year 10 it becomes obvious, that about 14% of the AAA have defaulted over this 10 year period. Furthermore it also becomes obvious, that the corresponding matrix has some flaws since there are no AA and A bonds which have after 10 years a AAA rating in contrast to some BB bonds. The reason for this is certainly the fact that the mitigation probabilities to AAA are very small and therefore not enough transitions have been observed over the observation time. In consequence the transition matrix contradicts expectations.

Finally it needs to be remarked that normally a change in rating implies a change in credit spread and hence implies a gain or a loss. In order to illustrate this, we use the same example from above and assume that the credit spread for a AA bond corresponds to 75 bp, the one for a A bond to 125 bp and the one for a BBB bond to 200 bp. In this case we have the following situation:

t	Coupon	Principal	Total CF	PV AA 4.75%	PV A 5.25%	PV BBB 6.00%
0			0	–	–	–
1	4000		4000	3818.61	3800.47	3773.58
2	4000		4000	3645.45	3610.90	3559.98
3	4000		4000	3480.14	3430.78	3358.47
4	4000		4000	3322.33	3259.65	3168.37
5	4000	100000	104000	82463.76	80523.53	77714.84
Total	20000	100000	120000	**96730.32**	**94625.35**	**91575.27**
Difference				*2104.97*	0	*-3050.07*

Also here the duration approximation can be used to estimate the impact of a change in credit spread. If we want to estimate the impact of an increase of the credit spread by 75 bp (eg A $\rightsquigarrow$ BBB) we have $d^{mod}(\mathcal{B}) \times 0.75\% \times 94625.35 = 4.40 \times 0.75\% \times 94625.35 \approx 3122$, and also here, this approximation is of acceptable quality.

Finally we want to have a look at the valuation using a Markov model. The corresponding formulae can be found in appendix B. Above we have seen the transition probabilities and also the interest used for discounting is obvious. Hence we still need to define the corresponding benefits. Since the contractual terms are honoured in all cases where the bond is not in default, we have for all states in $S^* = \{AAA, AA, \dots C\}$ the following:

$$a_{ij}^{\text{Post}}(t) = \begin{cases} 0 & \text{if} \quad k = 0 \\ i & \text{if} \ 0 < k < n \\ 1+i & \text{if} \quad k = n \\ 0 & \text{if} \quad k > n \end{cases} \qquad \forall (i,j) \in S^* \times S^*.$$

Now what happens for transitions $i \rightsquigarrow D$ with $i \in S^*$. In this case the investor gets back the value of the remaining cash flows $\times(1 - LGD)$, where LGD denotes the *loss given default*. A loss given default of 65% means that you get back 35000 EUR back in case a bond with a value of 100000 EUR. For the example a simplified calculation of the remaining value of the bond was used, in the sense that the remaining value of the bond was taken undiscounted. In the concrete example below we have chosen a loss given default of 65 % and an interest rate of 3%.

	AAA	AA	A	BBB	BB	B	C	D
0	926111	819694	809241	755500	661762	575428	538721	0
1	936413	831033	821436	768823	674901	582776	541098	0
2	946761	843599	834953	784101	691281	593642	546108	0
3	956986	857477	849866	801507	711386	608958	554563	0
4	966877	872758	866253	821224	735744	629936	567639	0
5	976175	889540	884193	843447	764914	658154	587107	0
6	984565	907929	903766	868382	799460	695660	615791	0
7	991660	928040	925052	896252	839910	745080	658450	0
8	996993	949997	948133	927296	886697	809687	723574	0
9	1000000	973934	973090	961780	940069	893334	827069	0
10	1000000	1000000	1000000	1000000	1000000	1000000	1000000	0

The really interesting thing about the Markov model is the fact that you can not only calculate the expected values, but also higher moments and the complete probability distribution function (see [Kol10]).

2.4 Shares

In the previous section we have looked at bonds. Another important asset class are shares. Here one invests in a company and the share price reflects the value of the company. An example could be a share of an insurance company, of a utility company etc. A share has two economic aspects which need to be distinguished. On one hand there is the amount of dividends a company pays and on the other hand there is the value of the share.

As with bonds each share has a nominal value, for example CHF 50. At each moment in time, this share has a market value, which might for example be CHF 320. The dividend yield of 10% would result in a dividend payment of CHF 5. This means that the dividend payment in relation to its market value is 1.56%.

Shares can be valued according to book value and market value principles. Both of these valuations are conceptually easier than bonds. Assume that we have bought a share for a price of EUR 50 per unit and we have 10000 units. Furthermore assume that the unit price has changed upwards and we want to see how different valuation principles look like. The book value principle requests that the share is valued at purchase price, or lower if the current market price is below the purchase price ("lower book or market"). There are several variations to this principle and the concrete set up depends on the legislation of the country. There is a stronger form of the "lower book or market" principle, where the book value has to be written down, once the market value falls below the book value and there is no possibility to write it up again. However, there are countries where the minimum condition does not apply immediately and the value of the shares can be smoothed over time unless there is a permanent impairment.

On the other hand the market value valuation of a share is rather easy. One takes the last paid price of the instrument. This is however not always as easy as it might seem. If you have an illiquid share where there are only few transactions. In this case the last transaction might be quite old and the last share might have been traded some weeks ago. Besides, market prices of such shares move normally more erratic than the paid prices of more liquid shares, and hence more care is needed for the valuation of them.

Now lets have a look at the example mentioned above. The table below illustrates the values of the shares for different unit prices:

Unit Price	Purchase Price	Number of Units	Market Value	Book Value
40	50	10000	400000	400000
50	50	10000	500000	500000
60	50	10000	600000	500000

It is obvious that the book value moves like the market value once the share price has fallen below the purchase price. On the other hand it remains constant above the purchase price. The difference between market value and book value is called *revaluation reserve*. In case of a unit price 60, it is 100000. In the past some companies had considerable revaluation reserves in equities and also in properties (which are valued according book values in a similar way). If the company wanted to access a part of this reserve, it had to sell and potentially re-buy the instrument in order to *realise* the gain.

In a next step we want to look at the risks of a share. Normally one assumes that the share price (for one unit) S_t moves according to a geometric Brownian motion. This implies that for any given time interval Δt we have the following

$$\log\left(\frac{S_{t+\Delta t}}{S_t}\right) \sim \mathcal{N}(\mu, \Delta t \times \sigma),$$

where $\mathcal{N}(\mu, \sigma)$ denotes the normal distribution with expected value μ and standard deviation σ. The parameter σ is the standard deviation for a one year time horizon and it is called *volatility*. For typical equity indexes (eg a normalised basket of equities) the expected yield is in the area of 7 to 8%. The volatility depends on the market sentiment and ranges in normal markets between 15 to 25%. If there is a distressed market, the so called spot volatility can be well above 40%.

Based on the normal distribution assumption we can easily estimate the risk of a share. We know the probability distribution function for a random variable $X \sim \mathcal{N}(0, \sigma)$:

x	$P[\frac{X}{\sigma} < x]$
1.000	84.134%
1.281	90.000%
1.644	95.000%
2.326	99.000%
2.575	99.500%
3.090	99.900%

Based on this, it is possible to form some rules of thumb. If we want to estimate the potential loss in a one in 200 year event we have to look at the 99.5% confidence level and we get 2.57. Assume now we want to estimate roughly the risk we are running for our 10000 shares at a market value of CHF 60 assuming a volatility of $\sigma = 18\%$. We get $10000 \times 2.57 \times 60 \times 0.18 = 277560$, which represents about 46% of the market value at this time. It is important to note that the rule of thumb uses an assumption, namely that the stock returns are log normally distributed. Since this is a model, it is only an approximation, and hence we might underestimate the true risk. Assume for sake of simplicity that we have observed a volatility of 18% over the last year. Our risk for a holding period of 1 year would be 277560. But a crisis might happen and the volatility could spike. In this case we would most likely underestimate the risk.

Before leaving shares we need to have a look at how volatility impacts the share price. Figure 2.3 shows a possible trajectory (share price development) over one year assuming three different levels of volatility 15%, 30% and 45%. We assume a drift $\mu = 7\%$ and a purchase price of CHF 500000.

2.5 Other Assets

This section will not go into the same detail as in the two previous section. Besides bonds and equities the most relevant asset classes an insurance company invests into are as follows:

Cash In order to be able to pay the claims due and because of regular premium income, insurance companies invest on a short time horizon into cash. This is a very risk adverse investment, where interest risk is normally virtually absent.

Properties In the past insurance companies have heavily invested in properties in order to safeguard the policy holder's assets. Properties are in a lot of regimes accounted at depreciated purchase price. This means over time of 20 to 30 years the purchase price is amortised.

Mortgages A lot of people finance their properties by mortgages and in some countries it is a custom to buy an insurance policy which serves as collateral for

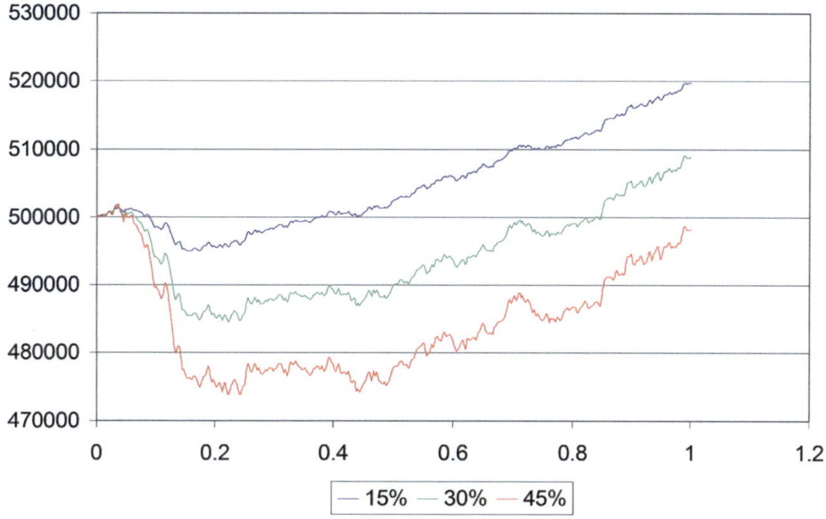

Fig. 2.3 Possible Trajectories of Shares

the mortgage. As a consequence insurance companies have built up mortgage portfolios.

Hedge Funds and Private Equities These two asset classes of alternative assets are normally valued like a share. Please note however that from a risk point of view they may behave completely differently and hence it is very dangerous to use the same models.

Commodities Commodities such as oil, precious metals etc are not an often used asset class for insurance companies. In some countries they are not allowed for covering policyholder funds.

2.6 Insurance Liabilities

In this book it is not possible to describe and value all possible insurance liabilities and hence we should to focus on the most important life insurance liabilities. We can distinguish between insurance liabilities where the policyholder assumes all risk and consequently invests in funds (see also appendix D). Here the value is normally quite clear and so we can focus on life insurance forms with investment guarantees. In this case the majority of the investment risk is born by the insurance company. From a conceptual point of view life insurance cover behaves very similar to a bond. In principle one agrees some payments, which have to be weighted with the corresponding probabilities. In the following we want to introduce the corresponding concepts.

Insurance liabilities can be valued according to a book value or a market value principle. In the first case future cash flows are discounted using discount rates based on the technical interest rate i. In Europe this rate is determined in a prudent way and should according to the 3rd life insurance directive normally not exceeding 60% of the yield of governance bonds. So if we assume that governance bonds in EUR yields 4%, the maximal technical interest rate would be 2.4%. In reality the rule is interpreted in a somewhat more ingenious way and one looks for example at rolling averages of yields of government bonds. Based on the technical interest rate a payment of 1 due in one year is discounted with $v = \frac{1}{1+i}$. So here the book value approach yields to higher liabilities representing a prudent valuation approach.

In this section we will also focus on the market valuation on a best estimate basis. This is the first step to determine the market value of an insurance liability. We assume however that the insurance cash flows are certain. Since there is in reality a risk involved, it will be necessary to revisit the concept of market values for liabilities later (Section 3.3.1 and appendix C). We will see there how risk enters in the valuation and how we can use this knowledge for risk adjusted performance metrics.

2.6.1 Life Insurance Model

In order to model a life insurance policy we consider a person aged x and denote by T the future life span and we remark that actually one would have to denote it $T(x)$ since it is dependent on the age x. The cumulative probability density function of T is

$$G(t) = P[T \leq t], \tag{2.1}$$

and we assume that there exists a probability density function for T. Hence we can write:

$$g(t)dt = P[t < T < t + dt]. \tag{2.2}$$

In order to do life insurance mathematics it is useful the define the following standard quantities:

$$\begin{aligned}
{}_tq_x &:= G(t), && (2.3)\\
{}_tp_x &:= 1 - G(t), && (2.4)\\
{}_{s|t}q_x &:= P[s < T < s + t] && (2.5)\\
&= G(t+s) - G(s) = {}_{s+t}q_x - {}_sq_x, && (2.6)\\
\overset{\circ}{e}_x &:= \mathbb{E}[T(x)] = \int_0^\infty tg(t)dt && (2.7)
\end{aligned}$$

$$= \int_0^\infty (1 - G(t))dt = \int_0^\infty {}_tp_x dt, \tag{2.8}$$

and we remark that $\overset{\circ}{e}_x$ is the expected future life span of a person aged x. We also remark that $q_x := {}_1q_x$ and $p_x := {}_1p_x$. Based on the above definitions we get the following equations:

$$\begin{aligned}
{}_tq_{x+s} &= G_{x+s}(t) &(2.9)\\
&= P[T(x+s) < t] &(2.10)\\
&= P[T(x) \le s+t | T(x) > s] = \frac{G(s+t) - G(s)}{1 - G(s)}, &(2.11)\\
{}_tp_{x+s} &= P[T \ge s+t | T > s] = \frac{1 - G(s+t)}{1 - G(s)}, &(2.12)\\
{}_{s+t}p_x &= 1 - G(s+t) &(2.13)\\
&= (1 - G(s))\frac{1 - G(s+t)}{1 - G(s)} = {}_sp_x\, {}_tp_{x+s}, &(2.14)\\
{}_{s|t}q_x &= G(s+t) - G(s) &(2.15)\\
&= (1 - G(s))\frac{G(s+t) - G(s)}{1 - G(s)} = {}_sp_x\, {}_tq_{x+s}. &(2.16)
\end{aligned}$$

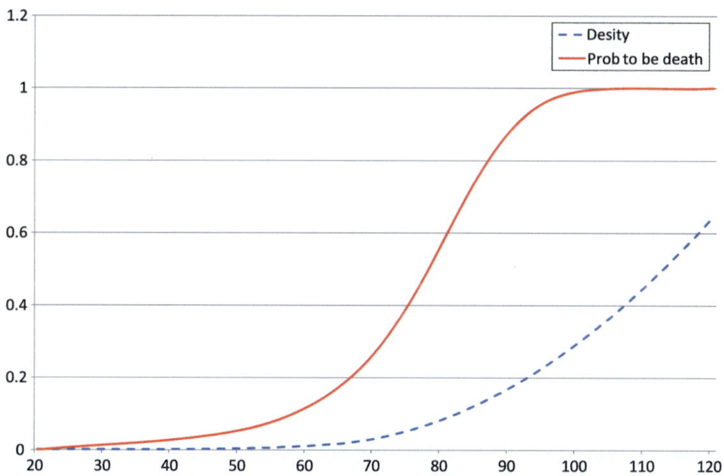

Fig. 2.4 Probability density function for the future life span and hazard rate

In order to calculate the quantities introduced above one normally uses mortality tables. Based on equation (2.14) we get the following

$$ {}_tp_x = \prod_{k=0}^{k<t} p_{x+k} = \prod_{k=0}^{k<t} (1 - q_{x+k}) \text{ for } t \in \mathbb{N}. $$

In order to simplify, one uses $K = \max\{k \in \mathbb{N}_0 : k \leq T\}$, the number of completely lived years before death.

2.6.2 Capital Insurance

Capital insurance is an insurance cover where there exists only one payment from the insurer during the contract and we distinguish between the following types of cover:

Term Insurance and Whole Life Insurance: In case of death a lump sum becomes due. The present value of this insurance type is denoted by $A^1_{x:\overline{n}|}$ if the cover is provided for n years (eg a term insurance for a 45 year old person with a cover period of 10 years is denoted by $A^1_{45:\overline{10}|}$). A whole life insurance is an insurance where $n = \infty$ and we denote its present value by $A_x = A^1_{x:\overline{\infty}|}$.

Pure Endowment: In case a person reaches a certain age (eg 65) a lump sum becomes due. The present value of this type of insurance is denoted by $A_{x:\overline{n}|}^{\;\;\,1}$.

Endowment: Combination of the two types above, eg. if the person dies before the age 65 a lump sum becomes due at the moment of death, otherwise the person receives the lump sum at 65. The present value of this insurance is denoted by $A_{x:\overline{n}|}$. So the present value of the benefits to be paid for a 35 year old person with maturity at age 65 is denoted by $A_{35:\overline{30}|}$.

In order to value a life insurance policy we need to know its value. In the normal life insurance model one expects lump sums in case of death to become due at the end of the year.

Whole Life Insurance

In case of death a payment of 1 is due, we have the following for the present value of the benefits as a random variable:

$$ Z := v^{K+1}, \tag{2.17} $$

where $K = 0, 1, 2, \ldots$. In case of a market consistent valuation Z reads as follows:

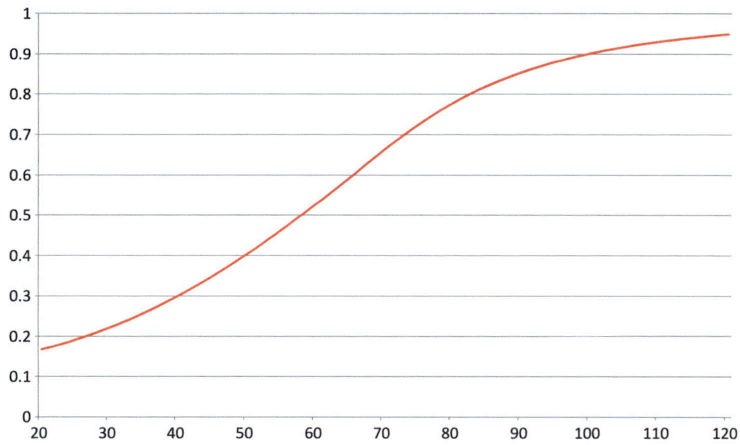

Fig. 2.5 Value of the benefits of a whole life insurance

$$Z := \pi(\mathcal{Z}_{(K+1)}), \tag{2.18}$$

Z takes values v, v^2, v^3, ... and $P[Z = v^{k+1}] = P[K = k] = {}_kp_x q_{x+k}$. Hence we get the following for the book value

$$A_x = \mathbb{E}[Z] = \mathbb{E}\left[v^{K+1}\right] = \sum_{k=0}^{\infty} v^{k+1} {}_kp_x q_{x+k} \tag{2.19}$$

and

$$A_x = \mathbb{E}[Z] = \mathbb{E}\left[\pi(\mathcal{Z}_{(K+1)})\right] = \sum_{k=0}^{\infty} \pi(\mathcal{Z}_{(k+1)}) {}_kp_x q_{x+k} \tag{2.20}$$

for the market consistent valuation of the expected cash flows. In a next step we can calculate the variance of Z as follows:

$$Var[Z] = \mathbb{E}\left[Z^2\right] - \mathbb{E}[Z]^2. \tag{2.21}$$

Term Insurance

The calculation is performed completely analogous: if the person dies within the contractual term (eg within n years) a capital 1 becomes due. In consequence we

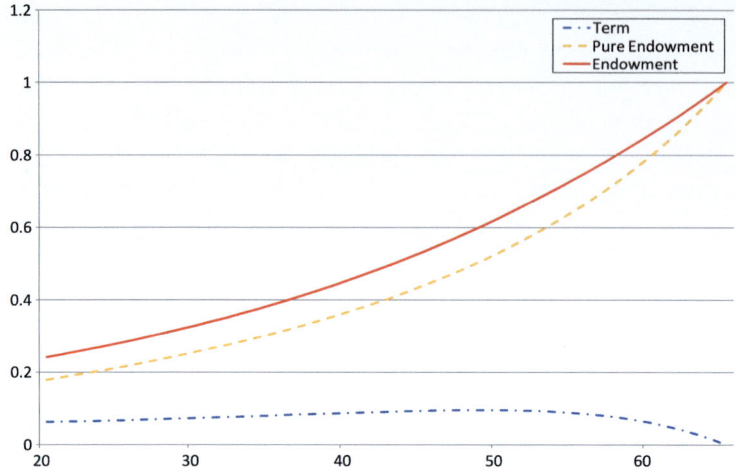

Fig. 2.6 Value of a term insurance

get the following random variable for the present value of the insurance liability:

$$Z = \begin{cases} v^{K+1}, & \text{for } K = 0, \ldots, n-1 \\ 0, & \text{otherwise} \end{cases} \tag{2.22}$$

and hence we have the following for the book value valuation:

$$A^{1}_{x:\overline{n}|} = \sum_{k=0}^{n-1} v^{k+1} {}_k p_x q_{x+k}.$$

For market values of the expected cash flows we have:

$$Z = \begin{cases} \pi(\mathcal{Z}_{(K+1)}), & \text{if } K = 0, \ldots, n-1 \\ 0, & \text{otherwise} \end{cases} \tag{2.23}$$

and hence

$$A^{1}_{x:\overline{n}|} = \sum_{k=0}^{n-1} \pi(\mathcal{Z}_{(k+1)}) {}_k p_x q_{x+k}.$$

2.6.3 Pure Endowment

The calculation is completely analogous. The only difference is the definition of the contractual payment stream.

$$Z = \begin{cases} 0, & \text{if } K = 0, 1, \ldots, n-1 \\ v^n, & \text{if } K = n, n+1, \ldots \end{cases} \tag{2.24}$$

and

$$\begin{aligned} A_{x:\overset{\;1}{\overline{n|}}} &= \sum_{k=0}^{\infty} Z(k) P[K = k] \\ &= \sum_{k=n}^{\infty} v^n P[K = k] = v^n P[K \geq n] \\ &= v^n (1 - P[K < n]) = v^n (1 - {}_nq_x) = v^n \, {}_np_x. \end{aligned}$$

For the market value of the expected cash flows we have:

$$A_{x:\overset{\;1}{\overline{n|}}} = \mathcal{Z}_{(n)} \, {}_np_x.$$

Endowment Insurance

Since an endowment is the sum of a term and a pure endowment insurance the arguments above apply mutatis mutandis and we get:

$$A_{x:\overline{n|}} = A^1_{x:\overline{n|}} + A_{x:\overset{\;1}{\overline{n|}}}.$$

2.6.4 Annuities

As with capital insurance there exist a variety of different annuity covers and we need to focus on some particularly important ones:

Immediate annuity: This is an annuity where the insured person receives at the beginning of every year an annuity 1 until death. The present value of such an annuity is denoted by $\ddot{a}_x$.

Deferred annuity: Here the payment starts in the future, but otherwise it is the same as above. For the present value of the deferred annuity we use ${}_n\,\ddot{a}_x$, where n stands for the number of years for which the annuity is deferred. So ${}_{30}\,\ddot{a}_{35}$ is a deferred annuity of a 35 year old person which is deferred by 30 years and hence starts at the age of 65.

Temporary annuity: This is the type of payment stream which is used to model a regular premium payment, starting immediately until death or when a certain term is reached.

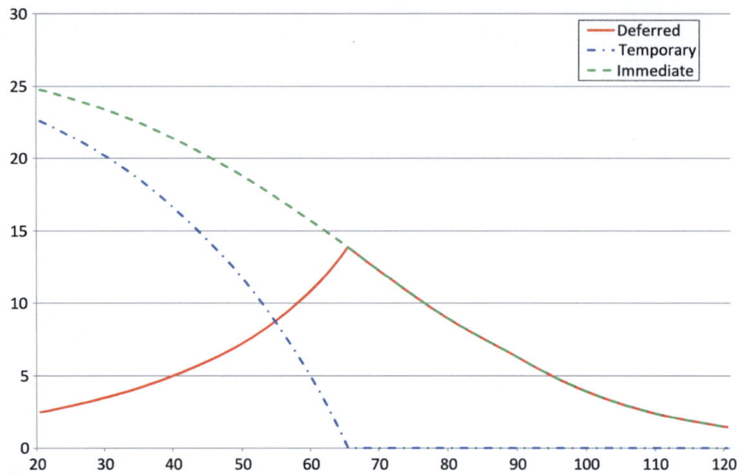

Fig. 2.7 Value of different annuity covers

In order to evaluate the value of an annuity we need in a first step to define the corresponding present value as a random variable Y.

$$Y = 1 + v + v^2 + \ldots + v^K = \ddot{a}_{\overline{K+1}|} \tag{2.25}$$

and we know that $P[Y = \ddot{a}_{\overline{k+1}|}] = P[K = k] = {}_k p_x q_{x+k}$. Hence we can calculate the expected present value for the book valuation as follows:

$$\ddot{a}_x = \mathbb{E}[Y] = \sum_{k=0}^{\infty} \ddot{a}_{\overline{k+1}|} {}_k p_x q_{x+k}. \tag{2.26}$$

There is also a second possibility where we interpret an annuity as a portfolio of pure endowment policies and hence we can write:

$$Y = \sum_{k=0}^{\infty} v^k \chi_{\{K \geq k\}}. \tag{2.27}$$

Here the present value can be calculated as follows:

$$\ddot{a}_x = \mathbb{E}[Y] = \sum_{k=0}^{\infty} v^k P[K \geq k] = \sum_{k=0}^{\infty} v^k {}_k p_x.$$

We remark that for the market value of expected cash-flows we need to replace v^k by $\mathcal{Z}_{(k)}$. In a next step it makes sense to indicate the relationship between capital insurance and annuities. We know the following relation:

$$Y = 1 + v + v^2 + \ldots + v^K = \frac{1 - v^{K+1}}{1 - v} = \frac{1 - v^{K+1}}{d},$$

which is valid as a random variable (please note that $d := \frac{1}{1-v}$). By applying the expected value operator we get the following useful relationship:

$$\begin{aligned}\ddot{a}_x &= \mathbb{E}[Y] = \mathbb{E}\left[\frac{1-Z}{d}\right] = \frac{1}{d} - \frac{\mathbb{E}[Z]}{d} \\ &= \frac{1 - A_x}{d},\end{aligned}$$

or in terms of an actuary

$$1 = d\ddot{a}_x + A_x. \tag{2.28}$$

By means of the above equation we can also calculate the corresponding variances as follows:

$$Var[Y] = Var\left[\frac{1-Z}{d}\right] = \frac{1}{d^2} Var[Z]. \tag{2.29}$$

Please note that the above relationship is not true for the market consistent present value of expected cash-flows since we normally do not have a flat yield curve.

2.6.5 Cash Flows and Valuation

After this introduction to life insurance mathematics we want now to look at an example and we will focus on an immediate annuity for a 65 year old man. We assume that the annuity is paid in yearly instalments of EUR 12000. Furthermore we want to look at a valuation as of 29.12.2006 for the market values and a technical interest rate of $i = 2.5\%$ for the calculation of book values. So the annuity can be characterised as follows:

ANNUITY	
Yearly payment - prenummerando	12000
Age	65
Currency	EUR
Valuation Date	29.12.2006
Technical Interest Rate	2.5%
Mortality Table	Swiss Males

Based on the above assumptions we get the following results:

Age	${}_t p_x$	Annuity Nominal	Annuity Risk Ajd.	Price $i = 2.5\%$	Price $\mathcal{Z}_{(t)}$	Value Book	Value Market
65	1.0000	12000	12000.00	1.0000	1.0000	12000.00	12000.00
66	0.9852	12000	11822.40	0.9756	0.9627	11534.04	11381.52
67	0.9692	12000	11630.87	0.9518	0.9266	11070.43	10778.13
68	0.9517	12000	11421.52	0.9285	0.8919	10606.01	10187.71
69	0.9331	12000	11197.65	0.9059	0.8577	10144.52	9605.12
70	0.9132	12000	10959.14	0.8838	0.8251	9686.29	9043.21
71	0.8918	12000	10702.70	0.8622	0.7935	9228.90	8492.71
72	0.8688	12000	10426.57	0.8412	0.7628	8771.51	7954.28
73	0.8441	12000	10129.93	0.8207	0.7335	8314.11	7430.54
74	0.8180	12000	9816.72	0.8007	0.7059	7860.52	6929.83
75	0.7900	12000	9481.08	0.7811	0.6782	7406.61	6430.32
76	0.7603	12000	9124.03	0.7621	0.6512	6953.83	5942.15
77	0.7288	12000	8746.38	0.7435	0.6251	6503.42	5468.02
78	0.6957	12000	8348.95	0.7254	0.5999	6056.49	5008.88
79	0.6610	12000	7932.33	0.7077	0.5755	5613.93	4565.42
80	0.6247	12000	7497.17	0.6904	0.5519	5176.53	4138.19
81	0.5870	12000	7044.19	0.6736	0.5297	4745.14	3731.38
82	0.5478	12000	6574.20	0.6571	0.5082	4320.53	3341.35
83	0.5073	12000	6088.43	0.6411	0.4875	3903.69	2968.54
84	0.4657	12000	5588.69	0.6255	0.4676	3495.88	2613.50
85	0.4231	12000	5077.49	0.6102	0.4484	3098.64	2276.94
...							
90	0.2098	12000	2517.90	0.5393	0.3649	1358.13	918.94
...							
100	0.0031	12000	37.95	0.4213	0.2471	15.99	9.37
...							
Total		552000.00	216736.59			**170312.90**	**149811.51**

The similarity of the above table with defaultable bonds is because of their intrinsic structural proximity. We can in particular observe that the reserve hold as book values is some 20000 EUR higher. One way to calculate the so called mathematical reserves V_x is based on the approach outlined above. There is also another efficient way to recursively calculate the reserves, based on the so called Thiele's difference equation. To this end we denote by $(a_k)_{k \in \mathbb{N}_0}$ the annuity vector to be paid at age k and by $(d_k)_{k \in \mathbb{N}_0}$ the death benefit at age k. Furthermore we denote by v the one year discount rate, which is in the market value context the corresponding discount rate based on the forward rate for the corresponding time. Equipped with this notation we get the following backwards recursion:

$$V_x = a_x + q_x \times v \times b_x + p_x \times v \times V_{x+1}.$$

For a proof of the above recursion we refer to appendix B.

2.6.6 Primer on Life Insurance Risks

There are typically two different risks which impact the level of reserves. The level of the interest rate, just in the same spirit as with bonds. Hence we do not show the corresponding example here and remark that the duration of the annuity above is about 8.38. A parallel shift of the interest rate by 1% changes the corresponding reserve by about 8.4%.

The second risk is the one related to the mortality risk. The table below shows the sensitivity in respect to mortality and we assume that the stressed mortality with respect to a parameter α is defined as $q_x(\alpha) = \alpha \times q_x$. Based on this we get the following:

Age α	$_t p_x$ 90%	$_t p_x$ 100%	$_t p_x$ 110%	MV 90%	MV 100%	MV 110%
65	1.0000	1.0000	1.0000	12000.00	12000.00	12000.00
66	0.9866	0.9852	0.9837	11398.62	11381.52	11364.42
67	0.9722	0.9692	0.9661	10812.09	10778.13	10744.22
68	0.9565	0.9517	0.9470	10238.55	10187.71	10137.04
69	0.9396	0.9331	0.9266	9672.35	9605.12	9538.24
70	0.9216	0.9132	0.9049	9126.33	9043.21	8960.70
71	0.9022	0.8918	0.8816	8591.31	8492.71	8395.06
72	0.8812	0.8688	0.8566	8067.93	7954.28	7842.00
73	0.8587	0.8441	0.8298	7558.78	7430.54	7304.20
74	0.8348	0.8180	0.8015	7071.92	6929.83	6790.26
75	0.8091	0.7900	0.7714	6585.40	6430.32	6278.51
...						
80	0.6553	0.6247	0.5954	4340.88	4138.19	3944.25
...						
85	0.4621	0.4231	0.3871	2486.94	2276.94	2083.60
...						
90	0.2469	0.2098	0.1779	1081.75	918.94	779.31
...						
100	0.0063	0.0031	0.0015	18.71	9.37	4.51
...						
Total				**154327.11**	**149811.51**	**145675.21**
Difference				*4515.60*		*-4136.30*

2.7 Shareholders Equity and Capital

After having discussed all the other assets and liabilities we will devote a short section to the shareholder equity or capital. As we have seen before there are different ways on how the different assets and liabilities are accounted. In principle shareholders equity is the difference between all assets and all liabilities other than the shareholder's equity. There might also be some adjustments for taxes not yet paid. The shareholder's equity represents the worth the shareholder has in the entity, and the purpose of the equity is twofold. It gives the shareholder the right to get a return on the capital he has invested. Moreover the shareholder equity or capital acts as a buffer in case of an adverse market development.

Since we want to illustrate this we will look first at a balance sheet based on book values and afterwards at the same balance sheet based on market values. In order to simplify we assume that we are living in a country with no taxes. Please note that we denote the mathematical reserves with MR. The insurance company we are looking at has the following balance sheet:

Balance sheet	Book A	Book L	Market A	Market L	
Cash	6200	47100	6200	48513	MR
Bonds	35700	**2200**	37842	**3569**	SHE
Shares	4400		4800		
Properties	1100		1300		
Loans	1400		1400		
Alternatives	500		540		
Total	49300	49300	52082	52082	

From this balance sheet it becomes obvious that there are revaluation reserves in both bonds and shares and that the mathematical reserves are (not taking the interest rate effect into account) about 5% too high. For the example we assume a duration of the bonds of 6 and of the reserves of 8. This means that the company is suffering from an economical point of view in case of decreasing interest rates.

Another typical effect is the fact that the shareholder capital based on a book value approach is about 1500 less than if using an economic valuation. This means that the risk absorbing capacity of the company in nominal terms is higher if we look at a realistic valuation. We need to stress that the statutory accounts in continental Europe often use book value accounting. Hence the dividend paying capacity is based on the lower equity (eg 2200) and therefore it is of utmost importance to always keep also this number in mind. Since local insolvency laws are also based on the statutory shareholder equity, the company would be in deep trouble if this number reduces too much.

But now lets look at the following three scenarios:

- Drop in equity prices by 20%,
- Interest rate decrease by 1% and
- Interest rate increase by 1%.

In the first scenario we look at a 20% decrease in stock markets. It becomes obvious that the reduction in shareholder equity is proportionally smaller in the case of book value accounting, since the first hit is absorbed by the revaluation reserves which were present before the shock in the equities. Since the shock is bigger than the revaluation reserve, also in the book value accounting, the value of the shares had to be adjusted downwards in accordance with the "lower book or market" principle.

Balance sheet *Shares -20%*	Book A	Book L	Market A	Market L	
Cash	6200	47100	6200	48513	MR
Bonds	35700	**1640**	37842	**2609**	SHE
Shares	3840		3840		
Properties	1100		1300		
Loans	1400		1400		
Alternatives	500		540		
Total	48740	48740	51122	51122	

In the second scenario we see the impact of a reduction in interest rate levels by 1%. Due to the nature of the amortised cost method, the effect is not reflected in the accounts in the book value world. We also see a material deteriorisation in the marked to market balance sheet, as a consequence of the duration gap of 2. We can see that a further reduction of the interest rate levels could become dangerous for the insurance company. As remarked before book value accounting "is blind" with respect to the issue in case of low interest rate in its purest form. In order to compensate for it there are reserve adequacy tests where one tests, whether the earned interest rate yield is sufficient for financing the technical interest rate for the reserves. If this is not the case the reserves would have to grow commensurately, hereby reducing the statutory shareholder equity.

Balance sheet *Int. -1%*	Book A	Book L	Market A	Market L	
Cash	6200	47100	6200	52281	MR
Bonds	35700	**2200**	39984	**1943**	SHE
Shares	4400		4800		
Properties	1100		1300		
Loans	1400		1400		
Alternatives	500		540		
Total	49300	49300	54224	54224	

The third scenario is a twin of the second one, just with the opposite direction of the interest rate shift. Here the available economic capital grows as a consequence of the duration gap. At this point it is worth mentioning that there are also other accounting standards, which are not symmetrical in the sense that book and market value principles are not applied in a consistent manner. IFRS for example can foresee market value principles for assets and book value principles for liabilities. We can see the corresponding consequences by regrouping the values accordingly. On doing so we acknowledge that under IFRS the shareholder equity reduces in case of an increase in interest rates for the simple reason, that the bonds lose value which is not compensated by a decrease in the policyholder mathematical reserves. Comparing IFRS accounting standards with economic principles also highlights one of the important paradoxes, namely that the two accounting standards contradict and that one can optimise both of them at the same time only in a limited manner.

Balance sheet *Int. +1%*	Book **A**	Book **L**	Market **A**	Market **L**	
Cash	6200	47100	6200	44745	MR
Bonds	35700	**2200**	35700	**5195**	SHE
Shares	4400		4800		
Properties	1100		1300		
Loans	1400		1400		
Alternatives	500		540		
Total	49300	49300	49940	49940	

Chapter 3
Accounting Principles

In this rather short chapter we want to have a deeper look at the different accounting principles which are normally used. As seen above, technically speaking an accounting principle on a balance sheet (x) is a function which allocates to each asset and liability its value. Since there are different possibilities, it is not always easy to see the main differences between the different accounting standards. We want to have a look at the following accounting standards:

- Statutory accounting,
- IFRS and US GAAP accounting,
- Embedded value accounting,
- Economic balance sheet accounting.

M. Koller, *Life Insurance Risk Management Essentials*, EAA Series,
DOI 10.1007/978-3-642-20721-1_3, © Springer-Verlag Berlin Heidelberg 2011

In order to do this, we always need to look at the same balance sheet and we want to see what are the material differences between the different standards. It needs to be stressed, that it will be impossible to explain all the different aspects and hence this chapter cannot substitute the in depth study of the corresponding standards.

We will look at the following balance sheet, which we have introduced before:

Balance sheet	Book **A**	Book **L**	Market **A**	Market **L**	
Cash	6200	47100	6200	48513	MR
Bonds	35700	**2200**	37842	**3569**	SHE
Shares	4400		4800		
Properties	1100		1300		
Loans	1400		1400		
Alternatives	500		540		
Total	49300	49300	52082	52082	

3.1 Statutory Accounting

Statutory accounting is, in most cases, one of the most prudent forms of accounting and the focus of the standards are smooth and continuous profits, as long as there are no market disruptions. Implicitly one assumes that the assets are held for a long period and bonds in particular are held until maturity. In consequence, bonds are valued according to the *amortised cost method* and shares are accounted for at the lower of book value and market value.

There are different ways how this "lower book or market" principle can be applied such as:

- Whether the book value needs to be written down and hence the asset stays afterwards at this lower price in the books or not,
- Whether there is only a need for a write-down if the decrease in asset value is permanent,
- Whether one can use other hidden values to offset the negative movement,
- Whether the impairment needs to take place at once or can be dispersed over some years.

One particular consequence of this set of accounting rules is the fact that there are normally unrealised capital gains and losses, which in this world only materialise if the asset is sold or the asset defaults.

3.2 IFRS and US GAAP Accounting

Since the different statutory accounting rules can vary considerably, US GAAP and IFRS accounting standards have been introduced with the aim to make balance sheets and income statements more comparable. As a consequence, the corresponding accounting standards, together with the guidance notes are very large, since the aim is to cover all possibilities. For a beginner it is not always easy to understand what is happening and why.

One can, in principle, try with each accounting standard to optimise the usability of both balance sheet and income statement. In the first case the focus is a most accurate representation within the balance sheet. In the second case the focus is in having profit and loss accounts where one tries to get a profit and loss statement, which allows best to judge the quality of the earnings of the company. This approach is also known as a deferral and matching approach. Both of the above mentioned accounting standards follow this philosophy. When looking at the DAC the deferral and matching approach will become more evident.

On the asset side there are normally different possible choices depending on the intention of the company. Whereas this might help the individual company to show their performance in the way they believe it is most suitable, these choices are also one of the root causes for the opacity of these standards (the other being the high intrinsic complexity).

These choices work as follows: Each asset is classified into a category and it is accounted for accordingly. In order to avoid accounting arbitrage there are limitations in respect to a change in the accounting category. For bonds the possible categories are shown in the table below:

	B/S Treatment	P/L Treatment
Hold to Maturity	Amortised Cost	Amortised Cost
Available for Sales	Market Value	Amortised Cost
Trading	Market Value	Market Value

This means that bonds are treated completely differently depending on the classification. For shares there also exists two different classifications, as for the ones above with the difference that obviously "Hold to Maturity" does not make a lot of sense. In the case of shares an impairment provision needs to be taken if there is a permanent impairment. Impairment of an asset is given if its carrying amount exceeds its recoverable amount. Since IFRS is very restrictive one usually finds accurate definitions of such terms, as:

The recoverable amounts for the following types of intangible assets should be measured annually whether or not there is any indication that it may be impaired. In some cases, the most recent detailed calculation of recoverable amounts made in a preceding period may be used in the impairment test for that asset in the current period:

- An intangible asset with an indefinite useful life.
- An intangible asset not yet available for use.
- Goodwill acquired in a business combination.

Depending on the valuation used for assets there may occur (gross) unrealised capital gains which step are broken down in different parts, such as latent taxes, etc. So summarising the situation can be quite difficult.

One particular asset is the so called DAC asset, which is a direct consequence of the deferral and matching principle. In order to understand this, one needs to understand how an insurance policy is sold. If an insurance policy is sold with a regular premium of 6000, one could, for example, expect that commissions of 10000 are paid to the distributor. This would result in an accounting loss even if no mathematical reserve would have to be set up. Hence the more the company sells, the worse its profit. Since it is expected that this initial loss is compensated in later times, there are several attempts to present an accounting standard which takes care of this. One of the possibilities is the embedded value method which we will describe in the following section. The other idea is to assume that the paid commission can be financed and amortised with the future gains. In consequence one creates in a first step an "intangible" asset called DAC ("Deferred Acquisition Costs") and one amortises it over time. Also here there is quite some discretionary, in respect to the following:

- How much of the acquisition costs are deferred?
- At the first application of the standards, which portfolios are considered going back and how?
- Which is the amortisation pattern which is used?[1]

After having looked at the assets we want next to look at the liabilities. There are, most importantly, the mathematical reserves, which are accounted at book values. This incongruence between assets and liabilities leads to an artificial balance sheet volatility. Furthermore it is known that the statutory reserves are based on a prudent approach resulting in hidden reserves as a consequence of this conservatism. Another important liability are latent taxes and deferred policyholder participations. The first effect is a consequence of having unrealised capital gains on the assets which are accounted according to market values. Once one sells them the unrealised capital gains would represent true gains before taxes. Since these gains are then taxed a corresponding liability is set up. The same is true for the policyholder participation. For a country such as Germany where 90 % of the gross profits have to be given back to the policyholder, it is clear that out of 1000 unrealised capital gains,

[1] Under US GAAP there are three different ways to amortise DAC, proportional to premium (FAS 60) and proportional to expected profits (FAS 97 and 120).

the shareholder can, in normal circumstances, only expect 10% or equally 100 (before tax). As a consequence, the remaining 900 are deferred policy holder bonuses. Hence based on 1000 of unrealised capital gains we would have the following reparticipation assuming 34 % of tax:

Deferred Policyholder Bonus	900
Deferred Tax	34
Part of Shareholder Equity	66
Unrealised Capital Gains	1000

Please note that in reality the situation is still more complex since one would expect that a part of these gains are used for the accelerated amortisation of the DAC asset and other similar effects.

3.3 Embedded Value and Economic Accounting

As we have seen before there is an intrinsic problem in respect to statutory accounting in the sense that companies writing profitable new business tend to show bad statutory returns as a consequence of the so called new business strain, the effect that the commissions paid are higher than the premium received. Obviously, statutory accounting misses some part in the value, namely the future gains. The aim of the embedded value accounting is to include future gains in the balance sheet. This is done by recognising the so called *present value of future profits (PVFP)*. Also embedded value accounting has changed over time since at the beginning it was based on statutory profits emerging over time. One realised that the original embedded value methodology was not clear enough and introduced the european embedded value (EEV). After recognising this later one is not *risk neutral* and that higher equity backing ratios (eg investments in equities) always leads to a higher value, one reconsidered this position with the introduction of a *market consistent embedded value (MCEV)*. Obviously, this topic could fill a whole book and hence the introduction must remain superficial. Depending on the parameters and interpretations chosen, a market consistent embedded value can be considered as an economic measure.

3.3.1 Economic Valuation / Market Consistent Embedded Value

In contrast to the *traditional embedded value*, the economic valuation or the market consistent embedded value is based on modern valuation techniques such as arbitrage free pricing etc. The idea here is to base discounting on a risk free rate. Risk is considered by setting appropriate capital for the different points in time and by setting an adequate cost of capital. For a more rigid approach than the one presented in this section we refer to appendix C. There we will show how to calculate different replicating portfolios and will also present an abstract approach to ALM.

The most important additional insights which will be provided by Solvency II are *economic balance sheets*, in particular with respect to insurance liabilities. This means on the asset side that all unrealised capital gains and losses are taken into account in a transparent way. On the liability side the situation is somewhat different because there are no tradeable instruments which can be used to perfectly replicate the liabilities in order to determine their economic price. It is clear however, that this information is of the utmost importance for managing the risks and therefore usually a model approach is used to get a reasonable approximation of the market values for the insurance liabilities. First, one needs to calculate the expected present value of the future policyholder benefits, as seen before. On top of this amount one requires a so called *market value margin* (MVM). In order to calculate the expected present value of the future policyholder benefits, one needs to calculate the corresponding cash flows. If we assume that the expected cash flows are independent on asset returns, such as for guaranteed benefits (eg. annuities in payment), the expected present value of the cash flows $(CF_t)_{t\in\{0,1,2,\dots\}}$ can be calculated by

$$\mathbb{E}[PV] = \sum_{k=0}^{\infty} \pi_t(\mathcal{Z}_{(k)}) \times \mathbb{E}[CF_k],$$

where $\pi_t(\mathcal{Z}_{(k)})$ denotes the market price of a zero coupon bond with maturity k at balance sheet date t, keeping in mind that we have assumed stochastic independence of the expected cash flows from the financial variables. If this independence is not the case, such as in products with discretionary bonus benefits or with GMDB products in relation to unit linked policies we need to apply a more general definition of replicating portfolios, such as the one presented in appendix C. Whereas this calculation is quite straight forward for P&C insurance, it requires some additional considerations for life portfolios, where it is typically based on a policy-by-policy calculation. In contrast to usual actuarial practise where mathematical reserves are based on the assumption that there are no lapses, it is key within a realistic valuation to also consider this effect. In doing so, the duration of liabilities usually reduces considerably. This clearly shows the importance to consider this effect.

On top on the expected present value, it is necessary to calculate the market value margin. In order to understand this we need to acknowledge that the $\mathbb{E}[PV]$ does not take into account that, even for guaranteed liability cash flows with no link to the capital markets, the cash flows can fluctuate over time for example as a consequence of a pandemic. As a consequence the insurance company need to carry a certain amount of risk capital to absorb such shocks and it will not be willing to assume the insurance liabilities just for $\mathbb{E}[PV]$. This can be shown mathematically (in appendix C) using the utility assumption of the investor and the Jensen-inequality. The cost of capital approach, which is also more formally introduced in appendix C aims to provide for a proxy of the market value of insurance liabilities.

Another motivation to use the cost of capital approach is based on the assumption that the originating life insurance company becomes insolvent and has to be wound up. In this case the market value margin, calculated based on the cost of capital approach incentives another company to assume the corresponding insurance liabilities, since it can produce a higher yield on the capital invested on the run-off portfolio assumed. Hence the cost of capital approach provides a mechanism to ensure that insurance portfolios are assumed once a life insurance company becomes insolvent (but not bankrupt).

The *cost of capital approach (CoC)* requires, that the risk capital (RC_t) is projected into the future. In a second step the CoC equals the present value of the corresponding costs for the future periods:

$$MVM = CoC = \sum_{k=0}^{\infty} \beta \times RC_k \times \pi_t(\mathcal{Z}_{(k)}).$$

The parameter β corresponds to the unit cost of capital and is usually in the order between 2% and 6%, for regulatory purposes. In case of a given hurdle rate γ (eg $\gamma = 7\%$), β can be calculated by the formula $\beta = \gamma -$ riskfree for the corresponding period, neglecting for the moment the effect of taxation. It is important to note that the CoC approach has two additional benefits: it can quite easily verify the corresponding results and it avoids double counting of capital.

Similarly one can calculate the internal rate of return by this approach. Assume that $\beta = \gamma -$ riskfree is constant (for example by introducing a constant spread over risk free), then the calculation becomes still easier:

$$IRR = \frac{\mathbb{E}[PV]}{\sum_{k=0}^{\infty} RC_k \times \pi_t(\mathcal{Z}_{(k)})} + \text{riskfree}.$$

In case of a yield curve which is not flat, the IRR (eg γ) can be calculated by the following formula:

$$IRR = \frac{\mathbb{E}[PV] + \sum_{k=0}^{\infty} i_k \times RC_k \times \pi_t(\mathcal{Z}_{(k)})}{\sum_{k=0}^{\infty} RC_k \times \pi_t(\mathcal{Z}_{(k)})},$$

where i corresponds to the corresponding forward return on the capital, eg $i_k = \frac{\pi(\mathcal{Z}_{(k+1)})}{\pi(\mathcal{Z}_{(k)})} - 1$. It is obvious that in case of a flat yield curve the two formulae are equal.

Having stated the importance of basing the solvency regime on a reliable economic balance sheet, there is another important question relating to the market consistent valuation of liabilities. What is the value of the different policyholder options such as the possibility to surrender a policy or to take the capital or annuity in a pension scheme? It is clear that these implicit options can have a considerable value, but there are few reliable methods to value them which are generally accepted. Therefore, a pragmatic approach has to be taken. This means that only the most relevant policyholder options should be quantified. The most prominent example is the guaranteed unit linked insurance contract. Here the valuation of the corresponding put option on the fund is relatively easy to quantify based for example on the Black-Scholes formula and the corresponding risk management techniques (see also section C and example 57).

Finally we need to realise that in the real world there are additional constraints, which have an impact on the value of a portfolio or a product sold. The most relevant are listed below:

- Frictional costs and
- Taxes,

Frictional costs stem from the fact that the company needs to hold at a certain time the corresponding statutory reserves V_t for an underlying block of business. Given the fact that the *best estimate liabilities* $\mathbb{E}[PV]$ may be inferior, the company needs to hold this additional amount, resulting in the above mentioned (pure) frictional capital costs:

$$FCC^* = \sum_{k=0}^{\infty} \beta \{\max(0, V_k - \mathbb{E}[PV]_k)\} \times \pi_t(\mathcal{Z}_{(k)}),$$

where $\mathbb{E}[PV]_k$ denotes the expected present value of liabilities as seen at time k. Based on the fact that the risk capital also qualifies as capital to fill up missing reserves, the total frictional capital costs amount to:

$$FCC = \max(0, FCC^* - CoC).$$

With respect to taxes, all values need to be considered after tax. Whether a certain tax applies and to what extent depends heavily on the country. In the simplest setting, pre-tax values can simply be multiplied by $(1 - \text{taxrate})$.

3.3.2 Valuation Methodology Revisited

Risk management is based on a market consistent valuation of the insurance liabilities. The following effects have to be considered:

1. Expected present value of the cash flows using a risk free interest rate.
2. The market value margin (MVM) which compensates the buyer of the portfolio for the risk he assumes.
3. The present value of the frictional capital costs, which are essentially a consequence of higher external capital requirements than those on a pure economic basis. Typical examples are higher statutory reserves or solvency requirements.
4. Other cash flow streams which need to be valued, such as cash flow swaps induced by funds withheld, etc.

The following figure illustrates the different parts:

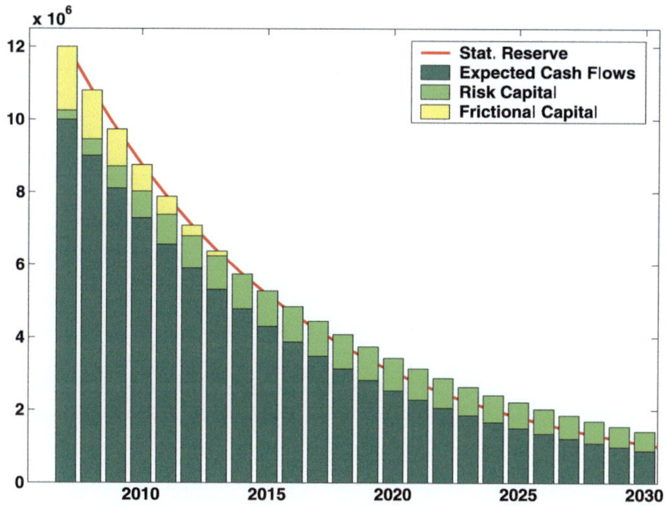

Best Estimate Liability (BEL)

Firstly, one needs to calculate the expected present value of the future policyholder benefits. On top of this amount one requires a so called market value margin MVM. In order to calculate the expected present value of the future policyholder benefits, one needs to calculate the corresponding cash flows. The expected present value of the cash flows $(CF_k)_{k\in\{0,1,2,\dots\}}$ is then calculated by

$$\mathbb{E}[PV] = \sum_{k=0}^{\infty} \pi_t(\mathcal{Z}_{(k)}) \times \mathbb{E}[CF_k],$$

where $\pi_t(\mathcal{Z}_{(k)})$ denotes the market price of a zero coupon bond with maturity k at balance sheet date t, assuming again that the cash flows are stochastically independent on the financial market variables. It has to be stressed that the consideration of lapses is key within a realistic valuation. Note that the calculation of cash flows including lapses can be done in a similar manner as in section 2.6. Moreover one can use the Markov chain life insurance model (see appendix B) by enlarging the state space, etc. by the state "lapse".

Market Value Margin/Cost of Capital

On top on the expected present value, it is necessary to calculate the market value margin. In a second step the CoC equals the present value of the corresponding costs for the future periods:

$$\begin{aligned} MVM = CoC &= \sum_{k=0}^{\infty} \gamma \times RC_k \times \pi_t(\mathcal{Z}_{(k)}) \\ &\quad - \sum_{k=0}^{\infty} \left(\frac{\pi_t(\mathcal{Z}_{(k-1)})}{\pi_t(\mathcal{Z}_{(k)})} - 1 \right) \times RC_k \times \pi_t(\mathcal{Z}_{(k)}) \\ &\quad \times (1 - \text{ Tax Rate}) \\ &= \sum_{k=0}^{\infty} \beta \times RC_k \times \pi_t(\mathcal{Z}_{(k)}). \end{aligned}$$

The parameter γ corresponds to the unit cost of capital and is usually in the order between 2% and 6%, for regulatory purposes. In case of a given hurdle rate γ (eg $\gamma = 13\%$), β can be calculated by the formula $\beta = \gamma -$ riskfree for the corresponding period, neglecting for the moment the effect of taxation.

Positions of a Fair Value Valuation

Position	Amount in USD	Relative Amount
Reserves in B/S	753400	19.78%
Present Value Premium	3055000	80.22%
Present Value Claims	-2945000	-77.33%
PV Exp - Internal	-50400	-1.32%
PV Exp - Overhead	-85730	-2.25%
PV Exp - Commissions	-332400	-8.72%
Subtotal	394800	10.36%
Market Value Margin	-97070	-2.54%
FCC	-33560	-0.88%
Funds Withheld	6281	0.16%
Tax	-127200	-3.33%
Total	**143200**	**3.76%**
PV Profit	284700	
PV Capital	2044000	
RoRAC		**13.93** %

3.4 Formulae

$$
\begin{aligned}
\mathbb{E}[PV] &= \sum_{k=0}^{\infty} \pi_t(\mathcal{Z}_{(k)}) \times \mathbb{E}[CF_k], \\
CoC &= \sum_{k=0}^{\infty} \beta \times RC_k \times \pi_t(\mathcal{Z}_{(k)}), \\
FCC^* &= \sum_{k=0}^{\infty} \beta \left\{ \max(0, V_k - \mathbb{E}[PV]_k) \right\} \times \pi_t(\mathcal{Z}_{(k)}), \\
FCC &= \max(0, FCC^* - CoC), \\
\text{Profit before Tax} &= \mathbb{E}[PV] - CoC - FCC, \\
\text{Profit after Tax} &= (1 - \text{taxrate}) \times \left\{ \mathbb{E}[PV] - CoC - FCC \right\}, \\
\beta &= \gamma - \text{riskfree for the corresponding period.}
\end{aligned}
$$

Note again that we have assumed here that the insurance cash flows are independent on the capital market variables, which is the case for guaranteed benefits, but *not* for discretionary policyholder participation and unit linked policies with guarantees. For a more rigid approach we refer to appendix C.

3.5 Examples

3.5.1 Annuity

We consider a real life annuity portfolio with a total face amount of about EUR 240 M p.a. In order to determine the expected cash flow and the replicating portfolio it is necessary to choose a mortality law for the description of the evolution of the mortality:

$$q_{x,t} = q_{x,t_0} \times \exp(-\lambda_x \times (t - t_0)).$$

Considering a x year old person, the present value of the annuity in payment of 1 EUR is given by

$$\ddot{a}_x = \sum_k {}_kp_x \times v^k.$$

This indicates that the expected cash-flow at time t equals ${}_tp_x$ and therefore the replicating portfolio corresponds to $\sum_t {}_tp_x \times \mathcal{Z}_{(t)}$, where $\mathcal{Z}_{(t)}$ represents an abstract basis for the corresponding zero coupon bonds. This policy has the following value at balance sheet date:

$$\mathbb{E}[PV] = \sum_k {}_kp_x \times \pi_t(\mathcal{Z}_{(k)}),$$

where $\pi_t(X)$ denotes the market price of the financial instrument X at time t.

It is now necessary to calculate the market value margin. In order to do so, one needs to determine the relevant risk factors together with their probability functions. In case of the annuity portfolio we assume longevity as main risk factor and assume that the mortality for future years follows the following law:

$$q_{x,t} = q_{x,t_0} \times \exp(-\lambda_x(\omega) \times (t - t_0)).$$

In this case we model the risk by replacing λ_x by $\lambda_x(\omega) = c(\omega) \times \lambda_x$. $c(\omega)$ corresponds to the relative change in mortality improvement in relation to the observed standard trend. In this case the present value of the loss equals the difference of the expected present values based on λ_x and $\lambda_x(\omega)$ respectively. By integrating over $d\omega$ one gets the desired result for the present value of the risk capital. Multiplying by the unit CoC results in the desired result.

The analysis is based on a real life annuity portfolio with reserves summing up to EUR 2.7 bn and annuities in payment of ca. EUR 240 M. We use a 99.5% shortfall as risk measure for the calculation of the CoC. At this point in time it is worth to mention the fact that

$$\sum_{k=0}^{\infty} \beta \times RC_k \times \pi_t(\mathcal{Z}_{(k)}) = \beta \times \sum_{k=0}^{\infty} RC_k \times \pi_t(\mathcal{Z}_{(k)}).$$

This means that it is possible to model the present value of the risk capital directly, which is done for this example. Furthermore the function:

$$\mathbb{N} \to \mathbb{R}, n \mapsto \frac{1}{\pi_t(\mathcal{Z}_{(n)})} \sum_{k=n}^{\infty} RC_k \times \pi_t(\mathcal{Z}_{(k)})$$

defines the required risk capital for the different periods. Figure 3.1 illustrates the replicating portfolio, on the one hand side with $c = 100\%$, and on the other with $c = 130\%$, showing the longer duration and hence the higher present value in the latter case.

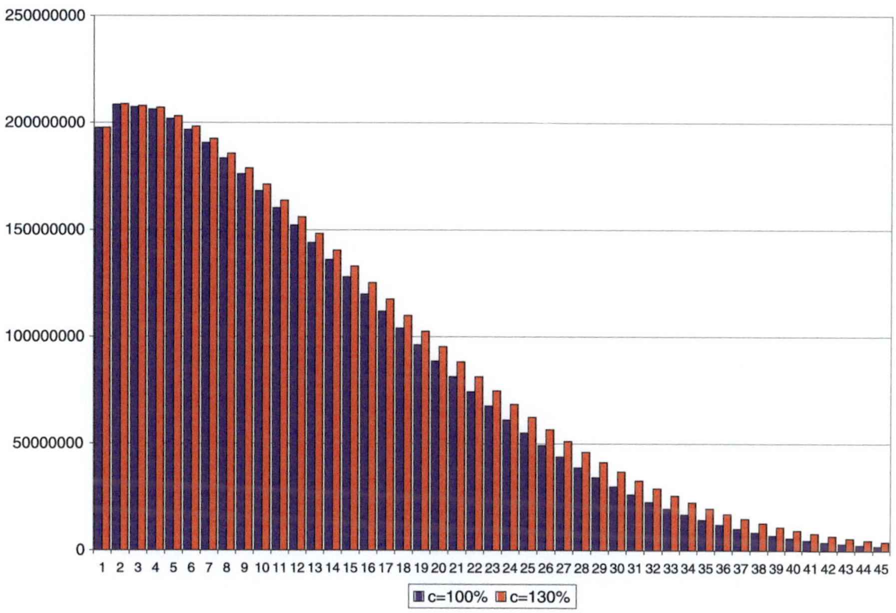

Fig. 3.1 Replicating Portfolio

By using the replicating portfolios as given in figure 3.1, the development of the required capital corresponds to figure 3.2, using a somewhat simplified version for the capital.

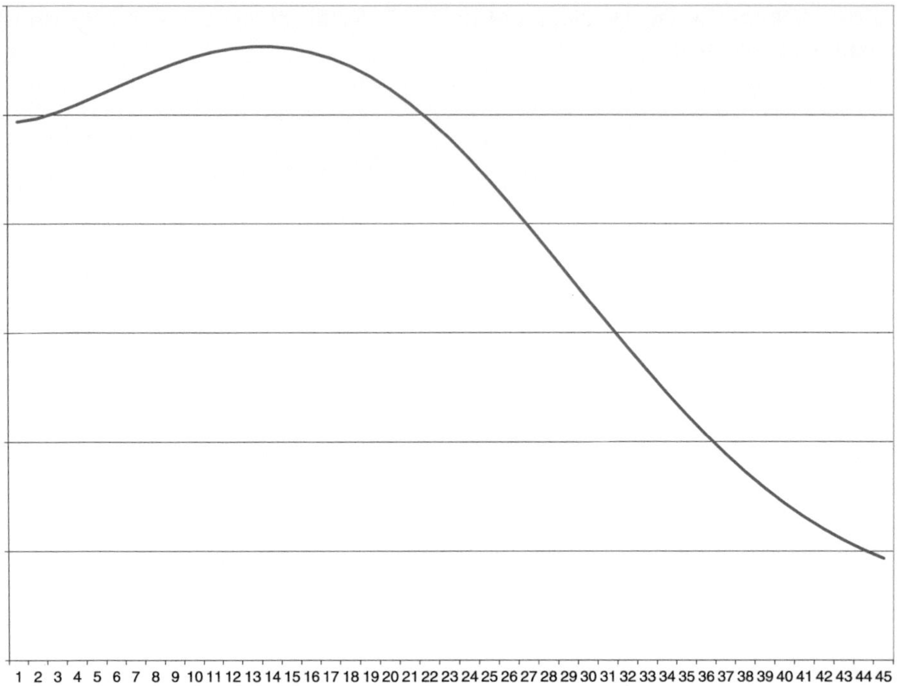

Fig. 3.2 Required capital over time

As a next step these calculations have to be done for the different $c(\omega)$, and weighted with the corresponding probabilities. The following table illustrates this:

$c(\omega)$	$P[c(\omega)]$	Loss($c(\omega)$) bn EUR	Contribution to TailVar bn EUR
1.0		0.000	
1.1		-0.032	
1.2		-0.065	
1.3		-0.099	
1.4		-0.133	
1.5		-0.169	
1.6		-0.205	
1.7		-0.242	
1.8	0.0005	-0.280	-0.028
1.9	0.0005	-0.319	-0.031
2.0	0.0004	-0.359	-0.028
2.5	0.0019	-0.572	-0.217
3.0	0.0017	-0.801	-0.272
Total	**0.0050**		**-0.578**

This indicates that the present value of the risk capital, calculated with the 99.5% TailVar, amounts to 0.578 bn EUR. For more details in respect of TailVar see chapter 4. Assuming that the statutory reserves or the price the company pays corresponds to 2.934 bn EUR we get the following:

		bn EUR	%
$+$	Statutory reserve	2.934	100.00%
$-$	$\mathbb{E}[PV]$	2.750	93.73%
$-$	CoC 13%	0.075	2.56%
$-$	Tax 25%	0.028	0.95%
$=$	**Profit Tax = 25%**	**0.080**	**2.72%**
IRR ca. 30%			

In this particular case the margins induced by the prudent mortality laws in the statutory reserves are partially offset by a low interest environment. It is however obvious that the statutory reserves carry about 4% of margin with respect to a market consistent valuation.

3.5.2 Capital Protection

Whereas we have considered in the first example an annuity portfolio, we now want to look at a life protection portfolio consisting of 100 $x = 30$ year old persons with a term of 30 years. The death benefit amounts (per policy) to 100000 EUR with a $\mathbb{E}[PV] = 14507$ EUR. In order to calculate the risk capital, we assume that the exogenous risk factors are types of pandemics, as follows:

$\theta =$ Relative q_x-level	Return period for θ	$F_{q_x\text{-level}}(\theta)$
1.0	0	0.000
1.1	10	0.975
1.2	20	0.976
1.3	30	0.978
1.4	40	0.980
1.5	50	0.981
2.0	100	0.990
2.5	175	0.994
3.0	250	0.996
4.0	500	0.998
10.0	1100	0.999
$20 + \epsilon$	∞	1.000

Based on this approach it is now possible to do a simulation by replacing the original q_x by a new random $q_x(\omega)$ given by $F_{q_x\text{-level}}(\theta)$. The following table summarises the main results of this simulation, where SaR stands for "Sum at Risk". The sum at risk is the amount of money the insurer loses for a certain policy in case the insured person dies. Hence it equals the sum insured minus the corresponding mathematical reserves hold in the balance sheet for this policy.

		Relative	in% SaR
Expected Value at level 100%	1407500		
Expected Addl Loss	43100	3.1%	0.43%
Sdtdev	97700	6.9%	0.97%
2.5 σ	244400	17.4%	2.44%
$F^{-1}(99\%)$	568800	40.4%	5.68%
$F^{-1}(99.6\%)$	720500	51.2%	7.20%
TVar(99%)	697100	49.5%	6.97%

This indicates that given a hurdle rate of 13% and using the VaR with respect to a return period of 250 years, the required single premium for this contract can be calculated as follows, neglecting the impact of taxes:

Item		Amount
$\mathbb{E}[PV]$		1450700
CoC @ 13%	$720,500 \times 13\%$	93600
Total		1544300

The following figures illustrate the effect of the 1918' influenza pandemic ('Spanish Flu').

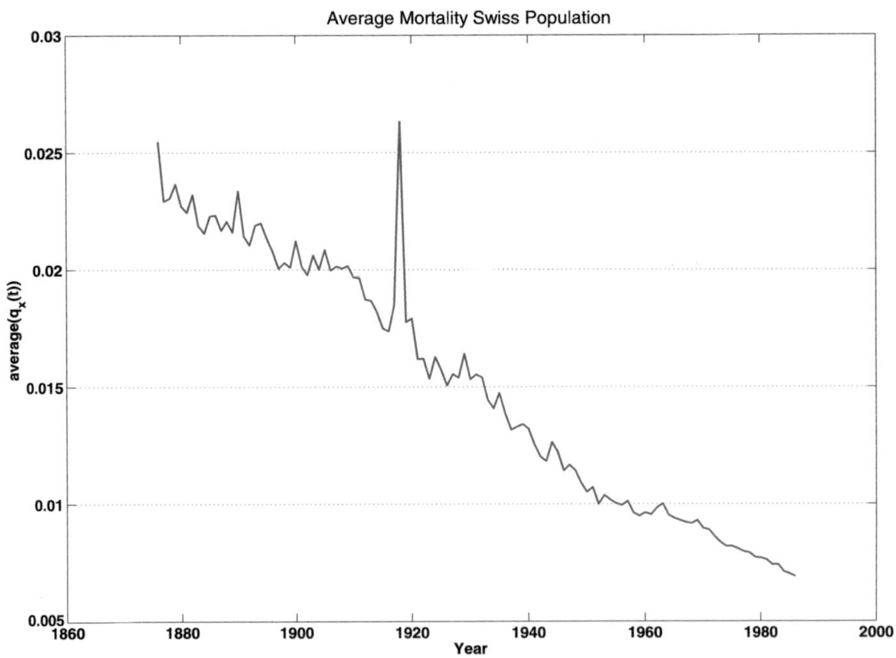

Fig. 3.3 Development of Mortality over Time

Figure 3.3 shows the change in average mortality over the years. It becomes obvious the average mortality for the year 1918, when the Spanish flu occurred equals roughly the one of 1860 and is much higher than the average mortality of neighbouring years. Figure 3.4 compares the mortality for the years 1908, 1918 and 1928. Interestingly the pandemic results in a much higher mortality for young people aged between 15 and 40. The older ages are relatively less affected.

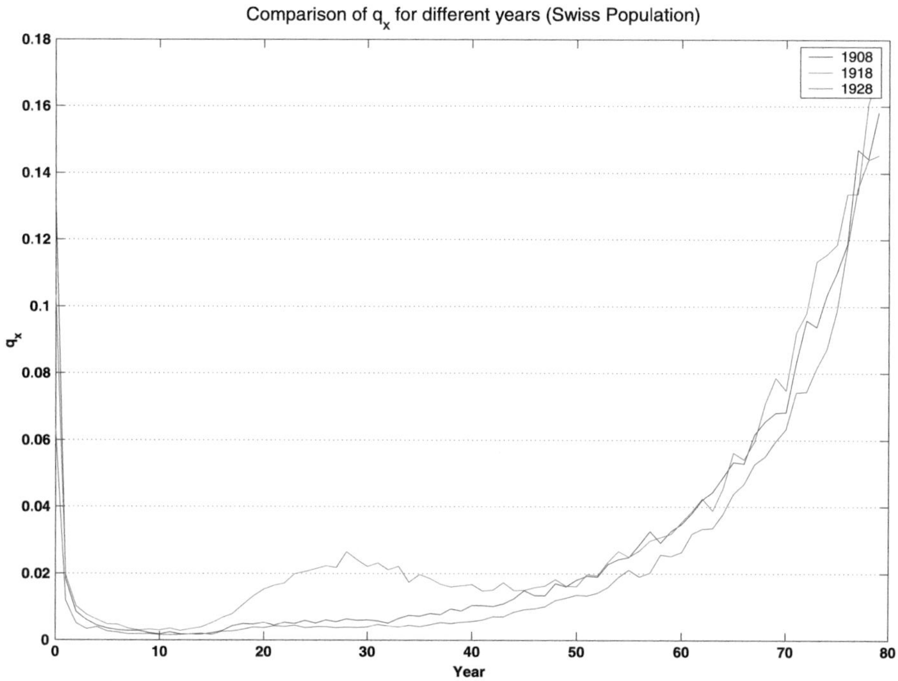

Fig. 3.4 Comparison of different years

Finally figure 3.5 shows the mortality per age-band and year. Again we see that the mortality for 30 year old people is considerably higher and equals the one of the 60 year old people. We observe at the same time that the main mortality improvement over this time span relates to the younger people and not to the very old.

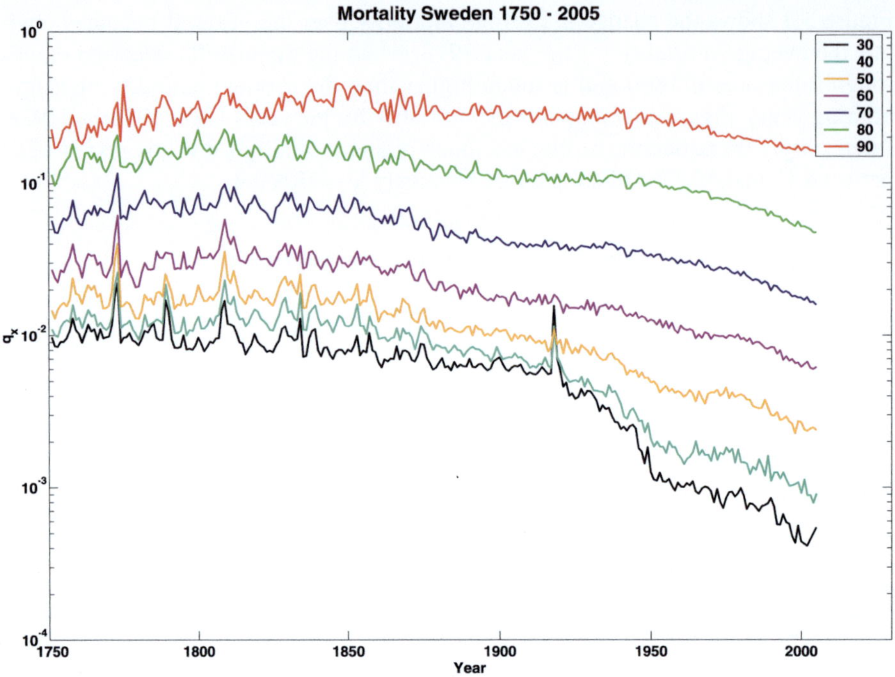

Fig. 3.5 Mortality by Ages for Sweden 1751 - 2005

Chapter 4
Risk Appetite and Tolerance

4.1 Risk Capacity and Risk Appetite

Risk appetite is a term that is frequently used throughout the risk management community. Recent changes in global regulations that encompass security and risk and control implications, have raised the awareness around the concept of risk appetite, particularly among the management team.

Also, the financial crisis of autumn 2008 has shown the critical importance of considering the possible risks which may be faced by financial service companies. This crisis involved several cases in which financial services companies, including insurers, suffered highly damaging losses from risks which they had not fully or correctly evaluated, or perhaps even not been aware of. Many of the failings of companies can

M. Koller, *Life Insurance Risk Management Essentials*, EAA Series,
DOI 10.1007/978-3-642-20721-1_4, © Springer-Verlag Berlin Heidelberg 2011

be attributed to the acceptance of excessive risks and the poor management of those risks or a lack of clarity around the level of compensation expected for risks taken. Also, in many cases, the link between risk and strategic planning or business decision making has been insufficient.

Risk capacity is defined by the available risk capital. From an economic viewpoint, this is defined as the adjusted difference between the market value of investments and the market value of liabilities (insurance liabilities and financial commitments). The risk appetite shall be commensurate with whichever risk the decision-makers (*risk owners*) are willing to assume. The risk appetite is measured in terms of the economic capital needed to cover a given risk exposure over a specified period of time and which must therefore be held in reserve. This capital must be sufficient with a high level of probability. Risk appetite must never exceed risk capacity.

Risk appetite, at the organisational level could be, in general terms, the amount of risk exposure or potential adverse impact from an event the organisation is willing to accept/retain.

With an increasing importance due to the latest events, risk appetite, tolerances, risk targets and limits are a critical element of prudent business management and an effective risk governance process and need to be more than a statement, but something you live every day by how decisions are made and companies are managed. The purpose of "defining risk appetite", whatever that may mean, is to control directly, or at least influence directly, how people make decisions on behalf of an organisation in the face of risk and uncertainty by specifying the importance of risk in some way. In establishing risk appetite, the picture of the whole strategy the risk management should follow is Fig. 4.1. In the concrete context it shows also the readiness of the organisation in implementing the respective steps, eg whether the step is implemented 25%, 50%, ... 100%.

Each one of the steps should be specified in the procedure at the organisational level, and needs to be analysed with a specific frequency to ensure the company is operating within the expectations of key stakeholders. Each step could broadly be defined as follows:

Risk Capacity Risk capacity could be defined as the maximum amount and type of risk a company is able to accept/retain in pursuit of its mission, vision, business objectives and value goals. It is directly related to an entity's capital and external stakeholder influences.

Risk Appetite Risk appetite, at the organisational level, is the amount of risk exposure, or potential adverse impact from an event, that the organisation is willing to accept/retain in pursuit of its mission, vision, business objectives and value goals.

The procedure for defining the risk appetite is called "limits system", which analyses the total exposure to the different risks and identifies "exposure limits". This means that for each risk it is possible to define the maximum threshold or expo-

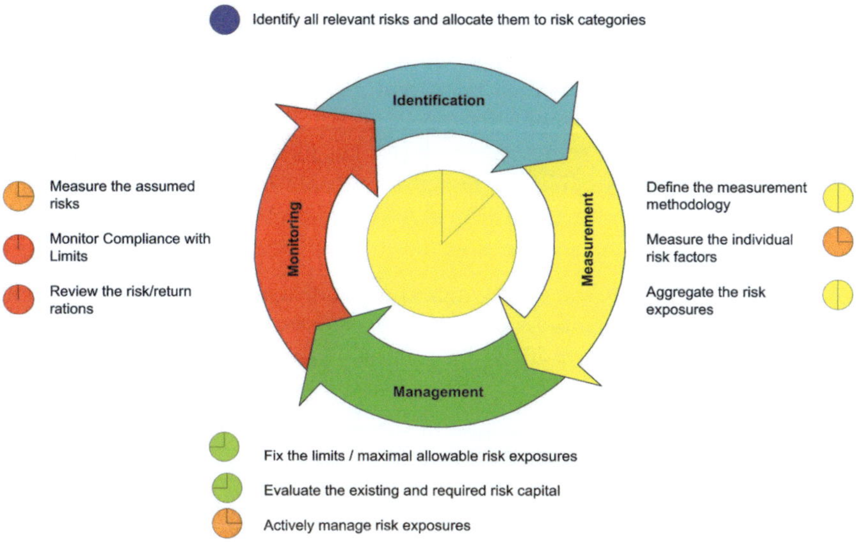

Fig. 4.1 Risk Identification Process

sure that the company is willing to accept and in a level above this limit the company needs to act to decrease the exposure.

The risk appetite, as we can see, is directly related to an entity's risk capacity as well as its culture, desired level of risk, capability and business strategy.

Entities often consider risk appetite both qualitatively and quantitatively. It is often expressed in acceptable/unacceptable outcomes focused in the downside risk, such as:

- Rating target of AA.
- No activity that will impair the ability to continue as a going concern.
- Defined probability of ruin at a specific confidence level (as 90%).

Typically, the level of a risk will be measured by the likelihood of it occurring and the financial impact if it does. We can:

- Capture expert's opinion of loss severities and frequencies.
- Calculating statistics for individual loss scenarios and the total losses an organisation could sustain as a result.

Most documents trying to define "risk appetite" say it is the amount of risk that "the organisation" is prepared to put up with. The idea is that it applies to the whole organisation. However, almost all practical applications of "risk appetite" involve multiple "appetites", though these may be intended to reflect an overall "appetite". "Risk appetites" in various forms are set out for the organisation as a whole, for sub-units of the organisation, for activities within an organisation, for types of risk, for individual risks, and even for liability products or asset portfolios. Sometimes a single "risk appetite" is laid down that is intended to be applied to all risks and sometimes each risk has its own individual appetite. All these are legitimate possibilities, though some only make sense for some kinds of decisions and there are some difficult issues to deal with in setting lower level appetites to reflect a higher level appetite.

Risk Tolerance Risk tolerance is typically a specific maximum applicable to each risk regarding the magnitude and the type of risks the organisation is willing to take in order to achieve its business strategy and objectives while operating within the broad risk appetite.

It should be set that the aggregation of risk tolerances ensures the organisation operates within the risk appetite.

Risk Tolerance can be expressed in terms of:

- Risk measures with a number (or category) that represents the quantity of risk, as the mathematically expected value of "impact", the total exposure, or VAR.
- Other variables that may be used are results that are not related to risk events. They may be solvency statements, results like profit or earnings, nominal measures as the amount of premiums written, or categories like high/medium/low applied to "impact" or to outcome levels.

Many constraints in "risk appetite statements" being written today do not mention risk at all. Instead, they are often rules on proxies for risk such as external conditions, activity levels, and results achieved. The reason for the use of this type of statement is that they are easy to analyse and reliable. For example, they may say that no more than 10% of investments will be in a certain currency, or that new offices will not be opened in countries with inflation above 4%, or that actual losses from operational incidents will not be more than 5% of revenues in any one month. All variables used should be carefully defined and their values, when used in decision making, should be made explicit, usually by being written down so that personal bias is harder to conceal. Historically, organisations typically consider risk appetite based on effects across the four general dimensions shown below by priority of use:

1. Statutory capital or level of surplus.
2. Credit ratings threshold.
3. Economic capital.
4. GAAP earnings.

Risk Target or Range The risk target is the optimal level of risk that the organisation desires to take to achieve its business strategy and objectives and to operate within its appetite/tolerance for risk. It is generally articulated as a range in the same units that risk tolerance is measured in.

The setting of risk targets should be based on the management's desired returns, the role of risk to achieve those returns (risk/return profile) and management's capability to manage each risk.

In setting specific risk targets, management aligns risk targets to ensure that it will meet both its strategic goals as well as operating within its risk appetite and tolerance.

Risk Limits The risk appetite is the amount of risk that it is willing to accept in pursuit of value. Risk appetite therefore reflects the desire to optimally exploit opportunities and minimise hazards to an acceptable level.

The actual implementation of the risk strategy is achieved through the fixing of permissible risk limits for the company as a whole. The corporate limits will be spread across the primary risk categories - separately for the markets, products, channels and functions. The risk committee delegates the limits to the next level of responsibility, documents the delegated limits and monitors compliance therewith. By doing this the risk committee ensures that it operates within the limits which have been granted by the group. In case of insufficient limits, the risk committee requests a higher capacity from group, if there is an appealing business case.

The risk committee shall be informed on a quarterly basis of the utilisation of limits by the company as a whole.

The risk limit is a threshold to control activities to ensure that variations from expected outcomes will be consistent with the risk target, but will not exceed the risk appetite/tolerance.

Limits are how the appetite/tolerance and the risk target are translated into practical constraints on business activity. For example, as mentioned in the risk appetite section above, each company has its own "exposure limits" for each of the risks, through the limits system.

Risk limits should be formulated so that on a probabilistic basis, considering utilisation, aggregation and correlation, they ensure that the organisation operates within its target range and does not exceed its aggregate risk tolerance.

The threshold can be set:

- By business unit.
- By individual risk exposure.
- Allowing for diversification.
- With controls and processes to maintain risk within risk appetite.

Once the risk appetite threshold has been breached, risk management treatments and business controls are implemented to bring the exposure level back to within the accepted range. However, we need to be aware that a threshold is a crude, all-or-nothing approach. No value for it is right in every situation and natural decision making can be disrupted by having to work with the threshold. For example, if an action results in a risk measure value that is above a limit then that action cannot be chosen. If the action results in a risk measure value that is below the limit then that action can be chosen. In other words, below the limit risk doesn't have any importance, but above it risk is of overwhelming importance and rules out the action. Although limits put a ceiling on risk taking they do not ensure that risks are properly weighed against rewards. If risk limits were the only way that risk was weighed in decision making then there would be no difference in risk terms between an action that was just within the limit and another that was well within the limit. To promote good risk reward decisions some more progressive weighting of risk can be used, or at least management need to be aware of the level in which each risk operates and analyse each situation carefully before making a decision.

Additional potential problems in defining thresholds could be:

- Setting levels is very difficult and they can often seem rather arbitrary, leading to problems getting people to take them seriously.
- The decisions to which the rules apply may be unclear.
- Breaking an overall "risk appetite" into smaller parts is difficult. If each individual element is constrained so that, in total, the overall appetite is not exceeded (in some sense) then the freedom to act given to individual elements has to be cut down. Many strategies that would be possible if risk was constrained only at the top level are blocked by having multiple lower level constraints. In effect, the breakdown into subsidiary risks and into activity levels can create an ever tightening straight jacket.

And also notice that risk appetite, tolerance and limits are not static. They must be updated with changes in strategy, the environment and market expectations. Ultimately, they should be a key element in driving risk taking and in turn in performance measurement.

4.2 Limit Systems

Limit systems are a way to express risk appetite. Normally risk appetite statements are given on a more global level and need to be broken down in a second step to actual limit systems which are more granular. As with risk appetite there are several possibilities on how to design a limit system and hence the example below is for illustration purpose only.

In this first section an overview is given in respect to the relative size of the underlying insurance company. Hence these are no limits but serve for comparing the corresponding risks.

In M EUR	Metric	BU
Local Cur		EUR
Total B/S	IFRS	53000
SH Equity	IFRS	1700

1 Risk Capital Limits

In M EUR	Metric	BU
Risk Capital Limit		
Market (ALM) Risk Cap	ICA	800
Credit Risk Capital	Group Method	200

In the above table the risk capital limit refers to the required capital which the company is willing to put at risk, as defined in chapter 9. The ALM risk capital and the credit risk capital refer to the required capital for the corresponding financial risk (see chapter 6).

2 Market Risk Limits

In M EUR	Metric	BU
Exposure Limits FX in%		
Local Currency	MV	100%
CHF	MV	15%
EUR	MV	100%
GBP	MV	20%
USD	MV	15%
Other FX	MV	5%
Total Equities	MV	500
Max. Single Stock Position	MV	50
Maximum interest rate sensitivity per 10 bps	Against guaranteed CF, scaled	300

All of the above quantities aim to steer the financial risk taking. One tries for example to limit the FX risk or also the maximal amount which can be invested in a single counter-party.

3. Credit Risk Limits

In M EUR	Metric	BU
Exposure Limits Credit		
Local Government	Nominal Value	unlimited
AAA Rating	Nominal Value	400
AA Rating	Nominal Value	200
A Rating	Nominal Value	100
BBB Rating	Nominal Value	50
Below BBB and NR	Nominal Value	25

4. Insurance Limits

In M EUR	Metric	BU
PV of Premium per Contract	EUR	15
Annuities	PV Annuities	5
Mortality	Sum at Risk	2
Disability	10 x annuity or lump sum	2
Stop Loss	Max Loss	15

The above table aims to limit some risks in relation to life insurance as outlined in chapter 7.

5. Other-Operational Limits

In M EUR	Metric	BU
Operational Limits	in M EUR	5
Capital Expenditure	in M EUR	2
Revenue Expenditure	in M EUR	2
Claims Settlement Authority	in M EUR	4
Bad debt write off p.a.	in M EUR	1
Asset Dispose	in M EUR	2
Reinsurance Commutation	in M EUR	4
Letters of Credit	in M EUR	4

4.3 Hedging Strategies and Response Strategies

We have seen in this chapter how risk appetite can be defined and we also know that there are the following principles for taking risks:

- Risk is rather limited than eliminated as responsible risk taking contributes to value creation. The approved risk appetite governs the level of risk an insurance company is willing to accept. Any risks outside of appetite will be proactively managed in a timely manner.

- Risks are only accepted where the required organisational capability, expertise and infrastructure to manage the risks are in place. In addition sufficient risk based capital buffers to withstand risks materialising even under extreme stressed conditions are required.
- In accepting risk we strive for capital efficiency and profitable growth.

The above means in particular that there is a need to reduce risks and have response strategies in place in order to bring risks outside risk appetite or risks violating some limits back into risk appetite.

The corresponding strategies for the reduction of financial risks are called *hedging strategies*. In the wider context also including all other types of risks one names them *response strategies*. The aim of this short section is to provide an overview what this could mean. For the (financial) assessment of the strategies we refer in particular to chapters 6 and 10. The objective of such strategies is to reduce the risk to an acceptable (agreed) level and to optimise the risk adjusted performance. In order to do that the different strategies are analysed and compared in order to choose an optimal one. An example can be found in section 12.4.

In the following we will show what a hedging or response strategy could mean for different risks an insurance company is facing:

Equity Risk: If an insurance company faces a too high equity exposure which could threaten it, there are different possible response strategies:

1. Do nothing (I will mention this only once ...).
2. Selling equities: This can be a lengthy process since a well diversified equity portfolio consists of may different equities. Furthermore not all equities are liquid enough to change the risk portfolio fast enough.
3. Using an overlay strategy. One possible choice is to sell futures in order to reduce the equity exposure. The selling of index futures can be performed very fast and hence it is possible to rapidly reduce the equity exposure. As with all proxy hedges there remains a *basis risk*. This means that the derisked equity portfolio (eg equities plus short future) will behave differently that the correspondingly reduced equity portfolio, since the actual equity portfolio may not fit the index chosen. Hence it is important in applying such hedges to see how close the hedge is to the actual portfolio and how big the deviation could be in an distressed environment.
4. Using derivative structure, such as buying an (index) put, or puts on individual equities. When using a put option on an index one also faces here a certain basis risk. Furthermore put options on individual equities tend to be quite costly, in particular if the underlying share is illiquid. The difference to the strategy using futures is that one pays for the options up front. For futures this is not the case. On the other hand futures require regular *margin calls*. This mechanism limits the counter-party exposure of the two parties engaging

in the future contract. Hence in case of short futures the insurance company has to pay cash to its counter-party (eg bank), when the stock market rises. There are two things to consider. On one hand is the basis risk. On the other hand the insurance company might be forced to sell equities in a rising market to pay the margin calls. For derivative structures it is worth to remark that there is also the possibility to do a self-financing derivative strategy. Hence one sells the upside (by selling a call) and uses the proceeds to buy the put. Such strategies could lead to the situation, where one acquires a downside protection at -15 % for the price of limiting its upside to +10 %.

Credit Risk: The question here is how to reduce credit risk; the possible choices are very similar to the one stated above with respect to equities. One can either sell the corresponding bonds directly or one can look for derivative structures (CDS) which mitigate the risk. The typical approach for a lot of insurance companies is to assume a very limited amount of credit risk and the tendency is then to sell the titles which are considered to carry excessive risk.

Interest Rate Risk: For interest rate risk we refer to chapter 6 and remark that the migration of interest rate risk can be performed either by directly selling and buying bonds, or by using *swaps* or *swaptions*. A Swap allows transforming the duration of a bond and it can be considered as an exchange of two different cash flow streams. Hence one can "change" a 3-year bond into a 10-year bond. A Swaption is an option which allows you to enter into a Swap contract at a predefined price.

Insurance Risk: The mitigation of insurance risk is done mainly via reinsurance contracts, where the insurer cedes a part of his risk to a reinsurer. There are different ways of doing this either by a *quota share*, where a certain percentage of the original risk is ceded to the reinsurer. A *non proportional treaty* is a reinsurance treaty where the cession is not linear (as with a quota share). A non proportional treaty would qualify in the context of financial risk as an option. A *stop loss treaty* is an example of such a non proportional contract. Here the reinsurer starts to pay if a certain threshold of the total loss is exceeded.

4.4 Introduction: Use of Capital

Economic capital is one of the cornerstones of risk management. It has, roughly speaking, the same purpose as a meter stick for an engineer: It serves to measure and compare different risks and to limit them. Hence the following tasks are performed using economic capital:

- Limit and control risks,
- Allocation of capital to different markets and different functions and lines of business,

- Measure risk adjusted profitability,
- For regulatory purposes and to define the risk appetite.

In order to understand the methodology outlined in this document we need to understand how such a model works and which are the generic steps to define it. In order to produce an economic capital model the following steps need to be performed. It should be stressed that this is not an easy task and that there exist models which can be used out of the box, such as the Swiss solvency test, JP Morgan Risk Metrics and others:

The following steps, which are explained in the sequel, are needed in order to calculate the required risk capital

- Definition of the risk factors,
- Definition of a probability density functions per risk factor,
- Definition of a valuation methodology,
- Definition of the joint distribution of all risk factors – diversification,
- Definition of risk measures,
- Definition of stress scenarios.

4.4.1 Definition of the Risk Factors Considered

The risk factors define a hierarchy of disjointed risks, which are modelled separately. Not all the risks will be quantified at the beginning. Normally, risk factors can be thought as a tree. Figure 1.5 shows the risk landscape, which can be used for a risk management framework.

In order to have a sufficiently granular risk map for measuring financial risk, the map as shown in figure 1.5 needs to be enhanced.

4.4.2 Probability Density Functions per Risk Factor

The probability density function for each risk factor describes mathematically its behaviour. In other words: We define these functions by how likely or unlikely a certain event is. As a first step this is done risk factor by risk factor. Next, the different risk factors are bound together. Then the question is whether the events are more likely to occur together or not. Is it, for example, more or less likely that the temperature drops in case of a thunder storm?

The modelling of the different probability density functions and the interaction of the risk factors is based on observable market data, e.g. one looks how the different risk factors have behaved in the past and assumes that this relationship is also valid for the future. Obviously this is a bold assumption and it is therefore important to acknowledge the limitations of *each* model. This however has to be put in relation to the use of a model in general. Assume you have a small lamp (model) in a dark night. Obviously this lamp cannot replace the sun (reality). Nevertheless nobody would go out and leave the lamp at home, only because it is not as bright as the sun ....

We have spoken before about linking different risk factors together in order to be able to observe whether two risk factors interact stronger or weaker. Technically speaking the benefit gained from two risk factors which level themselves out is called diversification or diversification benefit. By choosing a clever asset allocation one can "save" capital by relaying on this effect. *However* there is also a dark side, since it is known that diversification is normally collapsing in case of extreme market movements. Statistically there is a diversification benefit between equities and credit risk in normal circumstances. In September and October 2008 this diversification disappeared and equity and credit markets were highly correlated, resulting in higher losses than in a normal market.

In the above section we have seen that economic models shed some light on the risk characteristics of an insurance portfolio, but that it is dangerous to solely rely on this number. One practical method to know what might happen in extreme conditions is to use stress scenarios, which reflect such extreme states of the economy. With their use it is also possible to do "what if" calculations and regulators often expect companies to use them.

4.4.3 Valuation Methodology

After the definition of the various risks which are considered in the economic capital model in order to determine the required risk capital, it is necessary to know how much capital is available. This available risk capital serves as a buffer in order to absorb shocks which are induced by an adverse market movement. To determine shareholder capital and available risk capital there are several possibilities. We know statutory equity, IFRS equity, embedded value and realistic balance sheet equity. The choice of the type of accounting system to be used to determine the available resources is closely linked to the intrinsic methodology applied. Since economic capital models are usually based on economic principles, the normal approach is to use a so called economic or realistic balance sheet.

We assume here the familiarity with the IFRS accounting standards - at least on a very high level. This standard is - roughly speaking characterised by:

- Assets valued at market values,
- Liabilities valued at book (statutory) values,
- In order to partially compensate for this discrepancy, additional elements are used in IFRS such as DAC, shadow adjustments and goodwill.

As seen above IFRS is a hybrid accounting system somewhere between a traditional statutory balance sheet and a fully economic balance sheet. As such this balance sheet normally serves as a basis to determine the economic / realistic balance sheet used for economic capital purposes.

4.4.4 Risk Measures

Using a risk measure allows us to assign an amount of capital to the corresponding probability distribution if a ruin event occurs.

Without going into details there are two commonly used risk measures. One is the *value at risk* for a given confidence level and the other is the *Tail VaR* or *expected shortfall*. Both measures relate to a certain probability of occurrence. One normally speaks about a 99.6% VaR or a 99% TailVaR. This means that we consider events which happen on average every 250 years (eg $99.6\% = 1 - \frac{1}{250}$) or once in one hundred years (eg $99\% = 1 - \frac{1}{100}$) respectively. Since the occurrence of an event is linked to the time which elapses it is important to recognise the fact that a *VaR* or *TailVaR* is linked to a time span. In banking one uses, for example, a *daily VaR*, which means one looks at events which occur with a given probability within the next day. For insurance a yearly consideration is normal, hence we look a *99.6% yearly VaR*. In consequence these events can also be considered to have a 250 year return period. (Note that this concept is mainly used for natural perils in reinsurance.)

In order to better understand the two concepts – VaR and TailVaR – we want to show this based on an example and we assume that we have observed 1000 different observations of losses and we have sorted them according to their loss. The worst 12 losses were:

Number	Loss (in M EUR)
1000	210
999	175
998	150
997	145
996	140
995	130
994	125
993	120
992	115
991	112
990	110
989	105

We want now to determine the 99.6% VaR and the 99% Tail VaR. For the 99.6% VaR we look at the loss at position 997 (eg 4 events out of 1000) and find that the $VaR = 145$ M EUR. For the Tail VaR we have to look at the average loss, once we know that the loss is within the 1% worst. Hence we have to look at all losses between number 991 and 1000 and have to take the average. Hence we have:

$$\begin{aligned} TailVar &= \frac{210 + 175 + 150 + 145 + 140 + 130 + 125 + 120 + 115 + 112}{10} \\ &= 142.2. \end{aligned}$$

Since one can also plot the probability distribution, one can also locate the VaR and TailVaR within this graphic. Figure 4.2 show this relationship.

Stress Scenarios

We know that each economic capital model has limitations, in particular in case of extreme states of the market. This is also the case because there are not enough observations available to calibrate the model. Hence the use of a model which has been calibrated around the mean is dangerous, since the effective risk might be underestimated. In order to compensate for this effect it is common to use stress scenarios. Stress scenarios can be thought as instantaneous changes of the economy according to some predefined risks.

4.5 Risk Measures

Using a risk measure allows us to assign an amount of capital to the corresponding probability distribution if a ruin event occurs (a catastrophic scenario occurs when

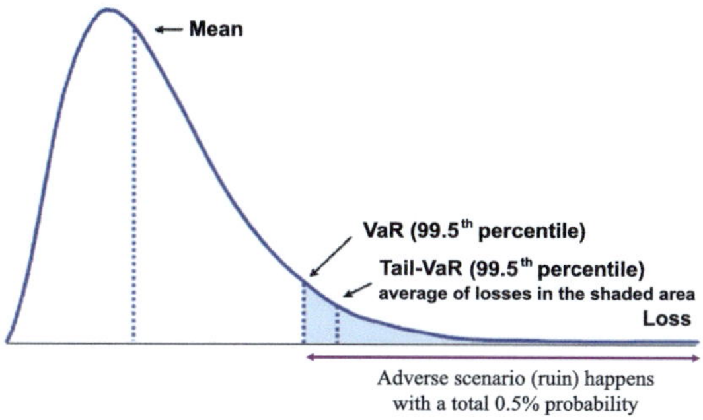

Fig. 4.2 Comparison VaR and TailVaR

the amount of admissible assets is lower than the total amount of technical provisions). Thus, a risk measure is what will allow calculation of the solvency margin and which will serve as a benchmark for the standard formula and the various internal models that companies may develop.

The chosen risk measure is a function $\rho : X \longmapsto \rho(X) \in \mathbb{R}$ where X is a real random variable which represents the possible loss for an insurer,

- $X > 0 \Longrightarrow$ loss,
- $X < 0 \Longrightarrow$ profit.

In the concrete set-up the possible loss X of an insurer is the difference in shareholder capital over a given time interval. A risk measure ρ is considered as "coherent" if it satisfies the four following axioms:

Monotonicity: $\rho(X) \leq \rho(Y)$ whenever $X \leq Y$
If the risk of a portfolio Y is greater than that of a portfolio X then the capital needed will be greater too.

Positive homogeneity: $\forall \alpha > 0 : \rho(\alpha \times X) = \alpha \times \rho(X)$
If each risk is multiplied by a factor α then the risk measure will also be.

Translational invariance: $\forall \alpha \in \mathbb{R}$: $\rho(X + \alpha) = \alpha + \rho(X)$

Sub-additivity: $\rho(X+Y) \leq \rho(X) + \rho(Y)$
Risk measure for two portfolios is lower than the sum of risk measures for these two portfolios. It represents the advantage of diversification effect.

4.5.1 Value at Risk (VaR)

This measure is much used in banks and insurance companies.

Definition: Given some confidence level $\alpha \in [0,1]$ the Value at Risk of a portfolio is the smallest number x such that the probability that the loss X does not exceed x is larger than $(1-\alpha)$. It represents the maximal potential loss accepted.

$$\mathrm{VaR(X)} = -\inf\{x \,|P[X \leq x] > (1-\alpha)\} \tag{4.1}$$

where X is a real random variable which represents the possible loss for an insurer.

In fact it is defined as a quantile α level: $VaR^{\alpha}(X) = Q^{\alpha}(X) = F_X^{-1}(1-\alpha)$ where F is the distribution function of the random value X continuous and purely increasing such as $F_X(x) = P[X \leq x] = 1-\alpha$.

The drawback of this measure is that VaR doesn't give any information about the tail of the distribution. Moreover it is not an "coherent"measure, indeed VaR is monotone, positively homogeneous and translationally invariant but not sub-additive (aggregating many risks can increase the risk and it is not really conservative).

Thus Artzner has developed the Tail Value at Risk as a risk measure which is more convenient than VaR especially in reinsurance since tail risks are covered. TVaR requires a less simulations to be estimated and to reaches stability faster than VaR.

4.5.2 Tail Value at Risk (TVaR)

Definition: Tail Value at Risk (TVaR) is the expected value of the loss in those cases where it exceeds the predefined confidence level. This measure is equal to the average loss a company will suffer in case of (extreme) situations where losses exceed a predefined threshold. Contrary to VaR, TVaR is "coherent"and considers the shape of the tail of the distribution.

$$\mathrm{TVaR}^{\alpha} = \frac{1}{1-\alpha}\int_{\alpha}^{1} VaR(X;\xi)d\xi \tag{4.2}$$

where X is a real random variable which represents the possible loss for an insurer.

We can define similar risk measures such as Conditional Tail Expectation (CTE), Expected Shortfall (ES) or Expected Tail Loss.

- Conditional Tail Expectation (CTE)
  CTE$^\alpha(X) = \mathbb{E}[X|X > VaR^\alpha(X)]$
  $\hookrightarrow$ CTE represents the mean value over VaR and satisfies the property of sub-additivity only in the case of a continuous distribution.
- Conditional Value at Risk (CVaR)
  CVaR$^\alpha(X) = \mathbb{E}[X - VaR^\alpha(X)|X > VaR^\alpha(X)]$
  CVaR$^\alpha(X) = \mathbb{E}[X|X > VaR^\alpha(X)] - VaR^\alpha(X)$
  CVaR$^\alpha(X) = CTE^\alpha(X) - VaR^\alpha(X)$
  $\hookrightarrow$ CVaR represents a weighted average between the value at risk and losses exceeding the value at risk
- Expected Shortfall (ES)
  ES$^\alpha(X) = \mathbb{E}[X|X \geq VaR^\alpha(X)]$
  $\hookrightarrow$ Expected shortfall is the conditional expectation of loss given that the loss is beyond the VaR level.

As we can see, in the graph above, a 99.5% TVaR gives the average of the highest 0.5% of losses. For this reason TVaR will be higher than the VaR estimate for the same percentile.
According to the CEIOPS[1], experience suggests that, on average, a 99% confidence level with a TailVaR risk measure may roughly be equivalent to a 99.6% confidence level with a VaR risk measure.

Afterwards, these risk measures will be used to determine our catastrophic scenario and consequently, the necessary amount of RAC.

4.5.3 Relationship Between Value at Risk and Expected Shortfall

We know that

$$\begin{aligned} \text{VaR}^\alpha(\text{X}) &= -\inf\left\{x\, |P[X \leq x] > (1-\alpha)\right\}, \\ ES^\alpha(X) &= \mathbb{E}[X|X \geq VaR^\alpha(X)], \end{aligned}$$

and we define

[1] www.ceiops.eu

$$e^{\alpha}(X) = \mathbb{E}[X - \mathrm{VaR}^{\alpha}(\mathrm{X}) \,|\, X \geq VaR^{\alpha}(X)]$$

the excess loss over the value at risk[2], and remark that $ES^{\alpha}(X) - VaR^{\alpha}(X) = e^{\alpha}(X)$. Hence this excess loss over VaR measures how much a possible loss "eats" in average more of the available capital in case a α–quantile event occurs. We remark that the $e^{\alpha}(X)$ is dependent on the distribution function of X. The example of section 4.4.4 show this additional "eating" of capital.

Hence the understanding of $e^{\alpha}(X)$ is essential when choosing a risk measure or defining risk appetite. In this section we want to have a quick look at the respective advantages and disadvantages of the two risk measures.

The main advantage of the VaR is that it is very well understood and widely used. It can be argued that this is the correct over all measure from a shareholders point of view if he wants to define how likely it is to lose all of its capital. If the idea behind measuring risk is however to consider a going concern and to limit the downside, the value at risk is dangerous since an additional amount of capital $e^{\alpha}(X)$ is needed to survive an α–quantile event.

It is therefore important to use the expected shortfall measure for the definition of risk appetite statements and for considerations where a going concern in an α–quantile event is envisaged. This is particularly true in case one wants to break down the total required risk capital in smaller pieces to define the risk appetite at a more granular level. A concrete example is a risk appetite statement of "We want to be able to continue operating normally and adhering to the one in 200 VaR after a one in 10 year event." In this example we need to hold an excess capital over the one in 200 Var of the 10 % expected shortfall (and *not* VaR) over the 99.5 % VaR, assuming the independence of the two events.

We finally remark that we obviously can stick to the VaR if we keep these facts in mind and have a view on the possible different outcomes by using VaR. From a theoretical point of view coherent risk measures have obviously its advantages when tying to analyse them.

Assume for the moment, that we use VaR as a risk measure, for example because the regulatory regime requests it and that in consequence the insurance company bases its target capital and its risk appetite on the α–quantile with respect to VaR. In case such a default occurs, in average, the amount $e^{\alpha}(X)$ is needed to run off the company in an orderly fashion. In consequence the average cost of default can be calculated as $\alpha \times e^{\alpha}(X)$.

We finally remark that the mean excess function for a random variable X is given by

$$e_X(x) = \mathbb{E}[X - x \,|\, X > x] = \frac{\int_x^{\infty} \{1 - F_X(\xi)\} d\xi}{1 - F_X(x)},$$

[2] It is important to remark, that $e^{\alpha}(X)$ has an interesting relationship to reinsurance, if we assume that $VaR^{\alpha}(X)$ is the retention of a reinsurance treaty, with no upper limit. In this case we can interpret $e^{\alpha}(X)$, as the expected loss of the reinsurer in case the retention limit is triggered.

and that we have $e^{\alpha}(X) = e_X(VaR^{\alpha}(X))$.

For the convenience of the reader we have listed some mean excess functions for different probability distributions in appendix A.3.

Before entering into a new topic, we want to have a look how VaR and TVaR concretely compare, assuming a standard $\mathcal{N}(0,1)$ normal distribution. We start at a confidence level α and calculate then VaR and TVaR. Moreover we also want to calculate the equivalent confidence level $\tilde{\alpha}$ with respect to VaR for a given TVaR at level α. The table below summarises the results and we remark that these numbers are heavily dependent on the distribution function chosen. For probability distribution functions with heavier tails, the difference between VaR and TVaR increases, and hence it is important to devote enough time and thought when choosing a suitable family of probability distributions for capital models.

Confidence Level	VaR	$e(x)$	TVaR	Equivalent VaR Level
0.84134	1.00000	0.52513	1.52513	0.93638
0.90000	1.28155	0.47343	1.75498	0.96036
0.95000	1.64485	0.41785	2.06271	0.98043
0.99000	2.32634	0.33886	2.66521	0.99615
0.99500	2.57582	0.31611	2.89194	0.99808
0.99900	3.09023	0.27685	3.36709	0.99962

4.5.4 What Is a Stress Scenarios Mathematically

A stress scenario, can be considered a stochastic process, associated with a point (Dirac) measure. For stock markets the drop of 30% can be considered as a stress scenario. At the beginning such a stress scenario is not linked with a probability. In order to use this methodology for determining a risk capital one needs to have a view on its probability.

The stress scenario is used to calculate the loss linked to it.

Chapter 5
Key Insurance Processes and Their Risks

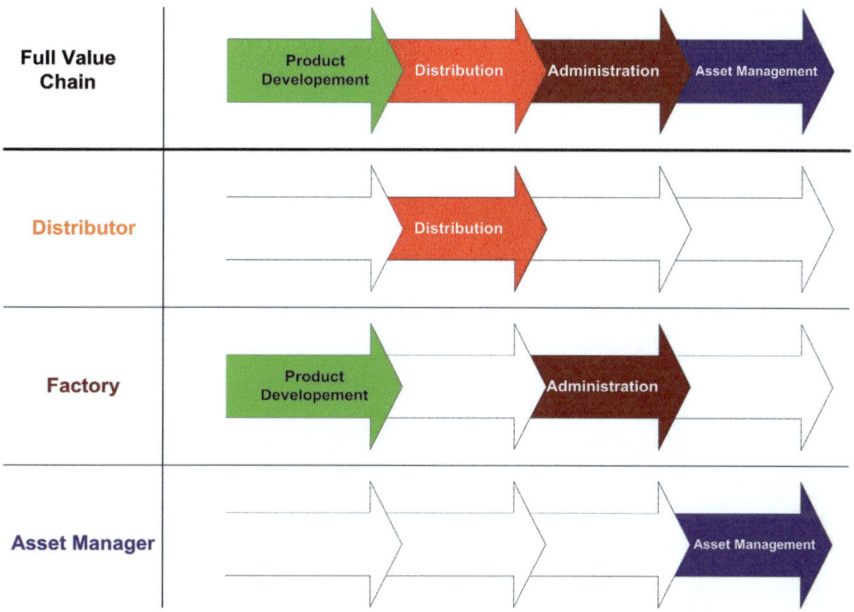

Fig. 5.1 Value Chain of an Insurance Company

The aim of this chapter is to highlight the main processes in an insurance company in order to decompose them. It is important to understand the value chain (see figure 5.1) of an insurance company and to see that each step of the value chain has its generic risks, which need a differentiated treatment. The main question here is, which part of a process has to be done by the first line of defence (line management) and which parts should be done by the second line of defence. As usual with all organisational set-ups there is no right or wrong, but there are different possibilities which could work, depending on the corresponding environment. In this sense the

M. Koller, *Life Insurance Risk Management Essentials*, EAA Series,
DOI 10.1007/978-3-642-20721-1_5,

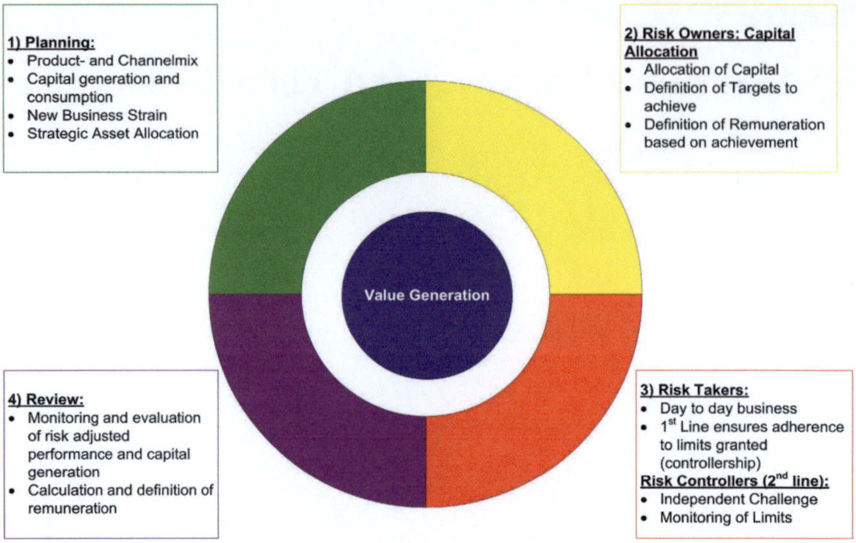

Fig. 5.2 Overview over Capital Allocation Process

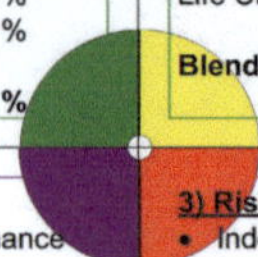

1) Planning:
• Opportunities:

Available Capital: 1'500

Type	Capital needed	Break-even return	Return offered
Reinsurance	100	13.0 %	15.0 %
Strat Project	150	11.0 %	12.0 %
GI	500	11.5 %	13.0 %
Life Trad	800	9.0 %	9.0 %
Life UL	400	9.0 %	11.0 %
Blended Total	**1'950**	**10.0 %**	**10.9 %**

2. Capital allocation:

Type	Capital allocated	Return (plan)	Hurdle for Bonus
Reinsurance	100	15.0 %	14.0 %
Strat Project	50	12.0 %	12.0 %
GI	350	13.0 %	11.0 %
Life Trad	700	9.0 %	8.5 %
Life UL	300	11.0 %	10.5 %
Blended Total	**1'500**	**10.8 %**	**10.0 %**

4) Review:
• Monitoring and evaluation of risk adjusted performance and capital generation
• Calculation and definition of remuneration

Type	Capital used	Return effective	Level Bonus
Reinsurance	99	15.5 %	14.0 %
Strat Project	65	12.0 %	12.0 %
GI	350	10.5 %	11.0 %
Life Trad	640	8.7 %	8.5 %
Life UL	300	12.0 %	10.5 %
Blended Total	**1'454**	**10.4 %**	**10.0 %**

3) Risk Taking:
• Independent Challenge
• Monitoring of Limits
• Redeployment of capital

Type	Capital limit	Capital used	Limit Headroom
Reinsurance	100	99	
Strat Project	50	70	
GI	350	360	
Life Trad	700	650	
Life UL	300	290	
Blended Total	**1'500**	**1469**	

Fig. 5.3 Example Capital Allocation Process

decomposition into processes and the split between the first and second line of defence need to be considered as examples and there may be good reasons to have a different organisational set up.

Before we start to decompose the different processes we need to first look at the different levels of intrinsic tension within an organisation. One could, in principle, define three different levels of independence.

1. No independence. The whole management is done within the first line of defence. One could, for example, consider sustainability as such an area. The reason for putting processes in this category might be the fact that the corresponding risks are less relevant or that the time until they materialise is longer.
2. Limited challenge from the second line of defence. In this category we may have risks which are more relevant than in the above category, be it in terms of timing or also in terms of severity. Here the challenging of the second line of defence is limited since there are no dedicated experts which could do this job, but there is rather a higher reliance on the first line of defence expertise. An example could be business protection.
3. Full independent challenge from the second line of defence. The corresponding risks are highly relevant for the company and in consequence the subject matter expertise in the second line of defence has the same level of professionalism as within the first line of defence. Risks which fall into this category are normally very important for the profitability and strategy of the company and normally include all types of financial risks such as market and ALM risks. In some legislation there is a need for some segregation of duties for example in doing independent valuation of OTC derivatives, etc.

Since it is obviously not possible to decompose all processes of an insurance company we limit ourselves to some key processes which normally form part of the third category.

Each of the process will be decomposed in the same way:

- Description of the process,
- Main risks,
- Organisational design,
- Key learnings.

5.1 Capital Allocation and Planning Process

Description of the Process: The capital allocation and planning process is the main economic process in an insurance company. All other economic processes are dependent on it. This process is described in figure 5.2 with an example in figure 5.3. It is particularly important to understand that such processes need to have feedback loops and hence this type of process never ends. Normally one

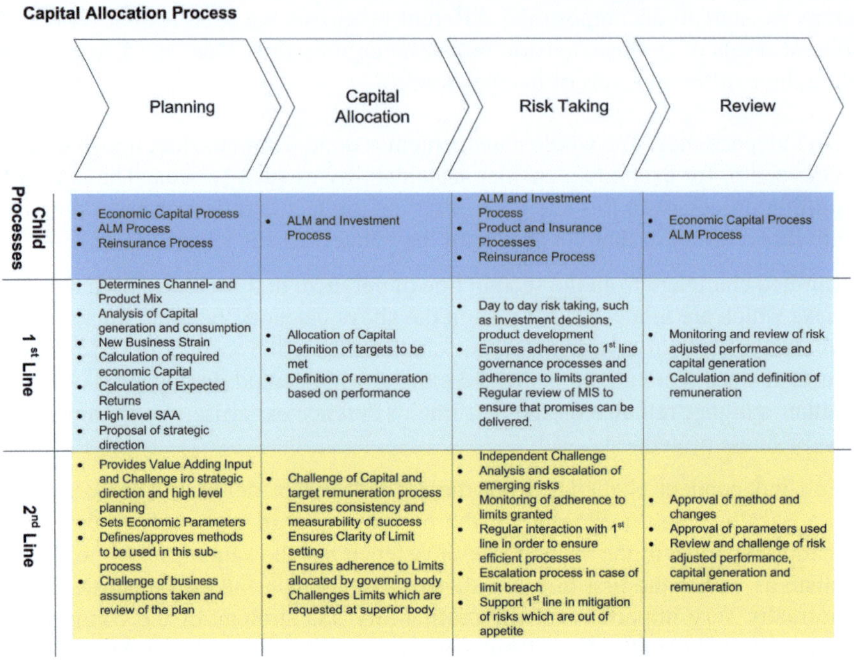

Fig. 5.4 Capital Allocation Process

starts with the amount of capital available and with a portfolio of business opportunities. Here the aim is to define an optimal product and channel mix which is in line with the strategic objectives and the risk appetite of the company. One has to weigh the capital consumption against new business strain and to determine the strategic asset allocation which fits best. For each business opportunity one has to have a view on the required capital and the corresponding expected returns. In a normal environment there is a scarcity of capital and it is not possible to execute all business opportunities. On the other hand there may be situations (for example in a reinsurance company when entering a softer market), that there are not enough business opportunities which yield the required return and hence it is not possible to deploy all capital. In these circumstances the company may also decide to give some capital back to its shareholders.

After gathering all the required information, one enters the next step of the process, where the risk owners decide how much capital is allocated to each of the business opportunities. This decision process is normally based on the pure economic facts but there are also some softer considerations, such as the strategic ambitions and considerations in respect of the insurance cycle. When allocating the capital, it is important to define the hurdles which need to be met in order that the corresponding managers get a bonus. It is important they are based on risk adjusted metrics. It is also here, where the risk appetite and the corresponding limits are defined and anchored.

After this second step, which normally happens before a new financial period, we enter into step number three, the risk taking. Here the line management aims to optimally deploy the allocated capital in order to generate superior returns and to meet the defined hurdle rates. At the same time they ensure that they operate within the risk limits granted to them. The second line of defence challenges key decisions taken by line management and independently monitors the adherence to the limits granted. In case of a limit violation the corresponding escalation mechanisms are indicated in order to bring the situation back within the agreed risk appetite. In some cases it will not be possible to deploy all capital or there is need for more capital for an interesting opportunity. In this case the risk owners reallocate capital as described in step number two.

At the end of the financial period, we enter into the review phase, where the actual achieved results are compared with the targets and where the risk adjusted returns are determined. As a consequence, the variable compensation of the line managers in function of their performance is determined.

Main Risks: Since this is the main process, we also face all possible risks, which are mainly under-performance due to wrong capital allocation, suboptimal business opportunities and the taking of undue risks, for example as a consequence of ill behaved remuneration structure and missing limit systems.

Organisational Design: Because this is a particular important process, it is necessary to have a clear segregation of duties between the first and second line of defence. Figure 5.4 shows a possible subdivision of the process. The main principle is that the persons who are remunerated according to metrics should not be able to determine how this is calculated. Therefore, it is one of the main tasks of the second line of defence to ensure the clarity in limits and bonus setting. They furthermore ensure an adequate challenge of the risk adjusted performance numbers in order to avoid self-fulfilling promises. It is key that the risk owners (who have the final responsibility for the business) define the risk appetite and the allocation of capital. Here the risk management function acts as an enabler providing value adding insights and challenges. There are another two important points. It is expected that the second line of defence also provides valuable input to the strategic dimension of the process. This is particularly evident since strategic errors have, in many cases, a large adverse effect. Finally it should be stressed that the second line of defence should not only check limits in order to fulfil the corresponding compliance requirements, but in case of a material limit breach or if the risk appetite is exceeded in a material manner, they should also provide help to bring things back within the risk appetite. This fact needs to be stressed, particularly in respect to financial risks, the corresponding know-how within the risk management function can be of considerable help.

Key Learnings: From this process it becomes particularly apparent that a clear vision of what needs to be done in the first and second line of defence is key. Furthermore, it is equally important to have the corresponding governance in place when making the abstract process living. This is done by exercising, which means that in case of governance committees, a frequency is required that is high enough so that people get used to the different concepts and tasks. It is also key to have reliable and robust processes and methods in place. Finally it is very important that one does this process with an action oriented focus in a structured manner.

5.2 Economic Capital Calculation Process

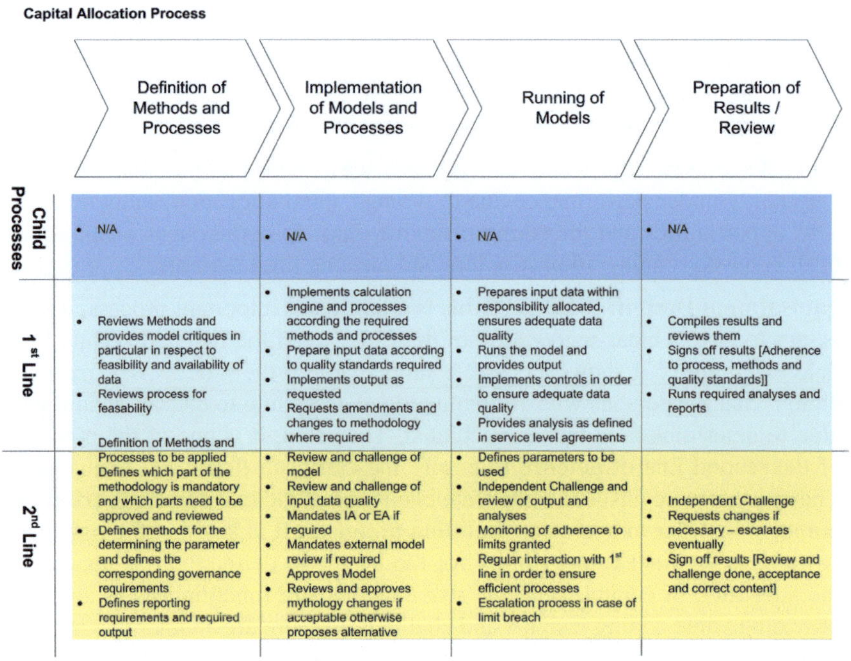

Fig. 5.5 Economic Capital Calculation Process

Description of the Process: The process of calculating the economic capital is split into two parts, the calculation of the available economic capital and the calculation of the required economic capital. Furthermore the process also assesses the risk adjusted profitability as mentioned in the capital allocation and planning process. From its characteristics this process is very technical and it is also very important. From a very high level point of view, it is the measuring process without which an objective assessment is impossible. As a consequence particular care needs to be applied to the split between first and second line of defence. It is also important to recognise that the separation is not trivial, since the design and the implementation of the corresponding models need to be split. In the same sense the people determining parameters need to be different from the people approving them. The split within this process is shown in figure 5.5.

Main Risks: As previously mentioned this process determines the "economic capital meter stick". There are two risks which could materialise if the process is not implemented correctly. The first is the risk that a wrong measurement either under or overstates capital, which leads to taking undue risks and suboptimal performance respectively. The second risk is in relation to the risk that the meter stick is adjusted only for the purpose of better returns during the assessment phase. Intrinsic to this risk is abuse of the management framework and the payment of undue bonuses.

Organisational Design: Since this process is the core of the economic risk measurement and valuation there are different stakeholders which participate in this process and two things need to be insured. On one hand we need to have enough tension in the system to get reliable and unbiased results. And on the other hand we need to have stable processes which ensure consistency with the financial accounts and so ensure the link to the business and finance. Obviously, also different design principles could be applied and the one proposed here (see Fig. 5.5) is aligned with Solvency II.

This has as a consequence that the risk people define the economic capital model and that the actuaries implement it and run it on a day-to-day basis. This means also that the actuaries propose parameters, which have to be signed off by the risk people and also the model and its results need to be signed off by them. Only by this separation of duties it is possible to have reliable checks and balances and to ensure that the system does not provide self-fulfilling promises.

Key Learnings: The main key learning here is that the economic capital models are very complex and it is necessary to have checks and balances intrinsic in this process in order to ensure adequate results. It is key to use the results in many areas of the business in order to embed economic thinking in the company and to meet the requirements of the "use test" imposed by Solvency II.

5.3 ALM and Bonus Setting Process

Description of the Process: If we go down one level from the capital allocation and controlling process, the next most important process is the ALM and bonus setting process. Before entering into details it is necessary to understand why we bind these two processes together. In a lot of cases ALM and bonus setting are considered as separate tasks. This can have devastating consequences. Some of them are explained in chapter 12.

The aim of this process is to steer the assets and liabilities in order to provide an optimal return for both shareholders and policyholders. Technically, the liabilities

ALM and Bonus setting process

	Planning and Analysis	Definition of SAA And Bonus principles	Risk Taking	Review
Child Processes	• Economic Capital Process	• Investment Process	• Investment Process • Product and Insurance Processes	• Economic Capital Process
1st Line	• Definition of business strategy • Definition of Shareholder return targets • Definition of PH return targets • Definition of risk appetite • Preparation of Liabilities (replicating portfolio) • Definition of possible investment and PH bonus strategies	• Allocation of Capital in asset and liability classes • Definition of SAA with limits and triggers • Definition of targets to be met • Definition of remuneration based on performance	• Day to day risk taking, such as investment decisions, product decisions • Ensures adherence to 1st line governance processes and adherence to limits granted • Regular review of MIS to ensure that promises can be delivered.	• Monitoring and review of risk adjusted performance and capital generation • Calculation and definition of remuneration
2nd Line	• Provides Value Adding Input and Challenge iro strategic direction and high level planning • Sets Economic Parameters • Defines/approves methods to be used in this sub-process • Challenge of business assumptions taken and review of the plan	• Challenge of Capital and target remuneration process • Ensures consistency and measurability of success • Ensures Clarity of Limit setting • Ensures adherence to Limits allocated by governing body • Challenges Limits which are requested at superior body • Challenge and value adding input iro SAA	• Independent Challenge • Analysis and escalation of emerging risks • Monitoring of adherence to limits granted • Regular interaction with 1st line in order to ensure efficient processes • Escalation process in case of limit breach • Support 1st line in mitigation of risks which are out of appetite	• Approval of method and changes • Approval of parameters used • Review and challenge of risk adjusted performance, capital generation and remuneration

Fig. 5.6 ALM and Bonus Setting Process

given by the insurance policies are compared with current assets. One has to do two things to define the bonus strategy and the target asset allocation optimising the relationship between assets and liabilities. During the year the investment strategy is implemented adhering to the limits allocated during the ALM process based on the capital allocated to this task. At the end of the year two things take place. Firstly the performance is measured and an excess generation of economic wealth is split between shareholders and policyholders. This process is called the bonus setting process.

Main Risks: There are many different risks inherent to this process. The main ones are a suboptimal investment strategy which either does not deliver the business objectives required or only meets the required returns by taking excessive investment risk. Another significant risk is the possible disjoint between the assets and the liabilities. This risk is normally manifested by granting too onerous bonus promises to the client, which cannot be achieved realistically without taking excessive investment risks. The crisis which hit a lot of insurers in the years 2001/02 is a consequence of this disjoint.

Organisational Design: Normally the issue is not that people are taking undue risks on purpose. The main issue is that they are not speaking to each other because either they do not understand the issues of the other key-player or because they speak a very different language. Furthermore the whole question around

ALM and bonus setting is not only highly complex from a technical point of view but also requires a deep understanding of insurance products and customer needs. As a consequence there are very few people who understand these relationships and the corresponding pitfalls in detail. Hence it is absolutely essential to work in a team, where the main functions that have a stake in this question are involved. This means that we need an ALCO with the following participants: distribution, products, finance, investment and risk.

Key Learnings: Since this process is highly complex it is necessary to work in an interdisciplinary team, where the different team members are able to understand the language of the other and where the whole process is based on a structured process supported by concise, action oriented and relevant management information. A lot of the corresponding material needed can be found in the following chapters.

It is clear that the ALM process and its decisions need to be broken down further in order to have a meaningful investment process. Without going into further details we refer to figure 5.7.

Investment process

	Planning and Analysis	Definition of TAA	Risk Taking	Review
Child Processes	• N/A	• N/A	• N/A	• N/A
1st Line	• Definition of investment strategy • Definition of asset return targets • Definition of risk appetite • Definition of possible tactical investment strategies	• Allocation of Capital in asset classes • Definition of TAA with limits and triggers • Definition of targets to be met • Definition of remuneration based on performance	• Day to day risk taking, such as investment decisions, product decisions • Ensures adherence to 1st line governance processes and adherence to limits granted • Regular review of MIS to ensure that promises can be delivered.	• Monitoring and review of risk adjusted performance and capital generation • Calculation and definition of remuneration
2nd Line	• Provides Value Adding Input and Challenge iro strategic direction and high level planning • Sets Economic Parameters • Defines/approves methods to be used in this sub-process • Challenge of business assumptions taken and review of the plan	• Challenge of Capital and target remuneration process • Ensures consistency and measurability of success • Ensures Clarity of Limit setting • Ensures adherence to Limits allocated by governing body • Challenges Limits which are requested at superior body • Challenge and value adding input iro TAA	• Independent Challenge • Analysis and escalation of emerging risks • Monitoring of adherence to limits granted • Regular interaction with 1st line in order to ensure efficient processes • Escalation process in case of limit breach • Support 1st line in mitigation of risks which are out of appetite	• Approval of method and changes • Approval of parameters used • Review and challenge of risk adjusted performance, capital generation and remuneration

Fig. 5.7 Investment Process

5.4 Product Design Process

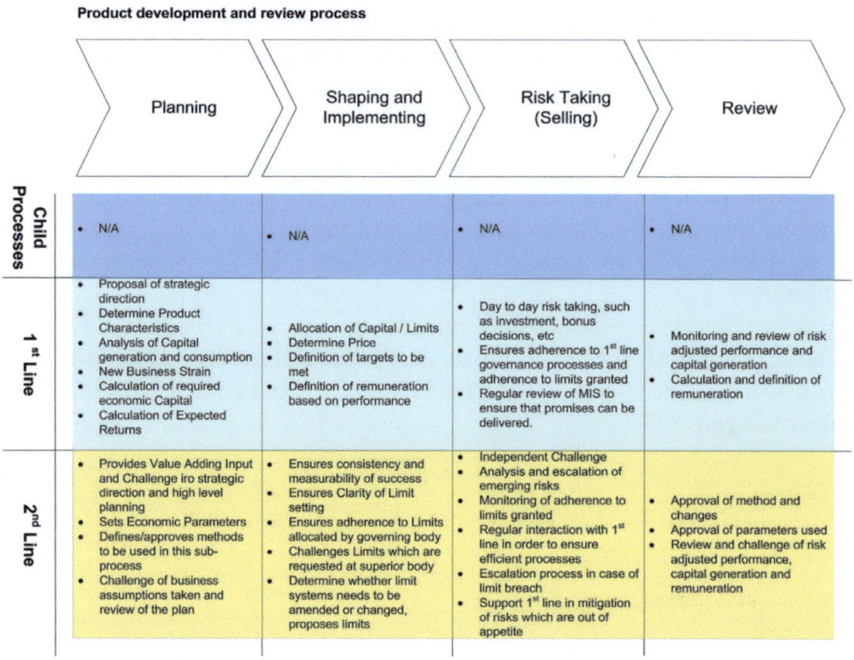

Fig. 5.8 Product Design Process

Description of the Process: As we will see in chapter 12 the product design process is another root cause of potentially big losses and problems in an insurance company. In contrast to investment risks, where gains and losses occur on a daily basis, losses as a consequence of a wrong product design can be considered as rare events with a high severity if they materialise. As the title says the aim of this process is to develop and introduce new or to change and improve existing products. Improvement in this context does not only mean increases in profitability, but also increases in sales volumes. Every new product starts with an idea, which can either come from looking outside the company, asking clients' demands, etc. Next, one has to shape the product and to look at the different angles, such as product design, product implementation, customer needs, asset liability management for the product, risks intrinsic to the product, pricing of the product etc. After having defined the corresponding boundaries one develops first a concept paper and next a technical documentation of the product, before implementing it. In parallel, the supporting material and training for the distribution channels is prepared in order to afterwards sell the product.

Main Risks: There are two main risks within the product design, namely a wrong structural design of the product and wrong pricing. Both effects can lead to either selling too few policies or too big losses. Chapter 12 explains some of the failures in relation to products leading to considerable losses for the insurance company.

Organisational Design: The root causes for ill behaved product designs are, from a structural point, almost the same as for the ALM and bonus setting process with the only difference being that the same people are not necessarily needed at the table. It is therefore important to create a product approval committee which overviews the issues in relation to products. Furthermore, we need adequately skilled people who are able to foresee issues such as the ones mentioned in chapter 12.

Key Learnings: As indicated before, the key learnings are very similar to the one of the ALM and bonus setting process.

Chapter 6
Financial Risks and Their Modelling

The aim of this chapter is to educate the readers, in order that they understand the basics of financial risk management and so that they can interpret the numbers within this report. For the underlying abstract valuation concept we refer to appendix C.

6.1 The Model Underlying Financial Risks

In order to develop a model for managing and measuring financial risks we have a look at the balance sheet, which have seen earlier in this book:

M. Koller, *Life Insurance Risk Management Essentials*, EAA Series,

DOI 10.1007/978-3-642-20721-1_6,

Balance sheet	Book A	Book L	Market A	Market A	
Cash	6200	47100	6200	48513	MR
Bonds	35700	**2200**	37842	**3569**	SHE
Shares	4400		4800		
Properties	1100		1300		
Loans	1400		1400		
Alternatives	500		540		
Total	49300	49300	52082	52082	

It is clear that we need to decouple the valuation π_t from the underlying asset. So formally the balance sheet consists of assets $(\mathcal{A}_i)_{i\in S_A}$ and Liabilities $(\mathcal{L}_i)_{i\in S_L}$ and we assume that both index sets S_A and S_L are finite. Now assume we have 1000 shares from HSBC. We could say that these 1000 shares are "one" asset. On the other hand we could model the same holding as holding 1000 pieces of the asset "1 HSBC share". Therefore we denote by $(\alpha_i)_{i\in S_A}$ and $(\lambda_i)_{i\in S_L}$ the number of units which we own at the certain point of time. Furthermore we want to separate the shareholder equity from the liabilities and we denote it $\mathcal{E}$.

If we write $\alpha_1 \mathcal{A}_1$ we assume that we are holding α_1 units of the asset $\mathcal{A}_1$. Hence our portfolio is an abstract finite dimensional linear vector space $\mathcal{Y} = \text{span}\{(\mathcal{A}_i)_{i\in S_A}, (\mathcal{L}_i)_{i\in S_L}, \mathcal{E}\}$. In this context our balance sheet is a point $x = \sum_{i\in S_A} \alpha_i \mathcal{A}_i + \sum_{i\in S_L} \lambda_i \mathcal{L}_i \in \mathcal{Y}$.

As seen before some assets and liabilities can be further decomposed in simpler assets and liabilities and hence we can find a suitable basis for the vector space $\mathcal{Y} = \text{span}\{e_1, \dots, e_n\}$, where $(e_k)_{k\in\mathbb{N}_n}$ is its basis, and we remark that we can also write our balance sheet as $x = \sum_{k\in\mathbb{N}_n} \gamma_k\, e_k$.

The idea to introduce $\mathcal{Y}$ is to have a normalised vector space. Assume for example that we hold some ordinary bonds. In this case we would use as $e_k = \mathcal{Z}_{(k)}$, the corresponding zero coupon bonds, etc.

We finally remark that the balance sheet $x \in \mathcal{Y}$ actually represents a random cash flow vector, and hence we strictly have x_t or $X_t(\omega) \in \mathcal{X}$ if we assume that the changes of the portfolio follow a stochastic process (cf. appendix D). For measuring the risk of the actual balance sheet it is normally sufficient to assume that $y \in \mathcal{Y}$ does not change.

Next we need to look at the second part, namely the valuation π_t, and we remark that:

- The valuation is dependent on time.
- We assume that the valuation is a linear functional $\pi_t : \mathcal{Y} \to \mathbb{R}$ which allocates to each asset its value (see also appendix C).
- A liability $\mathcal{L}$ is characterised by $\pi(\mathcal{L}) \leq 0$. In the same sense and asset has a positive value. As a consequence an $x \in \mathcal{Y}$ can in principle be both an asset or

a liability, depending on the economic environment and also depending on the valuation functional.

After having defined the different parts we need to have a closer look at what equity or capital ($\mathcal{E}$) means. In the context of the balance sheet we observe that the sum of the value of all assets equals the sum of the value of all liabilities (neglecting the sign). Hence we have the following:

$$\begin{aligned} x &= \sum_{i \in S_A} \alpha_i \mathcal{A}_i + \sum_{i \in S_L} \lambda_i \mathcal{L}_i + \mathcal{E} \in \mathcal{X}, \text{ and} \\ \pi(x) &= \pi\left(\sum_{i \in S_A} \alpha_i \mathcal{A}_i + \sum_{i \in S_L} \lambda_i \mathcal{L}_i + \mathcal{E} \right) \\ &= 0, \text{ and hence} \\ SHE = \pi(\mathcal{E}) &= -\pi\left(\sum_{i \in S_A} \alpha_i \mathcal{A}_i + \sum_{i \in S_L} \lambda_i \mathcal{L}_i \right). \end{aligned}$$

This means that we can always calculate the value of the shareholders' equity if we know the value of all other assets and liabilities.

Finally we want to show how to tackle the stochastic valuation functional π_t. Since we live in a linear vector space $\mathcal{Y}$ with a basis $(e_k)_{k \in \mathbb{N}_n}$, it is sufficient to define the price $\pi_t(e_k)$. The idea is to decouple the operator from the economy and the corresponding set up is to define the state of the economy by a stochastic process $(R_t)_{t \in \mathbb{R}} \in \mathbb{R}^m$. You could think that one of the components could be inflation, another could be the level of the 10 year interest rate, etc. In this setup we can define:

$$\pi_t(e_k) = f_k(R_t),$$

where $f_k : \mathbb{R}^m \to \mathbb{R}$ is a sufficiently regular function. If we assume for example that $R_t[10]$ is the interest rate for the 10 year bond, then we have (depending on our definition of π)

$$\pi_t(\mathcal{Z}_{(10)}) = (1 + R_t[10])^{-10}.$$

The idea of financial risk management is to assess and control the change of the value of the shareholder equity, e.g. the profit and loss induced by this change. If we assume for the moment that the time t is denoted in years, one is normally interested in the following quantity:

$$PL_T = (\pi_T(\mathcal{E}) - \pi_0(\mathcal{E})).$$

The loss which we encounter within the time interval $[0, T]$. Banks normally look at one week, eg $T = 1/52$, Solvency II looks at $T = 1$. One measures the risk, as indicated before based on the random variable PL_T.

Here again is a more formal environment: In order to assess the financial risk of an insurance company the following steps are needed.

1. Define the valuation methodology π_t,
2. Define (note this is a big model assumption) which stochastic process R_t models the economy,
3. Define the universe of all assets and liabilities $\mathcal{Y}$,
4. Define and calculate the functions $(f_k)_{k \in \mathbb{N}_n}$,
5. Analyse the possible balance sheets $x \in \mathcal{Y}$ and decompose each $\mathcal{A}_i$ and $\mathcal{L}_i$ into the basis $(e_k)_{k \in \mathbb{N}_n}$,
6. Define the risk measure to be used such as VaR, etc.,
7. Implement the model.

The implementation of the above steps in its purest form is very complex and therefore one normally has to make approximations.

6.2 Approximations

A common approximation starts with the simplification of the function f_k, by using a *Taylor approximation*. Since we are interested in

$$\begin{aligned} PL_T &= \pi_T(\mathcal{E}) - \pi_0(\mathcal{E}) \\ &= [\pi_T - \pi_0] \circ \left(\sum_{i \in S_A} \alpha_i \mathcal{A}_i + \sum_{i \in S_L} \lambda_i \mathcal{L}_i \right), \end{aligned}$$

we use the following first order Taylor approximation

$$\begin{aligned} \pi_T(e_k) - \pi_0(e_k) &= f_k(R_T) - f_k(R_0) \\ &\approx \nabla f_k(x) \|_{x=R_0} \times \Delta(R). \end{aligned}$$

If we apply this formula to all assets and liabilities we get a model where the gains and losses are linear in the risk factors R. If there is a balance sheet $x = \sum_{k \in \mathbb{N}_n} \gamma_k \, e_k$ we can obviously sum over the different e_k and we get the following approximation:

$$\begin{aligned}\pi_T(x) - \pi_0(x) &= \sum_{k \in \mathbb{N}_n} \gamma_k \times (f_k(R_T) - f_k(R_0)) \\ &\approx \delta^T \times \Delta(R),\end{aligned}$$

where

$$\delta = \sum_{k \in \mathbb{N}_n} \gamma_k \times \nabla f_k(x) \|_{x=R_0},$$

and where we denote with x^T the transposed of a matrix or vector.

Another simplification is to use a stochastic process, which is analytically easy to tackle. Both the risk metrics method and also the Swiss solvency test use a multi-dimensional normal distribution for $Z = \Delta R$.

Hence we have

$$Z \sim \mathcal{N}(0, \Sigma),$$

where we Σ denotes the *covariance matrix*. One can express this matrix by the standard deviation vector s for each of the risk factors and the correlation matrix ρ. In a first step we define the matrix $S = (v_i \times \delta_{ij})_{i,j}$. Furthermore we need to know that if $X_1 \sim \mathcal{N}(\mu_1, \Sigma_1)$ is a multidimensional normal distribution and A and b are a matrix and a vector, respectively, we then know that $X_2 := A \times X_1 + b \sim \mathcal{N}(\mu_1 + b, A \times \Sigma \times A^T)$. Using this formula we finally get the following relationship:

$$\Sigma = S \times \rho \times S,$$

keeping in mind that $S = S^T$.

If we use the two approximations, the calculation of the VaR at a level α (eg 99.5%) can be calculated as follows. In a first step we denote by

$$\zeta = F^{-1}_{\mathcal{N}(0,1)}(\alpha),$$

and we get in consequence:

$$\begin{aligned}VaR_{PL}(\alpha) &= F^{-1}_{\mathcal{N}(0,1)}(\alpha), \\ &= \zeta \times \sqrt{(s \times \delta)\rho(s \times \delta)^T}.\end{aligned}$$

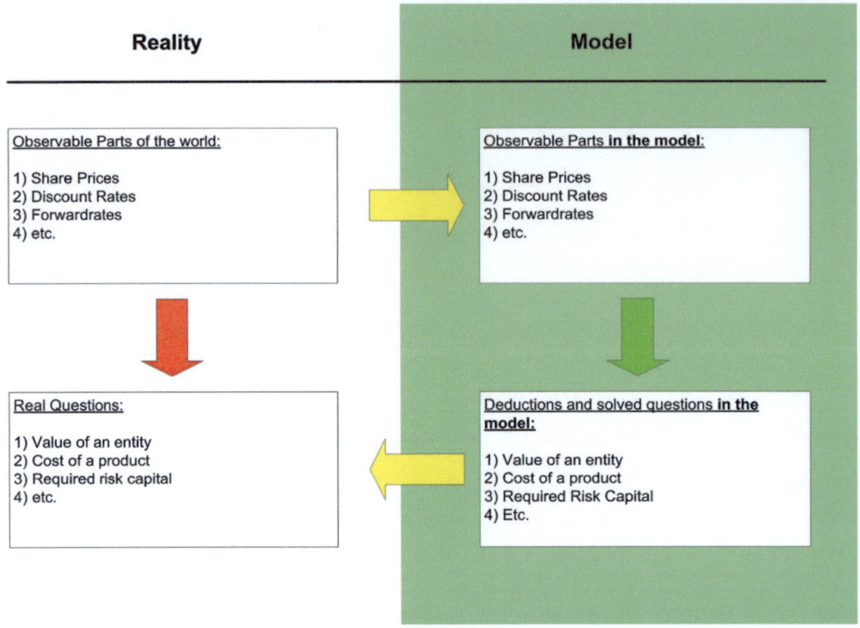

Fig. 6.1 Models and Model Risk

Hence the value at risk can easily be calculated using some simple matrix multiplications. The example which follows is based on these approximations.

At this point it is important to remark that every model has flaws and hence it is of utmost importance to understand the *limitations* of a model. The risk to choose a "wrong" or "inaccurate" model is called *model risk*. Here it important how a model is constructed. Figure 6.1 aims to show this. In principle there are the reality (left hand side of the figure) which one tries to model in order to answer "difficult" questions which can not be answered directly. In order to do that one creates a model (right hand side of the figure) and one should be able to answer the corresponding questions within the model. Next one translates the results back to reality and "hopes" that the diagram is commutative. From this point of view the model risk is the missing "approximate" commutativity of the model. As a corollary one needs to acknowledge that each model is suited and best adapted for a certain purpose and that it is dangerous to use the model outside that.

Another interesting aspect with respect to model risk is the fact that one can, from time to time, observe difficult and lengthy discussions between experts on which model is better. Such discussions can stem from the fact that these people do not distinguish between reality and the model and hence these discussions can end up in religion like beliefs.

In the same sense the results of every model depend on the parameters chosen. The risk of inaccurate model parameters is called *parameter risk*. An easy example is the equity volatility, which is for example used for the Black-Scholes model. The value of the corresponding options is heavily dependent of the volatility chosen. As remarked before the volatility for equity market indexes is normally in the region of say 17 %. In case of market disruptions this parameter can spike up to 30 % and above. Hence it is crucial to exactly know how the model behaves with respect to different parameters.

Finally it is worth noting that the distinguishing between model and parameter risk is not always clear.

6.3 Concrete Implementation

For the concrete implementation of a risk model for financial risk there are, in principle, the following three different approaches:

1. *Analytical approach*, such as the one used in the Swiss solvency test: Here the required risk capital is calculated based on a closed formula. The advantage here is the fast calculation times because this approach is only feasible for a limited class of model.
2. *Model based simulation* (aka Monte Carlo approach): One can, in principle, use whichever model is deemed to be adequate and one simulates the corresponding random variables. Here one can also use sophisticated methods to link variables together such as the copula method. This method is very flexible - for the price of having normally longer running times, since one requires normally a sizable amount of simulations in order to determine the tail probabilities with an adequate accuracy. Assume for example that we are interested in the 99.5% VaR. In this case we have only 500 simulations which are beyond this level for a sample of 100000.
3. *Historical Simulation*: In this case one uses past observed financial data to predict the future. The big advantage is the fact that we do not need to assume which is the correct distribution. In this class of methods we can either run through the past time series or one can use the boot-strapping method. The problem with this method is the fact that there are only quite short time series (say 50 years) for the underlying financial data. Since one is normally interested in rare events such as the one in a 200 year event one needs to amend this method correspondingly. Furthermore, one needs also to remark that the behaviour of some financial variables has changed considerably over the past 50 years, such as foreign exchange rates, which were fixed until the 1970s and are now floating.

As seen above there are different methods on which we can implement financial risk management. In this section we will have a closer look at the multi-normal model, as used in the analytic part of the Swiss solvency test. First we need to look at the risk factors used and then to calculate the risk capital for the balance sheet introduced above, based on a simplified model.

The Swiss solvency test uses the following risk factors:

- Zero coupon prices for CHF, EUR, GBP and USD, for 13 time buckets,
- Interest rate volatility,
- Credit Spreads for four different rating categories,
- Four different currencies vs CHF: EUR, GBP, USD and YEN,
- Seven equity indexes,
- Equity volatility,
- Real estate, hedge funds and private equity indexes,

each of which is modelled as a normal distribution. Before making a concrete example we want to have a look on how big the different quantities are. Since there are 81 risk factors, this would result in a 81x81 covariance matrix. In consequence we will have a look at a part of it. Firstly we want to look at the corresponding standard deviations (as of 31/12/08).

Risk Factor RF_i		Quantity	$\sigma(RF_i)$
EUR	1	bps	61.82
EUR	2	bps	72.08
EUR	3	bps	73.00
EUR	4	bps	73.12
EUR	5	bps	83.53
EUR	6	bps	70.43
EUR	7	bps	68.09
EUR	8	bps	65.93
EUR	9	bps	64.88
EUR	10	bps	63.54
EUR	15	bps	58.91
EUR	20	bps	60.94
EUR	30	bps	59.95
EURO STOXX		in%	18.78
Credit	AAA	bps	11.08
Credit	AA	bps	12.00
Credit	A	bps	23.80
Credit	BBB	bps	52.60

From the above table we see that the volatility for equities was about 19% and the standard deviation for spread risk increases if the credit quality deteriorates. In a next step we want to have a look at the (simplified) correlation matrix:

$\rho_{i,j}$	EUR 5	EUR 10	EUR 20	STOXX	AA	BBB
EUR 5	1.00	0.89	0.66	0.36	-0.14	-0.23
EUR 10		1.00	0.73	0.32	-0.17	-0.21
EUR 20			1.00	0.16	-0.09	-0.09
STOXX				1.00	-0.45	-0.50
AA					1.00	0.61
BBB						1.00

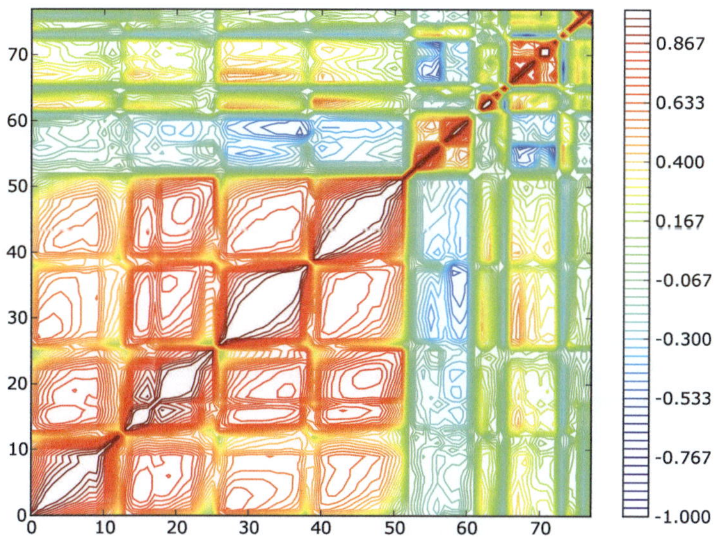

Fig. 6.2 Correlation Matrix

Looking at figure 6.2 we see clearly how the different risk factors are situated in the matrix. One sees the four times 13 risk factors relating to interest rates as first block, which is highly correlated between themselves and slightly less between different currencies. Afterwards one sees the correlation with the credit block, followed by the equity–like investments etc.

From the above table it becomes obvious that the credit spreads have a quite high negative correlation with stock market prices. This means that credit spreads increase normally at the same time when equity markets fall. One can observe that increasing stock market prices normally imply increasing interest rates. These two remarks are for example valid for the credit market crisis in 2008. Here we observed decreasing stock market indexes, reduced interest rate levels and increased credit spreads.

In order to make a concrete example based on the above data we need to base the calculation on a balance sheet and we assume:

Item	**EUR**	**Term**	**Rating**
MR	-10000	10	Government
Bond 1	5000	5	BBB
Bond 2	4000	20	AA
STOXX	2000		
Capital	1000		

So as a first step we need to calculate the sensitivities regarding our risk factor vector EUR 5,EUR 10, EUR 20, STOXX, AA, BBB. We assume for the sake of simplicity each of the bonds and the mathematical reserve (MR) is zero coupon bonds with the corresponding term. In this case the duration equals the term, as one can easily verify. Since the volatility for interest rates and credit spread movement is stated in bps, we also need to calculate the sensitivity of the corresponding values per bp.

Since the MR is considered as a $\mathcal{Z}_{(10)}$ there is only a sensitivity with respect to the EUR 10 year risk factor and an upward movement of 1% reduces the reserve by 10%, so from a capital point of view we have an entry of +1000. For 1 bp we hence have +10 and the sensitivity factor for this liability reads as $\delta_{x_1} = (0, 10, 0, 0, 0, 0)$. In the same sense we can calculate Bond 1 (δ_{x_2}) and Bond 2 (δ_{x_3}), remarking that both of them are sensitive also with respect to credit spreads we get

$$\begin{aligned}\delta_{x_2} &= (-2.5, 0, 0, 0, 0, -2.5),\\ \delta_{x_3} &= (0, 0, -8, 0, -8, 0).\end{aligned}$$

For the share we calculate the sensitivity for an increase of 1% and hence we get:

$$\delta_{x_4} = (0, 0, 0, 20, 0, 0),$$

and therefore we get for the total sensitivity:

$$\begin{aligned}\delta_{Tot} &= \sum_{k=1}^{4} \delta_{x_k}\\ &= (-2.5, +10, -8, +20, -8, -2.5).\end{aligned}$$

In a next step we need to calculate:

$$\begin{aligned}s \times \delta_{Tot} &= (83.53 \times -2.5, \ldots, 52.61 \times -2.5)\\ &= (-208.83, 635.42, -487.52, 375.70, -96.03, -131.51).\end{aligned}$$

Now we can calculate the standard deviation of the capital, considered as a random variable by:

$$\sigma = \sqrt{(s \times \delta)\rho(s \times \delta)^T}$$
$$= 546.6\text{M EUR.}$$

As a consequence of this the VaR for the 99.5% corresponds to $VaR_{99.5\%} = 2.57 \times 546.6 = 1404.7$ M EUR. If one further decomposes the VaR, one could look at pure interest rate VaR. In this case one would look at the corresponding δ:

$$s \times \delta_{Interest} = (-208.83, 635.42, -487.52, 0, 0, 0),$$

and we would get in the same way $VaR_{99.5\%}^{\text{Interest}} = 2.57 \times 487.9 = 1253.9$. This is the way how one determines which parts of the balance sheet contribute most to the risk. In the concrete example we get (in M EUR):

Item	Std Deviation	99.5% VaR
Bonds	487.9	1253.9
Equities	375.7	965.5
Credit	204.8	526.3
Simple Sum	1068.4	2745.7
Diversification	-521.8	-1341.0
Total	546.6	**1404.7**

Finally, we want to have a look at the accuracy of the linear approximation, which we have used. For shares there is nothing to do, since the change in value of the asset is linear. Therefore we want to look at the accuracy of the approximation for bonds. For simplicity we assume that the yield curve is flat at $i = 4\%$. In this context we have:

$$\pi_t(\mathcal{Z}_{(n)})[i + \Delta_i] - \pi_t(\mathcal{Z}_{(n)})[i] = (1 + i + \Delta_i)^{-n} - (1 + i)^{-n}.$$

We remark that the volatility of bonds is about 60 bps, therefore looking at the 99.5% (which is about $2.57 \times \sigma$) implies, looking at the precision of the approximation, at a shift of c 150 bps.

Term 20 yrs	True Change	Delta Approx.	Error
-300	0.3631	0.2738	-24.5%
-200	0.2165	0.1825	-15.7%
-150	0.1538	0.1369	-11.0%
-100	0.0972	0.0912	-6.1%
-50	0.0461	0.0456	-1.1%
50	-0.0417	-0.0456	9.3%
100	-0.0794	-0.0912	14.8%
150	-0.1136	-0.1369	20.4%
200	-0.1445	-0.1825	26.2%
300	-0.1979	-0.2738	38.3%

Consequently, we see that this approximation has some non negligible errors, which can be mitigated by adjusting the duration accordingly. Another source of such non-linearities are options, where a standard model is inadequate. Hence we want to have a look at a possible solution. In order to do that we need to go back to first principles, which define the model *before* approximations. In our case we assume that the risk factors are following a multi-normal distribution for $Z = \Delta R$. This means that for some of our assets $(\mathcal{A}_i)_{i \in S_A}$ or liabilities $(\mathcal{L}_i)_{i \in S_L}$ the linear approximation is inadequate. The method which we want to show here works in general and can either be applied to one or more of the underlying assets and liabilities. It works for example for plain-vanilla stock options, which can be valued using the:

The price for a *put*-option with payout $C(T, P) = \max(K - S; 0)$ at time t and strike price K and equity price S is given by:

$$\begin{aligned}
P &= K \times e^{-r \times T} \times \Phi(-d_2) - S_0 \times \Phi(-d_1), \\
d_1 &= \frac{\log(S_0/K) + (r + \sigma^2/2) \times T}{\sigma \times \sqrt{T}}, \\
d_2 &= d_1 - \sigma \times \sqrt{T}, \\
\Phi(x) &= \int_{-\infty}^{x} \frac{1}{\sqrt{2\pi}} \exp(\frac{-\zeta^2}{2}) d\zeta.
\end{aligned}$$

The risk factors which enter into the calculation of the price $\pi(\mathcal{A}) := P$ are the following:

- Share price S_t,
- Volatility of the share price σ, and
- The interest rate for the corresponding term r.

It needs to be stressed that the price of such an option is clearly not linear in the risk factors.

In order to assess the corresponding risk, one can for example use a partial or full simulation approach. In the first case the whole distribution is simulated with a sufficiently big sample and the effective change in capital is evaluated and recorded in order to determine the value of the chosen risk measure such as the VaR or TVaR. One can also use a partial simulation approach remarking that only nonlinear instruments need to be simulated. Here the question is how to "marry" the simulated and the analytical parts. One approach is to use control variables.

In order to describe this approach let's assume that we have one asset $\mathcal{A}$, which is not linearly dependent on the risk factor and assume for sake of simplicity that we have the following:

$$\tilde{\pi}_t(x) - \pi_0(x) = \delta^T \times \Delta(R), \text{ and}$$
$$\pi_t(x) - \pi_0(x) = f(\Delta(R)).$$

In the above we denote with $\tilde{\pi}_t(x)$ and $\pi_t(x)$, the approximated and the "true" change in value, respectively. The function f denotes the "true" change in value for a given $\Delta(R)$. So in order to do a partial simulation approach one needs to do the following:

1. Simulate n (say 50000) times the random variable R, resulting in a series of $(r_k)_{k=1,2,\ldots n}$.
2. Calculate the analytic value of the risk measure C_a^l for the linear approximation.
3. Calculate the simulated value of the risk measure C_r^l for the linear approximation, using the series $(r_k)_{k=1,2,\ldots n}$.
4. Calculate the simulated value of the risk measure C_r^f for the "true" value, using the series $(r_k)_{k=1,2,\ldots n}$.

As a consequence of the weak law of big numbers we have $C_a^l = \lim_{n\to\infty} C_r^l(n)$. Hence using the difference $C_a^l - C_r^l$ as a correction to C_r^f normally improves the quality of the approximation.

The table below show an example for the accuracy of the linear approximation in case of a plain vanilla put option. At time $t = 0$ we assume a stock price of 1000 and we consider a strike for the put option at 900. For this example a sample size of 50000 has been chosen and analysed for the first 500, the first 1000 samples etc.

Sample size	Linear Value	Linear Error	BS - Price Value	BS - Price Error
500	97680	+0.7%	159238	+1.5%
1000	97911	+0.9%	160029	+2.0%
2000	97908	+0.9%	160017	+2.0%
5000	96412	-0.6%	154571	-1.4%
10000	97911	+0.9%	160029	+2.0%
20000	96672	-0.4%	157896	+0.7%
50000	97038	ref	156831	ref

What can be seen from the above example is that the linear model converges much faster and in this case the value of the put-option to the company holding it is *underestimated.*

6.4 Interpreting the Results

This section provides a reporting template which can be used for financial risk management. This template risk report is subdivided into the following parts:

Summary: The aim of this section is to provide a concise summary. In order to get a high level view on the duration gap between assets and liabilities, the corresponding durations are calculated. Furthermore we see the impact of an increase in interest rates of 10 bps and an increase of 1% in equity prices, separately for assets and liabilities only and combined. After these deterministic measures we see some important key measures in terms of VaR, for both a one in ten year (1:10) and a one in 250 year (1:250) event. Here we look at combinations of risk factors. Namely we look separately at equities, bonds, surrenders and the total. This total VaR needs to be compared with the available capital (Market Value). Finally, also the Tail VaR or Expected Shortfall (ES) is shown. The figure underneath shows the required capital for different return periods (separately for assets and for the total). The two red balls represent the VaR in a 1:10 and a 1:250 year event and these numbers reconcile to the table.

Decomposition of VaR: In order to better understand where capital is consumed the total VaR is further decomposed into its components. It is possible to see which parts of the assets and liabilities consume the majority of the capital. In the concrete example we can see that most of the total required capital of 3216 M EUR is consumed by credit risk (2042 M EUR). Furthermore it becomes obvious that the pure ALM risk (in terms of interest rates) is quite small with 764 M EUR. Finally we see that equity risk and hedge funds account still with 764 M EUR and 464 M EUR respectively. At the bottom of the page we see that the surrender risk amounts to 736 M EUR.

Individual Capital Assessment (ICA): In order to be able to compare the model with the regulatory standard ICA model the corresponding results have been included in this section. It needs to be stressed that the ICA model covers more risk factors, but is not as granular for the market risk factors as the own model.

Scenarios: In this section some scenarios are shown and in particular, how the company balance sheet would look after such an event. Section 6.4.2 shows the main characteristics of the scenarios used. The balance sheet items for each of the scenarios follow a typical IFRS balance sheet. The figures underneath the table show the change in shareholders equity and the decomposition of the balance sheet post stress respectively.

Stress Tests: The section stress tests is thought to represent some additional scenarios as described above. The only difference is the fact that here the scenarios are shown in a summarised version and are based on group requirements. The scenarios currently used have been defined by FSA[1] and are quite self-explanatory.

Limits: This section aims to show the limits currently in place to limit the ALM risk. The table shows the limits currently in place: The target which is limited, the threshold, the current level and the headroom. The program has been built up in such a way that every number which is checked against a limit is either printed in green (e.g. within limit) or red (e.g. limit breach).

6.4.1 Notation

In this section the main elements in respect of notation are documented.

VaR	Value at risk, eg the Loss which occurs according to a certain probability. In the analysis a 99.6% VaR is used. This means that the loss represents a loss which in the long run is expected to occur every 250 years. It needs to be noted at this point of time, that analytical models tend to underestimate such losses since the risk factors have been modelled as normally distributed.
1:10	This symbol also relates to a VaR, in this case corresponding to the 90th quantile, e.g. once every 10 years.
Duration	The modified duration which represents the risk intrinsic to a bond portfolio
Sensi Bonds (+10 bps)	The change in value of a bond portfolio if the yield curve is shifted by 10 bps (= 0.1%).
ES 99%	The expected shortfall in a 1 in 100 year event is defined of the average loss looking at all events occurring less than once in 100 years. This measure is more sensitive in the tails than the VaR.
Intangibles	The intangibles in the balance sheet (eg goodwill etc.) In case of an impairment of participation the model reduces the intangibles in a first step.
MR	mathematical reserves for traditional business. They are moving in line with the interest rates.
UDS	Undistributed surplus.
Tax	Taxes and deferred tax assets and liabilities are not modelled.
SHE	Market Value Shareholder funds. This corresponds in principle to the corresponding MCEV.
Δ SHE	Change in SHE in case of a certain scenario.

[1] Financial Services Authority in the UK.

- The figure *Distribution of Losses* shows the probability density function of the losses. The two red circles represent the VaR 1:10 and in respect of the 99.5% quantile.
- The figure *Cash Flow Profile* shows the inflows (red: bond payments and yellow: premiums) v.s. the outflows (blue: expected claims).
- The figure *Diversification* shows the diversification effect in relation to the main asset risks.
- The figure *Decomposition of required Capital* and *Credit Risk by Rating* show which risk and which credit risk absorbs most of the required capital.

6.4.2 Scenarios

In the following section the different used scenarios are defined in some greater detail:

	Credit	Yen	Depr.	FSA	Hard Land.	Depr. ii
i-rate 2 yrs (bps)	0	-399	-399	50	90	-220
i-rate 7 yrs (bps)	0	-316	-316	50	90	-220
i-rate 10 yrs (bps)	0	-323	-323	50	90	-220
i-rate 25 yrs (bps)	0	-303	-303	50	90	-220
Shares (%)	-18	-18	-65	-34	9	-32
Properties (%)	-5	-5	-55	-19	-28	-36
AA Credit Spread (bps)	103	0	103	50	40	110

6.4.3 What Can and What Cannot Be Done with This

As indicated above, a model is not reality and hence it is of utmost importance to recognise the limitations of such a model. In this section we try to show some of the limitations of the model currently used. One of the possible risks of this model is that it is overly simplistic.

From a high level point of view the main shortcomings of the model are:

The model is linear: The different risk factors enter linearly into the calculation of the loss. Therefore for options, the corresponding delta equivalent is used. In a next step such effects should be captured better.

The model uses a standard multi-normal distribution.

Management actions: Management actions are not taken into account.

Insurance and operational risks: The model purely focuses on ALM risk.

Dynamic Lapses: Dynamic lapses are also an area where the model used needs refinement.

6.5 Reporting Example

6.5.1 Summary

Assessment and key Figures	Data Quality
• Globally the company has currently not enough risk capital, from a purely economic perspective, to run the corresponding ALM risks, since the margins have become tighter due to the losses in the equity and corporate bond portfolio and widening of the credit spread. • The statutory reserve set up for GMDB at the year-end (31.12.2008) was 130m EUR whereas the more economic vision used in the MCEV calculation produced a value of 162m EUR. During the first quarter the statutory reserves increased to c230m EUR, affecting the IGD Solvency adversely by c82m EUR.	• Private Equities are currently within the category hedge funds • The current analysis is still in draft form and is based on data as of 31.12.2008.

Item	Assets	Liabilities	Total
Duration	5.81	5.12	
Sensi Bonds (+10bps)	-177.87	144.98	-32.88
Sensi Equities (+1%)	14.90	-0.21	14.69
1:10 Bonds	1505.30	1174.40	352.94
1:10 Equities	370.32	15.68	380.18
1:10 Total	1809.60	983.14	1600.50
VaR Bonds	3025.60	2360.40	709.39
VaR Sx	–	736.69	736.69
VaR Equities	744.32	31.51	764.13
VaR Total	3637.10	1976.00	3216.90
ES 99%			3098.60
Market Value	–	45343.00	**2780.00**

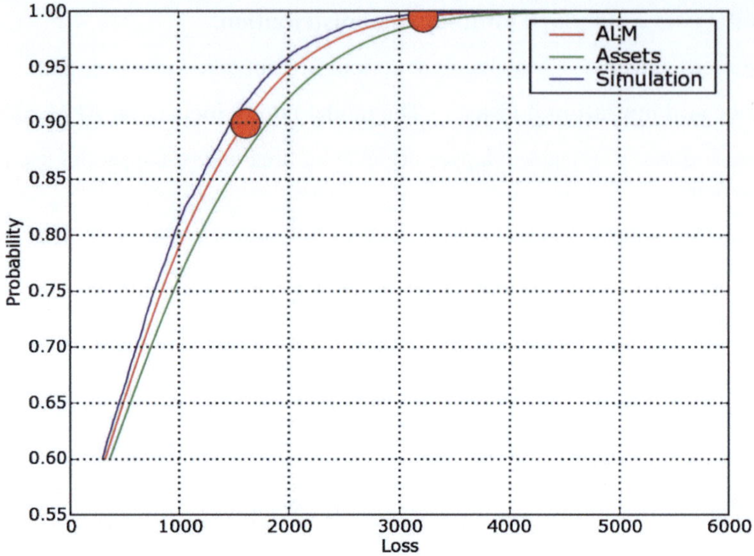

Fig. 6.3 Distribution of Losses

6.5.2 Decomposition of VaR

Assessment and key Figures	**Data Quality**
• The total required risk capital amounts to 3.2 bn EUR (pre Tax) and to c 2.1 bn EUR (post Tax), compared with a available risk capital of c 2.7 bn EUR (after Tax). The 2.1 bn EUR compare to 2.5 bn EUR for the YE2008 ICA calculation. The biggest difference is the fact the ALM Capital does not take into account risk other than market and surrender risks. The biggest additional contribution is the expense risk capital of c 0.6 bn EUR. Adding this to the ALM Capital the two numbers get closer with a difference below 0.1 bn EUR. Overall, both metrics result in similar numbers. • The ALM mismatch consumes about $\frac{1}{2}$ and the equities et al exposure ca. $\frac{1}{2}$ of the total risk capital. This indicates the company has a rather high risk in equities, private equities, properties and hedge funds. In particular, the capital needed for alternative investments is almost 20% of the total available risk capital. • Most of the ALM mismatch stems from the long duration liabilities which are not matched with corresponding assets.	• Replicating the portfolio for one major product line under review • Available risk capital not yet calculated and the current figure is based on an estimation. • GMDB exposure reflected via δ–equivalent for equities and volatility via θ. Interest rate sensitivity not yet reflected.

Item	Assets	Liabilities	Total
Market Value	–	45343.00	**2780.00**
Bonds EUR <3	48.26	52.35	11.78
Bonds EUR 3-7	616.54	454.94	162.66
Bonds EUR 8-12	1446.40	810.34	636.14
Bonds EUR 13-24	1007.60	925.96	109.24
Bonds EUR >25	26.64	263.45	236.81
Bonds EUR Total	3025.60	2360.40	709.39
Div. Ben.	-119.88	-146.66	-447.24
Bonds GBP <3	–	–	–
Bonds GBP 3-7	–	–	–
Bonds GBP 8-12	–	–	–
Bonds GBP 13-24	–	–	–
Bonds GBP >25	–	–	–
Bonds GBP Total	–	–	–
Div. Ben.	–	–	–
Bonds USD <3	–	–	–
Bonds USD 3-7	–	–	–
Bonds USD 8-12	–	–	–
Bonds USD 13-24	–	–	–
Bonds USD >25	–	–	–
Bonds USD Total	–	–	–
Div. Ben.	–	–	–
Bonds CHF <3	–	–	–
Bonds CHF 3-7	–	–	–
Bonds CHF 8-12	–	–	–
Bonds CHF 13-24	–	–	–
Bonds CHF >25	–	–	–
Bonds CHF Total	–	–	–
Div Ben.	–	–	–
All Bonds	**3025.60**	**2360.40**	**709.39**
Div. Ben.	–	–	–
Credit Risk	2042.20	–	2042.20
Shares MSCIEMU	744.32	–	744.32
Shares MSCICHF	–	–	–
Shares MSCIUK	–	–	–
Shares MSCIUS	–	–	–
All Shares	744.32	**31.51**	**764.13**
Div. Ben.	–	–	*-11.70*
FX GBP	–	–	–
FX USD	–	–	–
FX GBP	–	–	–
FX Total	**–**	**–**	**–**
Div. Ben.	–	–	–
Real Estate	131.20	9.45	121.75
Alternatives	464.46	–	464.46
Participations	191.90	–	191.90
Total	**3637.10**	**1976.00**	3216.90
Div. Ben.	*-920.34*	*-425.30*	*965.29*
Surrenders	–	736.69	736.69

6.5.3 Figures

Assessment and key Figures	Data Quality
• The duration of the bonds with 5.6 years is considerably shorter than the ones of the liabilities with 10.6 years. In part this is due to the special characteristics of a particular insurance portfolio and corresponding analysis are under way. • From a liquidity point of view the company has considerable amounts of bonds maturing within 1 year and 2 years leading to an excess liquidity of ca 2bn EUR and 1bn EUR respectively. • The table relating the shift in asset value for a shift in credit spreads shows clearly the high credit quality of the underlying assets corresponding to a average rating of slightly above AA	• Replicating portfolio for particular product line under review • Derivatives not yet reflected in analysis

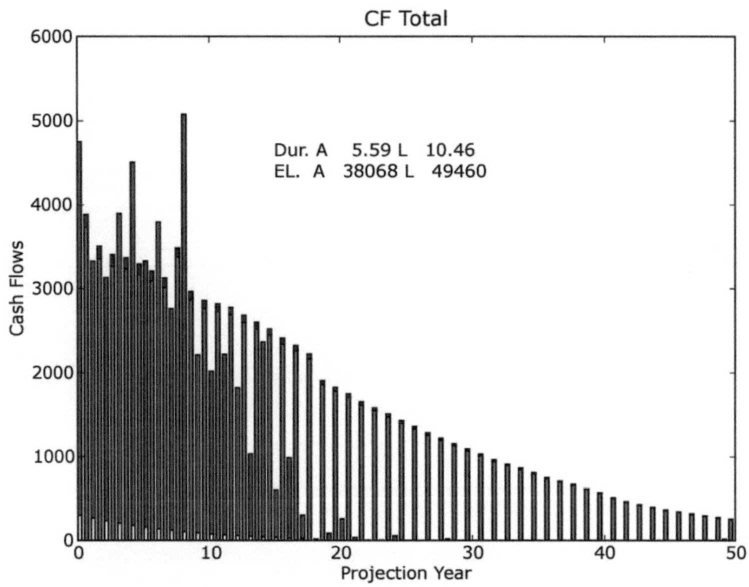

Fig. 6.4 Cash Flow Profile

Credit Quality	+10 bps Spread	Percentile	Δ Profit
EURO AAA	-89.13	1%	2905.30
EURO AA	-23.76	5%	2054.20
EURO A	-42.69	10%	1600.50
EURO BBB	-24.16	33%	538.04
USD AAA	–	66%	-537.70
USD AA	–	90%	-1600.50
USD A	–	95%	-2054.20
USD BBB	–	99%	-2905.30
Total	**-179.76**	99.5%	-3216.90

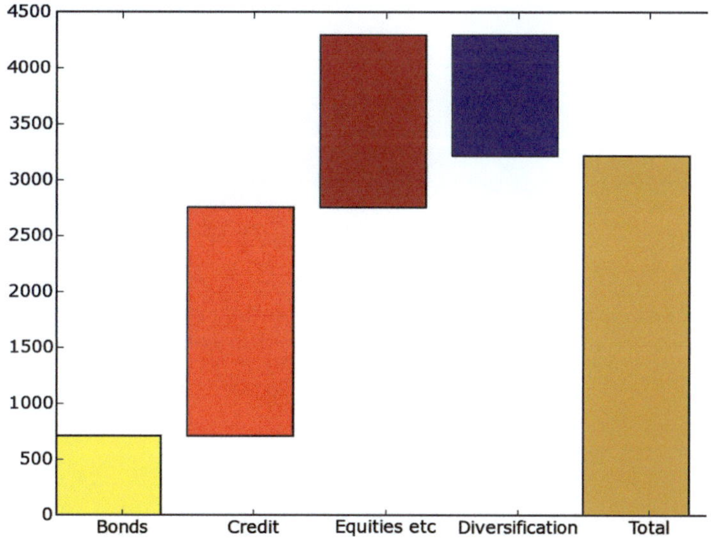

Fig. 6.5 Diversification

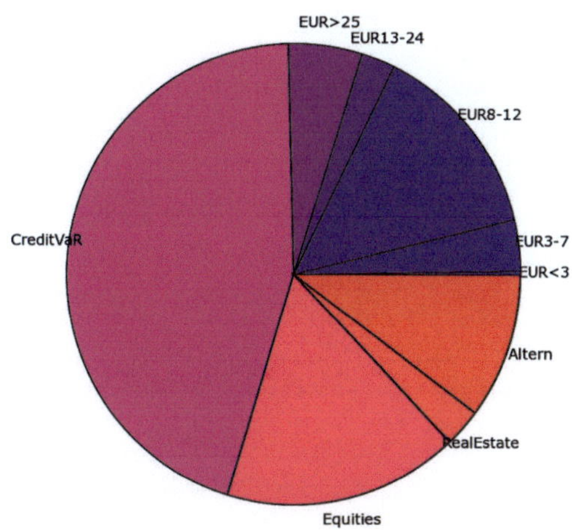

Fig. 6.6 Required Capital

6.5.4 Scenarios

Assessment and key Figures	**Data Quality**
• The main three scenarios consist of a widening of credit spreads by a further 50%, a falling of the interest rates to YEN levels and a global severe depression.	• The current analysis is work in progress and is based on data as of 30.12.08

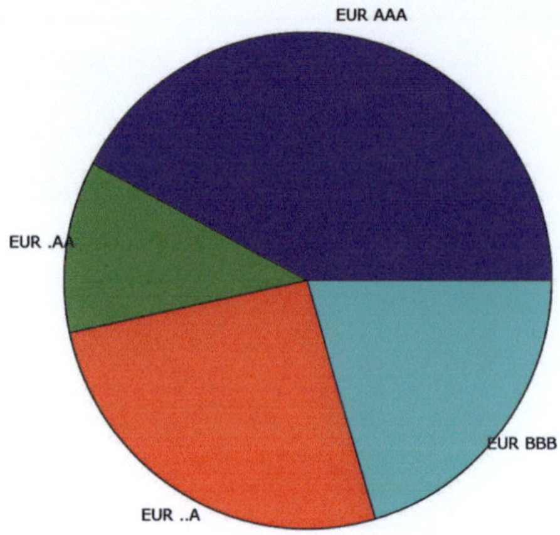

Fig. 6.7 Credit Risk by Rating

Item	Start	Y to D	Credit	YEN	Depression
Cash	6221.10	6221.10	6221.10	6187.00	6187.00
Bonds	35734.00	35734.00	33730.00	41567.00	39564.00
Shares	4468.30	4207.50	4207.50	4207.50	3499.50
Properties	1137.10	1081.70	1081.70	1081.70	528.15
Hedge Funds	595.70	521.24	521.24	521.24	59.57
Private Equity	64.00	56.00	12.80	56.00	6.40
Loans	1452.10	1452.10	1452.10	1452.10	1452.10
Unit Linked Assets	14330.00	14330.00	14330.00	14330.00	14330.00
Other Assets	3612.00	3612.00	3612.00	3612.00	3612.00
Intangibles	204.90	152.75	167.65	152.75	-63.30
MR	47066.00	47066.00	47326.00	51773.00	51773.00
Unit Linked Liabilities	14330.00	14330.00	14330.00	14330.00	14330.00
UDS	–	–	–	–	37.60
Debt	–	–	–	–	–
Deferred Tax	127.70	127.70	127.70	127.70	127.70
Other Lia	3801.90	3797.90	3797.90	3797.90	3758.00
SHE	2493.00	2046.20	-245.49	3139.00	-851.01
Δ SHE		**-446.82**	**-2738.50**	**645.97**	**-3344.00**

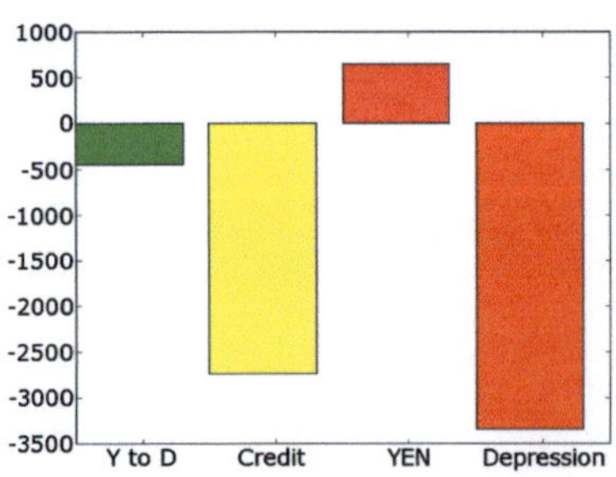

Fig. 6.8 Δ SHE for the scenarios

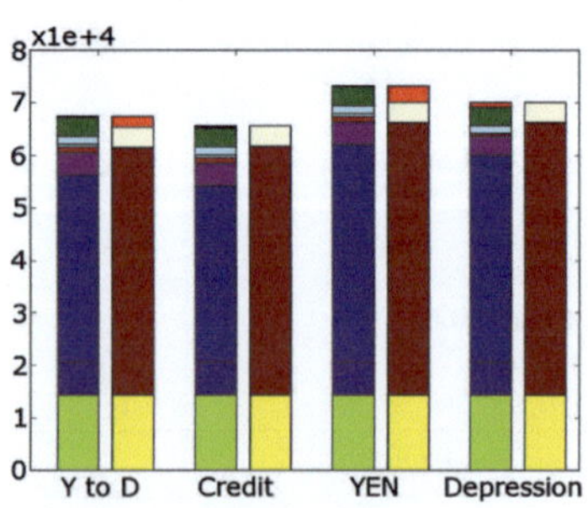

Fig. 6.9 BS for Scenarios

Year-to-date: The interest rates decreased by c 200 bps at the short and 30 bps at the long end. At the same time credit spreads widened between 90 bps (AAA) and 380 bps (BBB). Stock markets reduced by 35+%.
Credit Scenario: An additional spread widening of 50%.
YEN Scenario: Based on the current scenario, interest rates have been lowered to levels where the Yen was at the its deepest level
Depression: YEN interest rates, a 40% credit spread widening and a cumulative reduction of 65% for shares, PE, HF and properties.

6.5.5 ICA Capital

Assessment and key Figures

- The falls in available economic capital and the increase in capital requirements are largely driven by:
  Falls in equity markets and increases in credit risks have lead to minimum investment guarantees "biting", with a direct burn-through impact on shareholder assets.
  Saving products experienced significant erosion in value of future profits (VIF), with asset returns over the year close to minimum guarantees.
  Falls in the equity market meant unit linked contracts experienced an increase in Guaranteed Minimum Death Benefit (GMDB) risk.
- Changes in YE2008 stress methodology, in particular, the "softening" of equity and credit spread tests were key to keeping the funds from going into deficit on an economic basis.
- The company has completed an equity de-risking initiative, leading to a further fall in capital requirement for equity risk.
- The company believes a significant part of the credit spread widening is linked to liquidity premium and for some products, creates an artificial and unnecessarily high capital requirement.
- A separate exercise will need to be performed to quantify the reputational risk associated to structured products that have been sold with the underlying guarantees provided by third parties. The default risk is borne by the client but a reputational risk would remain with the company (total reserves 4,844m). This is not part of the YE08 SSTEC requirements, but will be investigated given the potentially material impact.

in M EUR	YE 2007	YE 2008	1Q2009	4Q2009
Available Economic Capital	3657	2780		
Reg. Capital Required	2083	2508		
Cover	176%	111%		
Diversification Benefit	39%	41%		
MV Assets	59390	56390		
MV Liabilities	54674	52558		

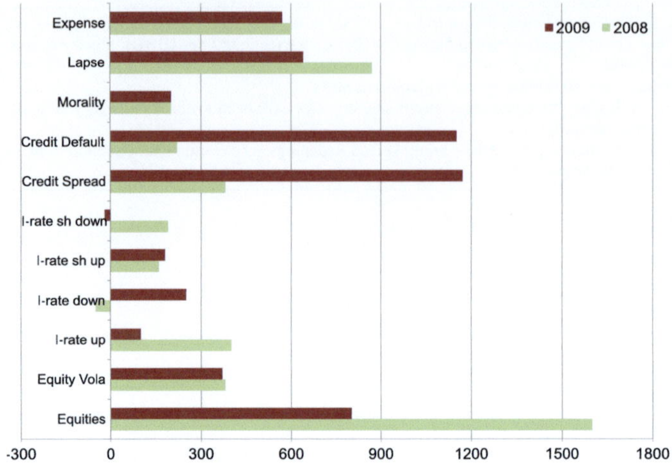

Fig. 6.10 Required ICA Capital by Risk

6.5.6 Stress Tests

Nr	Name	Equity	Δ Assets	Δ Lia	Δ Equity
	B/S	2493.00			
	Y to D	2046.20	-450.81	-3.98	-446.82
1	Equities -10%	1763.00	-603.96	126.01	-729.97
2	Equities -20%	1479.90	-757.11	256.01	-1013.10
3	Equities -40%	1043.60	-1063.40	386.01	-1449.40
4	Equity Vola + 10%	2044.10	-450.81	-1.88	-448.92
5	Property -7.5%	1908.00	-529.70	55.33	-585.03
6	Property -15%	1769.80	-608.58	114.65	-446.82
7	Property -30%	1558.40	-766.35	168.28	-934.64
8	I rates -50 bps	2210.60	438.52	720.94	-282.42
9	I rates -100 bps	2375.00	1327.80	1445.90	-118.01
10	I rates -200 bps	2703.80	3106.50	2895.70	210.79
11	I rates Twist: long down	1988.70	834.64	1339.00	-504.31
12	I rates Twist: long up	2102.20	-1750.40	-1359.60	-390.84
13	Cred spread +50 bps	1147.40	-1349.60	-3.98	-1345.60
14	Cred spread +100 bps	248.60	-2248.40	-3.98	-2244.40
15	Cred spread +200 bps	-1549.00	-4046.00	-3.98	-4042.00
16	FSA	530.28	-2703.00	-740.28	-1962.70
17	FSA Hard Landing	1287.10	-2533.00	-1327.00	-1205.90
18	FSA Depression 2010	206.99	875.04	3161.10	-2286.00

6.5.7 Limits

In this section the various limits are checked:

Limit	Threshold	Actual	Headroom
VaR Equity Asset (20%) ◇	409.24	744.32	-335.08
Total VaR (80%) ◇	1636.90	3216.90	-1580.00
Base Point Sensitivity	87.98	-32.88	55.10
Alternatives VaR (10%) ◇	204.62	464.46	-259.84
Credit VaR (20%) ◇	409.24	2042.20	-1633.00
Credit Scenario SHE > 0	0.00	3139.00	3139.00
Yen Scenario SHE > 0	0.00	-245.49	245.49
Combined 2 Scenario SHE > 0	0.00	1287.10	1287.10
Depression Scenario SHE > 0	0.00	-851.01	851.01
Properties VaR (10%)	204.62	131.20	73.42

6.6 Summary Reporting Example

Figure 6.11 provides an example of a summary on a page for the financial risk a company is facing. The aim is to be concise and also action oriented. Hence the table envisages the following entries:

Name The name of the risk is indicated together with a measure for its size, such as the amount of assets affected, a risk measure etc.

Risk Category The risk category aims to indicate which type of risk is described, such as credit risk, liquidity risk, ...

Risk The risk is described in a concise manner in order that a knowledgeable third party can understand, what the risk is.

Actions This one is the most important column, since the mitigation actions performed and planned are described. This helps to see the development with respect to the corresponding risk.

Remarks Here additional information needed to better understand the issue is documented.

The principle for the writing of such reportings must be *relevant, concise and action oriented.*

Name	Risk Category	Risk	Actions	Remarks
Credit Defaults	Credit Risk	The company had and has substantial credit exposure, in particular towards the banking sector and might suffer credit defaults	• Reduction in holdings in Italian and Greek government bonds • Reduction (€ 100 M) of counterparty exposure to X • Regular exposure control and reporting	The counterparty exposure vis-à-vis our strategic partners is a particular concern due to its strategic rationale.
Hedge Funds	Market Risk	An unclear asset investment rationale and risk controlling was found at the beginning of this year. As a consequence the company faces a variety of risks – exposure € c500M	• Decision to exit hedge funds • So far € 70M have been sold	Due to the introduction of gates and side pockets within the hedge funds the reduction of hedge fund investments will take some time
Equity and GMDB	Market Risk	As a consequence of the credit crises in 2008, the market value of equities dropped considerably together with a spike in equity-volatility. As a consequence the company was and still is vunerable against equity market movements.	• Reduction of € >3bn so far • Hedging of Equity exposure with derivatives • Mitigation actions with respect to GMDB exposure examined • Continuous monitoring	• Equity exposure (€ c900M) protected with put option strategy maturing Mar 2010 • GMDB reserve € 160M exposed towards volatility increase
Interest Rate	Market Risk	Since many of the bond portfolios covering insurance liabilities are not perfectly matched, the company is vunerable against interest rate movements. Most liability portfolios are of a shorter duration than the assets.	• Cash flow analysis performed for major blocks of business. As a consequence the corresponding risk for life portfolios is adequate • GI portfolios were short duration in liabilities and a reduction in asset duration is under way	Regular ALM risk controlling to be built up
Lapse Risk	Insurance Risk	As a consequence of the financial crisis, which now also effects the "real" economy, a lot of our policyholders lapsed their policies, with a potentially adverse impact on our franchise value	• Actions are under way to safeguard the franchise value of the company • The distribution channels have been asked to report back on the steps undertaken so far • Regular reporting of lapse experience	A side effect of the financial crisis is the fact that some of our clients switch their unit linked policies into traditional ones, which are less profitable

Fig. 6.11 Financial Risk Reporting

Chapter 7
Insurance Risks

7.1 Method for Allocation of Capital

The aim of this chapter is to introduce a very concrete risk capital model for life insurance risks and should help to understand the approach which needs to be taken and the respective necessary steps. It is important to understand that there are many other risks that need to be analysed and modelled, such as all GI risks and for life insurance also disability and in particular the lapse risk.

M. Koller, *Life Insurance Risk Management Essentials*, EAA Series,
DOI 10.1007/978-3-642-20721-1_7,

7.1.1 Steps Required

The following steps are needed in order to calculate the required risk capital for life risk

- Definition of the risk factors,
- Definition of a probability density functions per risk factor,
- Definition of a valuation methodology,
- Definition of the joint distribution of all risk factors – diversification,
- Definition of risk measures,
- Definition of the concept of stress scenarios.

7.1.2 Probability Density Functions per Risk Factor

Each of the risk factors needs to be described in terms of stochastic processes and corresponding probability density function.

7.1.3 Diversification

After the definition of the individual probability density function, it is necessary to define the joint distribution of all individual random variables. First, we will use a very simple model and assume that the present values of the corresponding losses are linked by a covariance matrix. Whilst not being the most elaborate method, this approach is pragmatic enough to capture the most relevant interactions.

7.2 Stochastic Models Used

7.2.1 Mortality

Historical statistics show that mortality has generally been improving for many decades. However, in contrast to the longevity, the major risk for mortality, from an insurer's point of view, is a pandemic-like event which could cause an exceptional mortality shock.

A pandemic is an epidemic that spreads worldwide or at least across a large region. A worldwide pandemic is recognised to be "virulent and contagious with high rates of illness and deaths, as well as significant social and economic disruption."

According to an article written by the *British Columbia Pandemic Influenza Advisory Committee (BCPIAC)*, several medical experts think that the threat of a severe "influenza" pandemic is to be feared.
Indeed, not only have past events shown that influenza pandemic strikes about three times a century (e.g. the Spanish Flu (1918 - 1919), the Asian Flu (1957 - 1958) and the Hong Kong Flu (1968 - 1969)), but all the factors contributing to the risk are in place.

In light of this, it is necessary to develop a model integrating a shock to evaluate the amount of capital to hold. As for the longevity model, we will take the future estimation of mortality rate ($q_{x,t}$) as an input.

Model Description

The following model for the mortality process assumes a mortality shock. The best estimate mortality is given by $q_{x,t}$. The shock modelled as a percentage of $q_{x,t}$ is simply given by a Bernoulli random variable ($\mathcal{B}(\gamma)$, with γ the frequency of the shock) multiplied by a log-normal random variable (severity of the shock):

$$Q^i_{x,t}(\omega) = q_{x,t} \times [1 + I_t(\omega) \times R_t(\omega)] + \epsilon^i_{x,t}(\omega).$$

With,

$$\begin{aligned}
(I_t)_{t\in\mathbb{N}} &\sim \mathcal{B}(\gamma) && \text{i.i.d.,}\\
(R_t) &\sim \text{Severity distribution, eg log-normal,}\\
R_t &= \exp(Y),\\
Y &\sim \mathcal{N}(\tilde{\mu}, \tilde{\sigma}),\\
\epsilon^i_{x,t} &\sim F && \text{i.i.d. with } \mathbb{E}(F) = 0,\\
(R, I, \epsilon) &\quad \text{Independent.}
\end{aligned}$$

We will first concentrate our efforts on determining the parameters of the $(R_t)_{t\in\mathbb{N}}$ and $(I_t)_{t\in\mathbb{N}}$ distribution, the aim of which is to simulate the non diversifying risk of mortality. The $(\epsilon)^i_{x,t}$ vector since it represents the diversifying risk.

Determination of $\gamma, \tilde{\mu}, \tilde{\sigma}$

We use the probability of death from the Human Mortality Database (HMD)[1] to fit the parameters needed. We used data from different countries: Sweden, England and France, in order to obtain greater information on the stability of the parameters.

To illustrate the data we used, here are the evolution of the probability of death at age 40 for the 3 countries, see figures 7.1, 7.2 and 7.3.

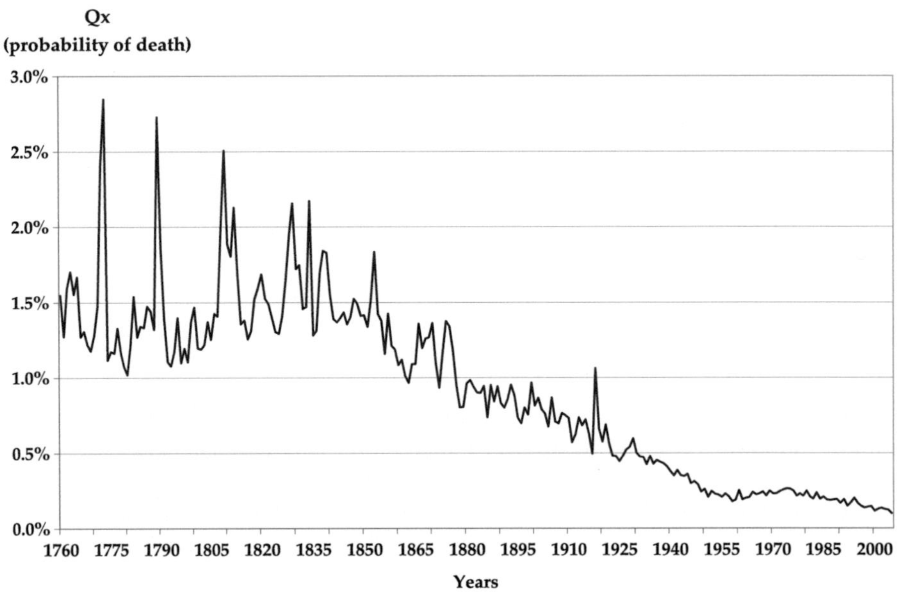

Fig. 7.1 Evolution of death rate at age 40 for males from Sweden

Estimation of Shocks in Past Data

As a first step, we tried to identify the shocks in mortality rate and express it as a percentage of the 10 years moving average q_x. Concretely, in total, 6 series of historical death rate (one for males and one for females for each of the following countries: France, England and Sweden) were at our disposal in a period of time $[T_1, T_2]$ (see values in table 7.1):

$(\hat{q}_{x,t})_{t\in[T_1;T_2];x\in[0;110]}$

Let us define, for $t \in [T_1 + 5; T_2 - 5]$ the following kernel estimator:

[1] www.mortality.org

Fig. 7.2 Evolution of death rate at age 40 for males from England

Table 7.1 Value of T_1 and T_2 by country

Country	T_1	T_2
France	1899	2004
England	1841	2003
Sweden	1751	2005

$$\tilde{q}_{x,t} = \frac{\sum_{i=-5}^{5} \hat{q}_{x,t+i}}{11}$$

$$\text{Shock}_{x,t} = \frac{\hat{q}_{x,t}}{\tilde{q}_{x,t}} - 1$$

In a next step we study the series of $\text{Shock}_{x,t}$ in order to find in past data which percentage of the average q_x should correspond to a shock in mortality rate which happens once in 250 years (99.6% percentile). Results per age band are shown in table 7.2.

These results show that if we look at the past data, the 1/250 catastrophic scenario should be based on a shock in mortality about 70% of the q_x for all combined ages, and about 150% for the specific age band [20; 40].

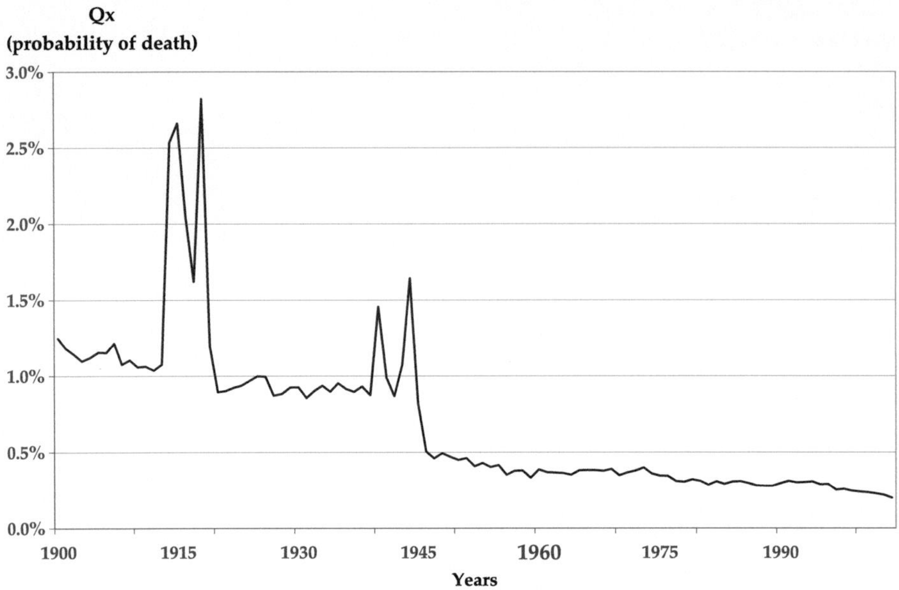

Fig. 7.3 Evolution of death rate at age 40 for males from France

Table 7.2 Historical mortality shocks (99.6% percentile of the $\mathrm{Shock}_{x,t}$ series by age band)

Age band	France M	France F	England M	England F	Sweden M	Sweden F
0-20	53%	55%	53%	35%	51%	49%
20-40	154%	135%	162%	51%	104%	87%
40-60	37%	34%	21%	19%	61%	54%
60-80	15%	18%	15%	15%	39%	31%
Total	68%	63%	62%	27%	60%	53%

Determination of γ

We keep studying the series of $\mathrm{Shock}_{x,t}$. We want to obtain the γ parameter. In table 7.3 we summarise frequencies of $\mathrm{Shock}_{x,t}$ which are greater than 20%.

Table 7.3 Historical mortality shocks frequency series by age band

Age band	France M	France F	England M	England F	Sweden M	Sweden F
0-20	7.3%	4.2%	3.9%	1.3%	7.8%	7.8%
20-40	8.3%	4.2%	5.2%	2%	8.3%	6.8%
40-60	3.1%	2.1%	1.3%	0.7%	5.7%	5.7%
60-80	0%	0%	0%	0%	2.6%	3.6%
Total	7.3%	4.2%	3.3%	0.7%	5.2%	4.2%

We can observe that for the age band 60-80 shocks frequencies are null for England and France. So we will skip this age band for the following part of the study.

For the risk capital in relation to mortality risk we define $\gamma^{\text{MORT}} := \gamma$.

Relationship Between $\text{Shock}_{x,t}$, I_t and R_t

Let us assume that:

$$\text{Shock}_{x,t} \sim I_t \times R_t.$$

We want to establish a link between the percentile from $\text{Shock}_{x,t}$ and the one from R_t. If ϕ is the 99.6% percentile from $\text{Shock}_{x,t}$, then we have:

$$P[\text{Shock}_{x,t} \geq \phi] = 0.004,$$
$$P[I_t = 1, R_t \geq \phi] = 0.004.$$

Then, because I_t and R_t are independent,

$$P[R_t \geq \phi] = \frac{0.004}{P[I_t = 1]}. \tag{7.1}$$

Then ϕ is the $1 - \frac{0.004}{P[I_t=1]}$ percentile from R_t.

Determination of $\tilde{\mu}$, $\tilde{\sigma}$

Let us introduce $F[\tilde{\mu}, \tilde{\sigma}]$ the cumulative density function of a log-normal distribution with parameters $\tilde{\mu}$ and $\tilde{\sigma}$:

Given relationship 7.1, we will be able to fit parameters $\tilde{\mu}$, $\tilde{\sigma}$ with the historical data. We define 2 sets of conditions to fit $\tilde{\mu}$ and $\tilde{\sigma}$:

$$F[\tilde{\mu}, \tilde{\sigma}]^{-1}(1 - \frac{0.004}{\gamma}) = \text{Shock}_{x,t}(99.6\%) \text{ and}$$
$$\exp\left(\tilde{\mu} + \frac{\tilde{\sigma}^2}{2}\right) = AV(\text{Shock}_{x,t}|\text{Shock}_{x,t} > 20\%).$$

Where $\text{Shock}_{x,t}(99.6\%)$ is the 99.6% percentile of the $\text{Shock}_{x,t}$ historical series, and $AV(\text{Shock}_{x,t}|\text{Shock}_{x,t} > 20\%)$ is the average of the $\text{Shock}_{x,t}$ historical series where $\text{Shock}_{x,t} > 20\%$ (see table 7.4).

Table 7.4 Historical mortality shocks average where the shock is greater than 20%

Age band	France M	France F	England M	England F	Sweden M	Sweden F
0-20	38%	42%	38%	37%	32%	34%
20-40	95%	86%	87%	46%	48%	40%
40-60	30%	32%	21%	22%	35%	37%
Total	46%	42%	47%	42%	39%	39%

We obtain the following set of parameters:

Table 7.5 Values for $\tilde{\mu}$

Age band	France M	France F	England M	England F	Sweden M	Sweden F
0-20	-5.41	-3.74	-3.46	-1.64	-5.49	-5.64
20-40	-4.59	-2.59	-2.84	-1.94	-4.59	-4.05
40-60	-3.36	-2.57	-2.08	-1.57	-4.22	-4.54
Total	-5.08	-3.41	-2.88	-1.1	-4.12	-3.69

Table 7.6 Values for $\tilde{\sigma}$

Age band	France M	France F	England M	England F	Sweden M	Sweden F
0-20	2.98	2.4	2.23	1.13	2.95	3.01
20-40	3.01	2.2	2.32	1.51	2.77	2.49
40-60	2.07	1.69	1.03	0.28	2.52	2.65
Total	2.93	2.25	2.06	0.68	2.52	2.34

At this stage, it is difficult to have a strong idea of which parameter we should use except that since most of our portfolio comes from UK, we should choose parameters for UK and for the age band concerned.

For Mortality we denote by $(\mu^{\text{MORT}}, \sigma^{\text{MORT}}) = (\tilde{\mu}, \tilde{\sigma})$.

7.2.2 Longevity

As with mortality, there are many different possibilities to model the longevity risk. We want to show one particular model in order to understand the different steps needed. For an overview of other models and approaches we refer to [PDHO09] and sources therein.

We use the following model for longevity. Assume the best estimate mortality is given by $q_{x,t}$. Then we use the following model for the mortality process:

$$Q^i_{x,t}(\omega) = q_{x,t} \times C_t(\omega) + \epsilon^i_{x,t}(\omega).$$

With,

$$\begin{aligned} C_t &= \exp(X_t) \times C_{t-1}, \\ (X_j)_{j\in\mathbb{N}} &\sim \mathcal{N}(\mu,\sigma) \qquad \text{i.i.d.}, \\ \epsilon^i_{x,t} &\sim F \qquad \text{i.i.d. with } \mathbb{E}(F)=0, \\ (C,\epsilon) &\quad \text{Independent}, \\ C_0 &= 1, \end{aligned}$$

where x represents the age of the insured, t the period in time we focus on.

This model aims to describe how longevity behaves around the projected expected values. We will concentrate our efforts on determining the parameters of the $(C_t)_{t\in\mathbb{N}}$ distribution the aim of which is to simulate the non diversifying risk of longevity. We will not work on the $(\epsilon)^i_{x,t}$ vector since it represents the diversifying risk.

This approach allows a best estimate mortality projection given by the common assumption:

$$q_{x,t} = q_{x,t_0} \times \exp\left(-\lambda_x \times \{t - t_0\}\right).$$

Comparison with the Lee-Carter Model

In 1992 Lee and Carter [LC92] developed an new method to forecast mortality and so to obtain prospective mortality tables. This stochastic methodology suggested a log-bilinear form for the central death rate $\mu_x(t)$ for age x at time t . It consists in decomposing the age-specific mortality in two components:

- A set of age-specific constants a_x, b_x,
- A time-varying index of mortality κ_t.

The model has the following form:

$$\ln(\mu_x(t)) = a_x + b_x \times \kappa_t + \epsilon_{x,t}.$$

Since this admits several solutions, restrictions (on the parameters) must be added to obtain an identifiable model:

$$\sum_{x_{min}}^{x_{max}} \beta_x = 1,$$

$$\sum_{t_{min}}^{t_{max}} \kappa_t = 0.$$

The signification of the parameters is described below:

a_x represents the average level of the $\ln(\mu_x(t))$ surface over time. Therefore, $\exp(a_x)$ is the general shape of mortality at age x over time. κ_t is an indicator of mortality evolution over time. It describes the change in overall mortality over time. b_x indicates the sensitivity of rates to the mortality evolution over time, for a given age x; so, mortality evolution at age x depends on the index of mortality κ_t.

To build prospective mortality tables, it is necessary to

- Estimate the parameters of the Lee-Carter Model.
  An optimal solution can be found with the method of least squares and is given by a singular value decomposition (SVD) (for the determination of parameters κ_t and b_x).
- Model the time-varying parameter κ_t describing mortality evolution. In fact it is necessary to specify the process shape of κ_t ($AR(p)$, $MA(q)$, $ARMA(p,q)$, $\ldots$) in order to make forecasts and thus, to extrapolate the trend.
  In this way it is possible to obtain predictions of the probabilities of death and interval forecasts by doing simulations.

Choice of the Model

Since there is no generally accepted model, we chose this one that could be considered as a simplified version of Lee-Carter as we can see if we express the 2 models in the same way:

$$\ln(\mu_x(t)) = [a_x + b_x \times \kappa_t] + \epsilon_{x,t} \quad \text{(Lee-Carter model)},$$

$$\ln(Q_{x,t}) = \ln(q_{x,t}) + \sum_{i=0}^{t} X_i \quad \text{(our model)}.$$

Indeed, both use a log-normal distribution to estimate the deviation around a global tendency. But we have also some major differences. Lee-Carter has a stochastic predictive part $(a_x + b_x \times \kappa_t)$ whereas our model separates this prediction work and assumes it is given by the $(q_{x,t})$. The variance of the stochastic part of our model $(\sum_{i=0}^{i=t} X_i)$ is growing with time.

Model Parameters

Since the model assumes that our best estimate mortality is given by $q_{x,t}$, we need the following condition:

$$E\left[\exp\left(X_t\right)\right] = 1, \tag{7.2}$$

since $\exp\left(X_t\right)$ follows a log-normal distribution. This implies:

$$E\left[\exp\left(X_t\right)\right] = \exp\left(\mu + \frac{\sigma^2}{2}\right). \tag{7.3}$$

Equations (7.2) and (7.3) yield to:

$$\mu = -\frac{\sigma^2}{2}.$$

So the only parameter for our model is given by $\sigma^{\text{LONG}} := \sigma$.

7.3 Concrete Example: An Annuity Portfolio

Annuity portfolios are normally re-insured in the form of a quota share or in form of a mortality swap. In both cases the premiums paid are compared with the annuities paid out. For an individual policy$k$ the payout is given by the following formula:

$$R_k(\omega, t) = \chi_{\{T_k \geq t\}} \times R_k,$$

where $T_k(\omega)$ denotes the (random) future life span of policy k and R_k the annuity to be paid per annum. As a consequence the price (BEL) of this annuity corresponds to

$$V_k(t) = \sum_{t=0}^{\infty} {}_tp_x \times R_k \times \pi(\mathcal{Z}_{(t)}),$$

denoting by $\pi(\mathcal{Z}_{(t)})$ the price of the t-year zero coupon bond at valuation date. The problem with the above mentioned cover is the fact that both the cover length and the potential loss for the reinsurer is unlimited, leading to relatively high prices. In contrast, the typical non-life reinsurance covers are limited in time and in maximal amounts. The aim is to analyse the applicability of the corresponding methods in mortality swaps.

The analogy of the instruments which we will describe in the financial world are most likely swaptions to the extent that there is a real swap in certain situations and that there is a limit in time and amount.

In order to illustrate this type of product, we will have a look at the different steps. The main ingredient is a clear definition of the risk covered and the corresponding cash flows. We remark that figure 7.3 shows the corresponding results by putting always 5 years together. Hence bucket number 1 corresponds to years 1 to 5, bucket number 2 to the years 6 to 10, etc.

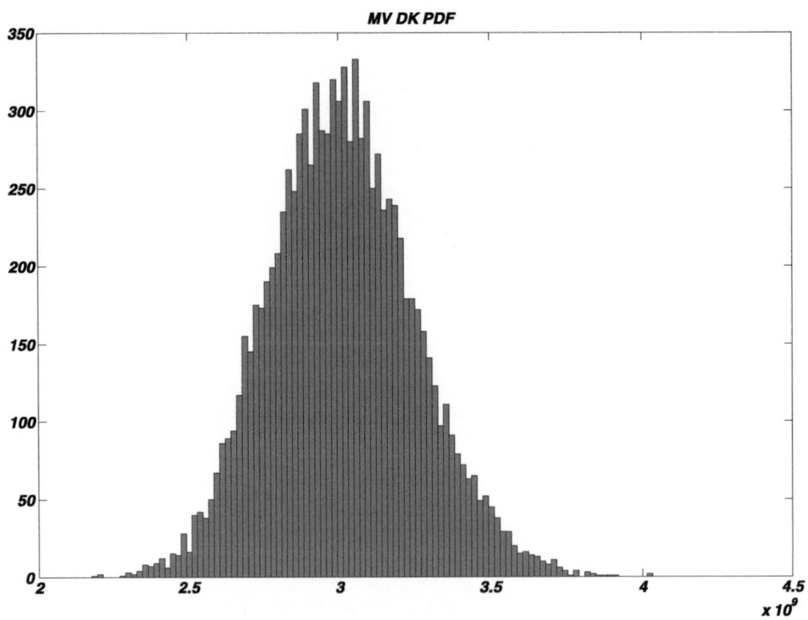

Fig. 7.4 Distribution of the present value of the future cash flows

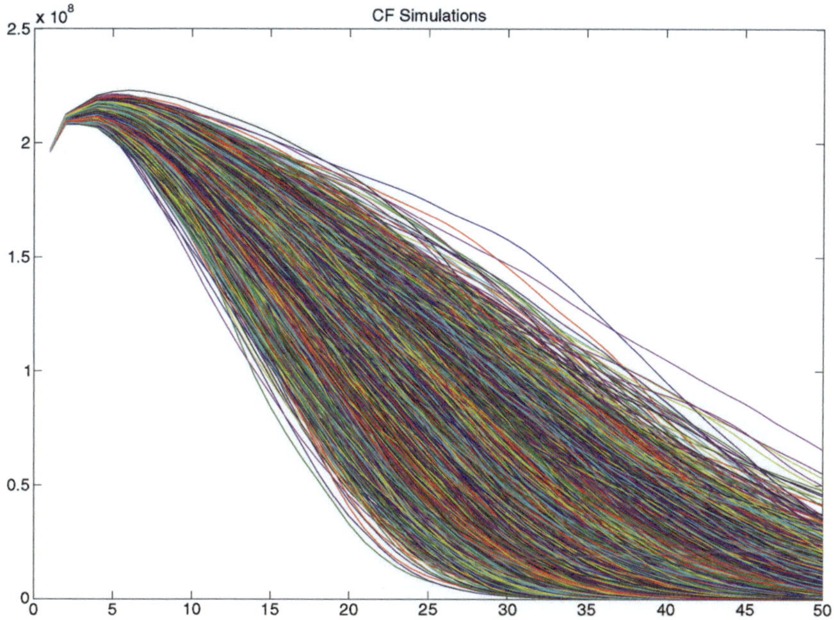

Fig. 7.5 Sample of CF Trajectories

7.3.1 Formulae

In principle, the usual calculations for expected values of cash flows CF_t apply, with exception that one needs to replace them by the correspondingly adjusted ones. Assume, for example, a layer between α and β within the time interval $[t_1, t_2]$. In this case we have

$$CF_t^* = \begin{cases} 0 & \text{if } t \notin [t_1, t_2], \\ \min(\beta - \alpha, \max(CF_t - \alpha, 0)) & \text{else.} \end{cases}$$

Normally one would agree that both α and β are defined relative to the expected value of the corresponding cash flows. For example $\alpha_t = 110\% \times \mathbb{E}[CF_t]$ and $\beta_t = 120\% \times \mathbb{E}[CF_t]$ The present values and premium need to be calculated by using a standard approach.

In the following we will use the Swiss mortality table ERM/F 2000 for the calculations. The mortality law for ERM/F 2000 can be defined as follows, with $C(t, \omega) = 1 \, \forall t$:

$$q_{x,t,\omega} = q_{x,t_0} \times \exp\{\lambda_x \times (t - t_0)\} \times C(t, \omega).$$

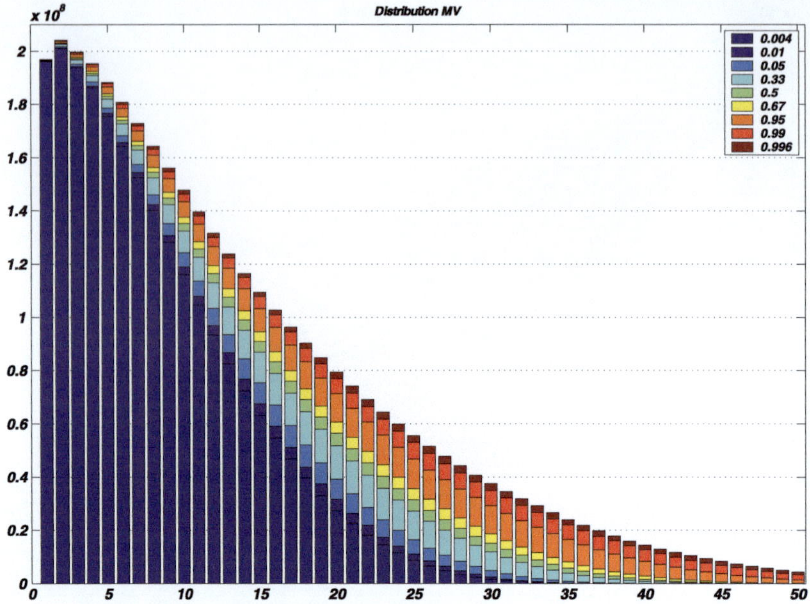

Fig. 7.6 Probability Density Function of the Present Values

The model applied for these calculations uses a non-trivial C: We denote by $(X_t)_{t\in\{0,1,2,\dots\}}$ iid $\mathcal{N}(\mu,\sigma)$ normal distributed random variables, and define

$$C(t) = \exp(\sum_{k=0}^{t} X_k).$$

Note that this model is very similar to the well known Lee Carter model, which is given by:

$$\begin{aligned} \log(q_{x,t}) &= a_x + b_x\, k_t + \epsilon_{x,t}, \\ k_{t+1} &= k_t + \mu + \sigma\, X_t, \\ (X_t)_{t\in\mathbb{N}} &\sim \mathcal{N}(0,1). \end{aligned}$$

Obviously the two models interrelate by

$$\begin{aligned} q_{x,t_0} &\mapsto a_x, \\ \lambda_x &\mapsto b_x. \end{aligned}$$

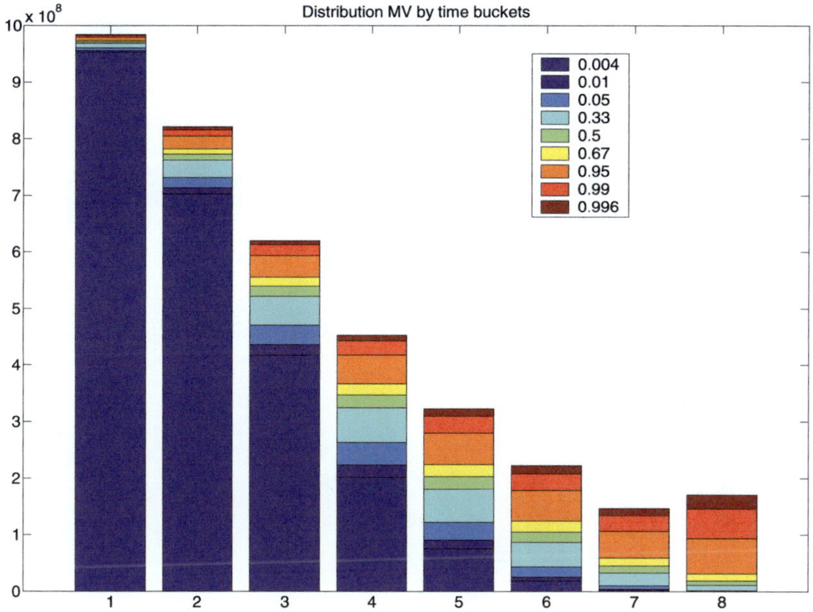

Fig. 7.7 Market Values by Time Buckets

The main difference between the two models is the different parameters for the fluctuation of the trend (relative to the model introduced above):

Model used	$X_k \sim \mathcal{N}(0, \sigma)$
Lee Carter	$X_k \sim \mathcal{N}(\lambda_x \times \mu, \lambda_x \times \sigma)$

7.3.2 Application to Insurance Linked Securities

The mortality swap is intrinsically linked to insurance linked securities (ILS) and therefore this section describes some of the above mentioned features. Firstly it needs to be noted that investors have the following criteria to invest in certain investments. The investments should satisfy as many of the following characteristics as possible:

- High return in relation to the corresponding risk,
- Clearly defined risks and the possibility to quantify the risk,
- Liquid secondary market in order to offset the risk in adverse situations,
- Short binding period, in case there is no liquid secondary market.

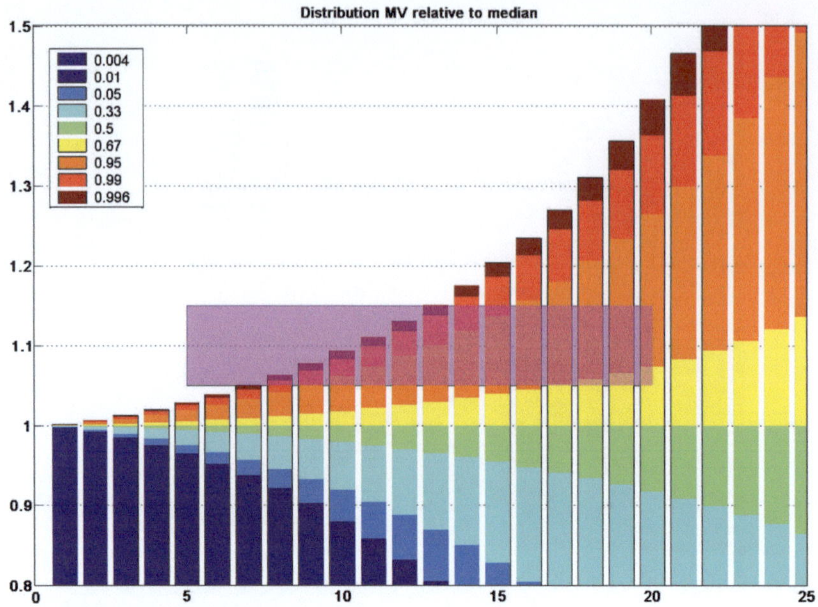

Fig. 7.8 Example of a covered layer

In case of longevity ILS the risk is clearly defined but difficult to quantify. The recent introduction of well defined indexes relating to these risks and the companion documents and methods (cf "JPMorgan LifeMetrics") help to close this gap. Until now there exists no liquid secondary market and therefore the binding period of the risk needs to be analysed further. The easiest thing would be to offer longevity ILS only for a limited period of time to the investor. This is clearly feasible as defined above, but has the drawback that neither the most relevant risk for the pension funds and primary life companies is considered. Furthermore, interestingly, non-correlated, investment opportunities would be left out. The idea is therefore to slice a given mortality portfolio in analogy to CDS constructions in bits which have different risk characteristics, ranking from Bond investments with a equivalent counter-party risk of "AAA" down to junk bond and an equity part. In contrast to the CDS the differentiation would be done over time. As the longevity risk increases with time, the risk related to the annuities which will be paid out in the near future corresponds to bonds with a higher rating and less return. The further away the annuity is paid, the higher risk and return for the corresponding bond. Figure 7.9 shows this.

Moreover the following table shows how the differences match to different investor characteristics:

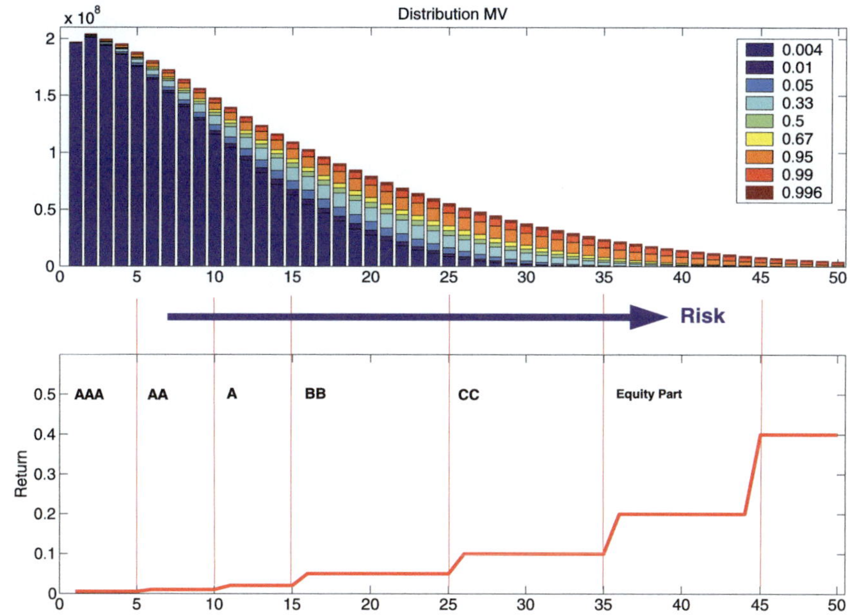

Fig. 7.9 CDS structuring for longevity ILS

Time Bucket	Rating idea	Investor Characteristics
$[0,5[$	AAA	Corporate Bonds
$[6,10[$	AA	Corporate Bonds / Reinsurance
$[11,15[$	A	Reinsurance
$[15,25[$	BB	Reinsurance / Hedge Funds
$[25,35[$	CC	Hedge Funds
$[35,45[$	Equity Part	Hedge Funds / Private Equity
$[46,\infty[$	Equity Part	Private Equity

After this abstract structuring there is a mechanism to calculate the profit and loss for each slice and time needs to be calculated. The profit and loss for a certain period is given by

$$L_t(\omega) = \{\, {}_tp_x - \chi_{T>t}\} \times \pi(\mathcal{Z}_{(t)}).$$

Therefore the calculation of the profit/loss in a certain point in time can be calculated if either $q_{x,t}$ or ${}_tp_{x_0,t_0}$ is given. In case of a slice the corresponding profits and losses are calculated by integrating over the exposure, eg the loss within time $[t_1, t_2]$ is given by:

$$L_{[t_1,t_2[}(\omega) = \sum_{\tau=t_1, x\in\mathbb{N}}^{t_2-1} R_{\tau,x} \times \{\, {}_\tau p_x - \chi_{T_x>\tau}\} \times \pi(\mathcal{Z}_{(\tau)}),$$

where $R_{x,t}$ denotes the annuity (notional amount) to be paid for the persons aged x at inception of the ILS at time τ.

We consider now the annuity part at time n with an expected annuity of A_n and a risk capital at that time, for example with respect to the 99.6 % Var of η_n of the expected annuity. For simplicity we assume $A_n = 1$. For this treaty the investor wants a return of κ_n on the capital invested. By $A_n(\omega)$ we actual annuity paid out at time n. We have the following relations, denoting with C_n the corresponding risk capital.

$$\begin{aligned} A_n &= \mathbb{E}[A_n(\omega)] \\ C_n &= F_{A_n}^{-1}(99.6\%) \end{aligned}$$

From the investor's point of view we have the following cash flows:

Time	Amount
$t = 0$	$-C_n + \xi$
$t = n$	$[C_n - A_n(\omega)]^+ + C_n \times \kappa_n$

where ξ denotes the risk premium for the investor. Because of equilibrium we have the following equation:

$$C_n - \xi = \pi(\mathcal{Z}_{(n)}) \times \left\{ \int [C_n - A_n(\omega)]^+ d\omega + C_n \times \kappa_n \right\}.$$

7.3.3 *Statistics*

Table used	ERM/F 2000 second order for q_x and trend
Date of Valuation	1.1.2006
Model for Morality law	Mortality improvements following cumulative log-normal path
μ	$-\frac{\sigma^2}{2}$
σ	10 %
Discount factors	Yield curve in Euro as of 29.12.2006
Annuities paid out p.a.	EUR 244.18 M
Proxy for stat. MR 3.5 %	EUR 3,164.16 M

```
MV DK Mean      3011.22 M     Stddev       239.47 M (7.95 %)
MV DK -.---- Min          2176.70 M     (  72.29 % of mean)
```

```
MV DK 0.0040 Quantile      2407.04 M    (  79.94 % of mean)
MV DK 0.0100 Quantile      2485.36 M    (  82.54 % of mean)
MV DK 0.0500 Quantile      2632.77 M    (  87.43 % of mean)
MV DK 0.3300 Quantile      2898.58 M    (  96.26 % of mean)
MV DK 0.5000 Quantile      3005.98 M    (  99.83 % of mean)
MV DK 0.6700 Quantile      3109.00 M    ( 103.25 % of mean)
MV DK 0.9500 Quantile      3420.31 M    ( 113.59 % of mean)
MV DK 0.9900 Quantile      3599.37 M    ( 119.53 % of mean)
MV DK 0.9960 Quantile      3691.62 M    ( 122.60 % of mean)
MV DK -.---- Max           4032.62 M    ( 133.92 % of mean)

Bucket [ 1, 5] DK Mean       970.83 M    Stddev         5.78 M (0.60 %)
Bucket [ 1, 5] DK -.---- Min            946.91 M    (  97.54 % of mean)
Bucket [ 1, 5] DK 0.0040 Quantile       953.44 M    (  98.21 % of mean)
Bucket [ 1, 5] DK 0.0100 Quantile       955.80 M    (  98.45 % of mean)
Bucket [ 1, 5] DK 0.0500 Quantile       960.67 M    (  98.95 % of mean)
Bucket [ 1, 5] DK 0.3300 Quantile       968.64 M    (  99.77 % of mean)
Bucket [ 1, 5] DK 0.5000 Quantile       971.19 M    ( 100.04 % of mean)
Bucket [ 1, 5] DK 0.6700 Quantile       973.63 M    ( 100.29 % of mean)
Bucket [ 1, 5] DK 0.9500 Quantile       979.67 M    ( 100.91 % of mean)
Bucket [ 1, 5] DK 0.9900 Quantile       982.73 M    ( 101.23 % of mean)
Bucket [ 1, 5] DK 0.9960 Quantile       984.00 M    ( 101.36 % of mean)
Bucket [ 1, 5] DK -.---- Max            989.74 M    ( 101.95 % of mean)

Bucket [ 6,10] DK Mean       771.25 M    Stddev        22.21 M (2.88 %)
Bucket [ 6,10] DK -.---- Min            662.71 M    (  85.93 % of mean)
Bucket [ 6,10] DK 0.0040 Quantile       703.04 M    (  91.16 % of mean)
Bucket [ 6,10] DK 0.0100 Quantile       713.85 M    (  92.56 % of mean)
Bucket [ 6,10] DK 0.0500 Quantile       731.78 M    (  94.88 % of mean)
Bucket [ 6,10] DK 0.3300 Quantile       762.71 M    (  98.89 % of mean)
Bucket [ 6,10] DK 0.5000 Quantile       773.09 M    ( 100.24 % of mean)
Bucket [ 6,10] DK 0.6700 Quantile       782.07 M    ( 101.40 % of mean)
Bucket [ 6,10] DK 0.9500 Quantile       804.71 M    ( 104.34 % of mean)
Bucket [ 6,10] DK 0.9900 Quantile       816.04 M    ( 105.81 % of mean)
Bucket [ 6,10] DK 0.9960 Quantile       821.40 M    ( 106.50 % of mean)
Bucket [ 6,10] DK -.---- Max            843.75 M    ( 109.40 % of mean)

Bucket [11,15] DK Mean       536.56 M    Stddev        37.94 M (7.07 %)
Bucket [11,15] DK -.---- Min            351.99 M    (  65.60 % of mean)
Bucket [11,15] DK 0.0040 Quantile       417.68 M    (  77.84 % of mean)
Bucket [11,15] DK 0.0100 Quantile       436.71 M    (  81.39 % of mean)
Bucket [11,15] DK 0.0500 Quantile       471.03 M    (  87.79 % of mean)
Bucket [11,15] DK 0.3300 Quantile       521.64 M    (  97.22 % of mean)
Bucket [11,15] DK 0.5000 Quantile       539.72 M    ( 100.59 % of mean)
Bucket [11,15] DK 0.6700 Quantile       555.49 M    ( 103.53 % of mean)
Bucket [11,15] DK 0.9500 Quantile       593.68 M    ( 110.65 % of mean)
Bucket [11,15] DK 0.9900 Quantile       613.08 M    ( 114.26 % of mean)
Bucket [11,15] DK 0.9960 Quantile       619.86 M    ( 115.52 % of mean)
Bucket [11,15] DK -.---- Max            657.53 M    ( 122.54 % of mean)

Bucket [16,20] DK Mean       344.11 M    Stddev        47.36 M(13.76 %)
Bucket [16,20] DK -.---- Min            135.87 M    (  39.49 % of mean)
Bucket [16,20] DK 0.0040 Quantile       202.06 M    (  58.72 % of mean)
Bucket [16,20] DK 0.0100 Quantile       223.62 M    (  64.98 % of mean)
Bucket [16,20] DK 0.0500 Quantile       263.32 M    (  76.52 % of mean)
Bucket [16,20] DK 0.3300 Quantile       324.41 M    (  94.28 % of mean)
Bucket [16,20] DK 0.5000 Quantile       347.04 M    ( 100.85 % of mean)
Bucket [16,20] DK 0.6700 Quantile       366.76 M    ( 106.58 % of mean)
Bucket [16,20] DK 0.9500 Quantile       417.42 M    ( 121.31 % of mean)
Bucket [16,20] DK 0.9900 Quantile       443.26 M    ( 128.82 % of mean)
Bucket [16,20] DK 0.9960 Quantile       453.25 M    ( 131.72 % of mean)
Bucket [16,20] DK -.---- Max            493.63 M    ( 143.45 % of mean)

Bucket [21,25] DK Mean       202.24 M    Stddev        47.95 M(23.71 %)
Bucket [21,25] DK -.---- Min             32.43 M    (  16.04 % of mean)
Bucket [21,25] DK 0.0040 Quantile        75.37 M    (  37.27 % of mean)
Bucket [21,25] DK 0.0100 Quantile        90.32 M    (  44.66 % of mean)
```

```
Bucket [21,25] DK 0.0500 Quantile     122.05 M    (  60.35 % of mean)
Bucket [21,25] DK 0.3300 Quantile     181.05 M    (  89.52 % of mean)
Bucket [21,25] DK 0.5000 Quantile     203.22 M    ( 100.49 % of mean)
Bucket [21,25] DK 0.6700 Quantile     224.45 M    ( 110.98 % of mean)
Bucket [21,25] DK 0.9500 Quantile     279.91 M    ( 138.41 % of mean)
Bucket [21,25] DK 0.9900 Quantile     310.11 M    ( 153.34 % of mean)
Bucket [21,25] DK 0.9960 Quantile     322.97 M    ( 159.70 % of mean)
Bucket [21,25] DK -.---- Max          372.89 M    ( 184.38 % of mean)

Bucket [26,30] DK Mean       106.99 M    Stddev      40.99 M(38.32 %)
Bucket [26,30] DK -.---- Min            5.61 M    (   5.24 % of mean)
Bucket [26,30] DK 0.0040 Quantile      18.68 M    (  17.46 % of mean)
Bucket [26,30] DK 0.0100 Quantile      25.57 M    (  23.90 % of mean)
Bucket [26,30] DK 0.0500 Quantile      43.53 M    (  40.69 % of mean)
Bucket [26,30] DK 0.3300 Quantile      86.64 M    (  80.99 % of mean)
Bucket [26,30] DK 0.5000 Quantile     104.68 M    (  97.85 % of mean)
Bucket [26,30] DK 0.6700 Quantile     124.07 M    ( 115.96 % of mean)
Bucket [26,30] DK 0.9500 Quantile     178.23 M    ( 166.59 % of mean)
Bucket [26,30] DK 0.9900 Quantile     208.37 M    ( 194.77 % of mean)
Bucket [26,30] DK 0.9960 Quantile     222.52 M    ( 207.99 % of mean)
Bucket [26,30] DK -.---- Max          278.46 M    ( 260.27 % of mean)

Bucket [31,35] DK Mean        49.80 M    Stddev      29.55 M(59.33 %)
Bucket [31,35] DK -.---- Min            0.65 M    (   1.30 % of mean)
Bucket [31,35] DK 0.0040 Quantile       2.66 M    (   5.34 % of mean)
Bucket [31,35] DK 0.0100 Quantile       4.20 M    (   8.43 % of mean)
Bucket [31,35] DK 0.0500 Quantile      10.34 M    (  20.76 % of mean)
Bucket [31,35] DK 0.3300 Quantile      32.84 M    (  65.94 % of mean)
Bucket [31,35] DK 0.5000 Quantile      45.17 M    (  90.70 % of mean)
Bucket [31,35] DK 0.6700 Quantile      59.07 M    ( 118.63 % of mean)
Bucket [31,35] DK 0.9500 Quantile     106.17 M    ( 213.20 % of mean)
Bucket [31,35] DK 0.9900 Quantile     134.08 M    ( 269.24 % of mean)
Bucket [31,35] DK 0.9960 Quantile     146.43 M    ( 294.05 % of mean)
Bucket [31,35] DK -.---- Max          198.37 M    ( 398.34 % of mean)

Bucket [36,50] DK Mean        29.45 M    Stddev    31.12 M(105.69 %)
Bucket [36,50] DK -.---- Min            0.11 M    (   0.36 % of mean)
Bucket [36,50] DK 0.0040 Quantile       0.30 M    (   1.01 % of mean)
Bucket [36,50] DK 0.0100 Quantile       0.50 M    (   1.69 % of mean)
Bucket [36,50] DK 0.0500 Quantile       1.67 M    (   5.67 % of mean)
Bucket [36,50] DK 0.3300 Quantile      11.08 M    (  37.62 % of mean)
Bucket [36,50] DK 0.5000 Quantile      18.93 M    (  64.29 % of mean)
Bucket [36,50] DK 0.6700 Quantile      31.22 M    ( 106.02 % of mean)
Bucket [36,50] DK 0.9500 Quantile      93.36 M    ( 317.04 % of mean)
Bucket [36,50] DK 0.9900 Quantile     145.94 M    ( 495.60 % of mean)
Bucket [36,50] DK 0.9960 Quantile     170.37 M    ( 578.55 % of mean)
Bucket [36,50] DK -.---- Max          273.39 M    ( 928.41 % of mean)
```

7.3.4 Allocation of Risk Capital in Proportion to Premium

Based on the simulation approach it is possible to allocate a risk capital per year in proportion of the annuity paid out, as follows. We know from the simulation the 99.6% quantile as a percentage of the best estimate annuity paid out.

t	η_t	t	η_t
1	100.18 %	26	186.87 %
2	100.67 %	27	197.50 %
3	101.30 %	28	210.19 %
4	102.06 %	29	222.65 %
5	102.95 %	30	238.05 %
6	104.03 %	31	255.49 %
7	105.20 %	32	276.07 %
8	106.59 %	33	299.82 %
9	108.13 %	34	324.58 %
10	109.82 %	35	349.16 %
11	111.60 %	36	383.51 %
12	113.62 %	37	422.22 %
13	115.77 %	38	462.09 %
14	118.26 %	39	503.20 %
15	121.21 %	40	560.08 %
16	124.48 %	41	619.86 %
17	128.09 %	42	693.29 %
18	132.12 %	43	770.48 %
19	136.80 %	44	859.48 %
20	142.08 %	45	945.10 %
21	147.77 %	46	1033.18 %
22	153.63 %	47	1149.51 %
23	160.64 %	48	1277.50 %
24	168.35 %	49	1382.71 %
25	176.94 %	50	1506.23 %

Assume that for the portfolio considered the best estimate annuities are given by $(R_t)_{t\in\mathbb{N}}$. In this case the present value of the necessary risk capital can be calculated by

$$C_0 = \sum_{\tau=0}^{\infty} R_\tau \; (\eta_\tau - 1) \; \pi(\mathcal{Z}_{(\tau)}) \text{ and,}$$

$$C_t = \frac{1}{\pi(\mathcal{Z}_{(t)})} \sum_{\tau=t}^{\infty} R_\tau \; (\eta_\tau - 1) \; \pi(\mathcal{Z}_{(\tau)}).$$

At this point there is still a need for a better explanation for the capital needed at a certain point in time, especially for annuities:

1. The determination of the capital C_0 can be put in a relation to the annuities paid, in the sense that the present value of the difference equals the risk capital. A re-

lated question is now whether there is an additional need for risk capital during the projection for $t = 1$ etc. This is not necessary as the following calculation shows: The economic risk at a given time t for an annuity 1 at inception is equivalent to the difference of the relevant payments:

$$L_t(\omega) = \{ \, _tp_x - \chi_{T>t}\} \times \pi(\mathcal{Z}_{(t)}).$$

Therefore the risk capital to be allocated at this time is the corresponding possible annuity in an adverse scenario. The sum over the different points in time results in the above mentioned C_0. Therefore the actual risk capital allocated at a certain point in time equals $\Gamma(L_t)$, the corresponding risk measure for the possible excess annuity to be paid.

2. The second question, which needs some explanation relates to the required capital C_1, but not as seen from time $t = 0$, but if we are at time $t = 1$ and know the development until then. In this situation the corresponding mortality processes have moved forward by one year and therefore, we need to take the *conditional* risk measure at time $t = 1$, as a formula:

$$\begin{aligned}\text{Risk Capital at time } 1 &= \Gamma\left[\frac{1}{\mathcal{Z}_{(1)}}\sum_{t=1}^{\infty} L_k \,\middle|\, \mathcal{F}_1\right]\\ &= \mathbb{E}\left[\frac{1}{\mathcal{Z}_{(1)}}\sum_{t=1}^{\infty} L_k \,\middle|\, \mathcal{F}_1, L > F_\alpha^{-1}(L)\right].\end{aligned}$$

in case of the 99 % shortfall as risk measure. This relates assuming some homogenity assumptions in the following formula in therms of η:

$$\begin{aligned}C_0 &= \sum_{\tau=0}^{\infty} R_\tau\ (\eta_\tau - 1)\ \pi(\mathcal{Z}_{(\tau)}) \text{ and,}\\ \tilde{C}_t &= \frac{1}{\pi(\mathcal{Z}_{(t)})}\sum_{\tau=t}^{\infty} R_\tau\ (\eta_{\tau-t} - 1)\ \pi(\mathcal{Z}_{(\tau)}).\end{aligned}$$

7.3.5 Reporting Templates

This section provides an easy reporting template in relation to insurance risks. The concrete example is quite detailed and was intended for use in a life reinsurance company as its core business. For a primary insurance company the reporting template might be too detailed and one would expect other dimensions which matter more such as the lapses per distribution channel. As one can see the numbers in the template are purely virtual. Most of the risk categories are self explaining with the exception of ILA, which stands for impaired life annuities, eg annuities sold to people with a short remaining life time, be it for their old age, or be it for the fact that they are ill.

a) By Risk Factor

	Var(1:75)	Var(1:250)	TVar(1:100)	Limit (1:250)
Mortality	25.0	50.0	60.0	250.0
Longevity	25.0	50.0	60.0	200.0
ILA	25.0	50.0	60.0	
Disability	25.0	50.0	60.0	
Lapses	25.0	50.0	60.0	
Other	25.0	50.0	60.0	
Simple Sum	150.0	300.0	360.0	
Diversification	-50.0	-100.0	-120.0	
Total	100.0	200.0	240.0	500.0

b) By Lines Of Business:

	Var(1:75)	Var(1:250)	TVar(1:100)	Limit (1:250)
TCI	25.0	50.0	60.0	
GMDB	25.0	50.0	60.0	150.0
Financing	25.0	50.0	60.0	
Longevity /MS	25.0	50.0	60.0	
ILA	25.0	50.0	60.0	
Other	...	...	...	
Simple Sum	150.0	300.0	360.0	
Diversification	-50.0	-100.0	-120.0	
Total	100.0	200.0	240.0	

c) Stress Scenarios:

Scenario	SC1	SC2	SC3
TCI	25.0	50.0	60.0
GMDB	25.0	50.0	60.0
Financing	25.0	50.0	60.0
Longevity /MS	25.0	50.0	60.0
ILA	25.0	50.0	60.0
Other	...	...	...
Total	100.0	200.0	240.0

SC1: Pandemic as seen in 1918
SC2: Global increase in life span by 3 yrs at age 65
SC3: Run of the bank due to economic distress

Chapter 8
Operational Risks

8.1 Introduction

Whereas it is quite easy to see the consequences of financial and insurance risks, operational risks are more opaque and less easy to quantify. First it is necessary to define, what operational risks are. Normally operational risks are the ones which are not financial or insurance risks. In order to be more concrete, the following list represents some important operational risks:

M. Koller, *Life Insurance Risk Management Essentials*, EAA Series, DOI 10.1007/978-3-642-20721-1_8, © Springer-Verlag Berlin Heidelberg 2011

Regulatory risk: Change in laws. Failure to comply with regulatory requirements.

Financial crime: Fraud, money laundry and others.

Tax: Risk paying too high taxes, risk to fail to comply with tax laws and regulation.

Reputational: Risk putting reputation and brand at danger.

IT and business protection: Risk of IT failure, risk of data loss, missing IT security, viruses.

GI claims: Bad claims settlement processes, leakage and claim fraud.

Name	Risk Category	Risk	Actions	Remarks
ALM and Investment Processes	Market Risk Credit Risk Derivatives	There is a decentralised risk taking with respect to ALM and Investment risk, which might result in suboptimal performance and corresponding losses	• Set up and staffing of a centralised financial and ALM risk management function • Centralisation of risk taking	In addition to the organisational changes, processes to monitor the exposure have been set up. They however need to be improved further
Reputational Risks wrt structured products	Reputational Customer	As a consequence of the credit crunch some structured products in Italy have defaulted. Our clients suffered the economic loss and we are exposed towards reputational and customer risk	• Customer care actions to mitigate the reputational risk (cost € 65 M) • Set up of clear requirements which need to be met in order to launch a new structured product • Stringent product approval process involving Products, Finance and Risk	At group level there is also a new product approval process to follow in relation to structured products
Change Programe	Project Risk	The aim of quantum is to mitigate some of the major strategic and operational risks such as being subscale in some markets. Like each change project there are some material intrinsic project and people risks. In addition there is the risk to be focussed towards inwards and to neglect the changing external environment	• Set up of a strong PMO • Rigorous project approval and monitor process • Dedicated risk manager for Quantum • High level involvement of the European Executive together with regular reporting	In order to get a holistic risk assessment there are regular structured interviews with the main stakeholders. Moreover the project managers of the individual projects are asked to regularly report on the respective risks.
Underperformance of distribution partners	Distribution Credit Risk	As a consequence of the economic downturn, we observe significant underperformance of our distribution partners.	• Review of the credit exposure with our partners • Increased focus on relationship management	

Fig. 8.1 Operational Risk Reporting

As one sees from the above table there are myriads of operational risks. One main difference from financial risk is the fact that the mitigation of such risks normally takes longer. Assume for example an outdated IT platform, where you have a risk of data loss and also the risk that the people knowing the platform die before it can be replaced, such as with the Y2K problem. At that point of time (around 1998/1999) there was a big fear that legacy IT platforms from the 70s were not able to cope with the year 2000. There was a huge request for COBOL programmers and some companies had to ask their retired workforce for support .... Obviously the redesign of such an IT platform, which could have been used for policy administration, cannot be done overnight. In contrast to setting up a plain vanilla equity hedge this can be done in very expedient way.

This example makes also obvious, that the risk profile of operational risks cannot be adjusted suddenly and in consequence it is important to weigh the expected benefits against the costs incurred. Assume for example that the company is eager to reduce the business protection risk to a very low level. This means a lengthy process and commensurate (big) costs. Also once these measures have been implemented, one has paid for them, and there is little sense to increase the risk appetite again.

But how are such risks managed? In most companies this is done by corresponding risk policies which define the minimal control standards to reduce the risk to an acceptable level. After that, the controls are checked regularly in order to ensure adherence to the risk policies. Furthermore one normally uses loss data bases in order to monitor the operational risk that occurred, in order to determine whether the measures taken are adequate.

Analogously the operational risk reporting focuses on operational risks which are likely or have crystallised. Figure 8.1 shows such a reporting, where it becomes obvious that the focus is on an actionable remediation plan.

8.2 Risk Appetite

As indicated above the risk appetite is normally given by the corresponding risk policy. Since there are so many operational risks we want to focus on one operational risk policy. In section 8.3 we want to have a look at a holistic risk control framework.

It is key that the requirements are clear and that the first line of defence is able to provide evidence that corresponding measures have been implemented.

We will now have a look at the business protection risk policy, which covers the following areas:

- Information security - a breach of the confidentiality, integrity or availability of information or the IT assets used to store, process or transmit such information.
- Physical security -staff, premises and assets may not be protected against internal or external threats.
- Incident and business continuity management - the inability to maintain business activities and operations following an incident that has a significant impact.

The BP policy could for example state the following risk appetite:

> Businesses must comply with the business protection minimum standards of control; namely, information security, physical security, incident and business continuity management, unless a waiver or exception has been granted.

After defining this abstract risk appetite it is in a next step necessary to define the minimal control standards, which in turn limit the risk. The following list shows some of the possible minimal standards:

IT Access

1. Access to applications and underlying system is appropriately restricted and configured (e.g. by implementation of identification and authentication mechanisms) to reduce the risk of unauthorised/inappropriate access.
2. Procedures exist to ensure that access is added, modified and deleted in a timely manner. Access is restricted in accordance with the role or function of the individual.
3. Procedures exist to ensure that user access to applications is regularly reviewed.
4. Controls are in place to prevent unauthorised remote access.
5. Procedures exist to ensure that privileged access is added, modified and deleted in a timely manner. Access is restricted in accordance with the role or function of the individual.
6. Technical protective measures are to be implemented on all IT systems, applications and infrastructure.

Physical security

1. Physical access to computer facilities and data centres are restricted.
2. Physical access to all buildings must be controlled.
3. Staff must be subject to employment screening.
4. Regular security inspections of all buildings must be carried out.

After the definition of the minimal standards it is now necessary to implement them accordingly.

8.3 Risk Measurement and Exposure

Financial risks are to be recorded and quantified on the basis of generally recognised methods and standards. Operational and strategic risks, on the other hand, may not be amenable to meaningful quantification. These risks are to be evaluated in a suitable manner.

The following requirements apply to the methods of measurement:

- It must be ensured that all risks are recorded and assessed.
- The risk positions can be considered in total and can also be aggregated with respect to other relevant criteria, such as markets, functions, products, etc.
- Risk measurement should be consistent, accurate and comprehensive.
- The risk measurement algorithms must be technically correct and relevant to the risk/business in question. The measurement methods are to be precisely implemented in relation to the algorithms used. Adequate data quality must be ensured for risk measurement purposes.
- The methods that are used in measuring risk must always, as a minimum, fulfil the regulatory requirements.

The individual risk exposures are summed and reported in the economic capital reporting at company level according to the risk category/type of risk. The individual risk totals are then consolidated to form the overall Risk.

8.4 Controlling Process in General and Embedding Risk Policies

This section provides a metric in order to evaluate the level of implementation of the risk management guidelines within the business. It is one of many possible ways to control implementation. First, in order to measure implementation and the progress over time, it is necessary to define the corresponding metrics. The basis for this measurement are the risk management guidelines, which have been defined by the company.

In a second step the relative importance of the guideline has to be defined. Furthermore, it is also important to recognise that in a diversified group there are companies of different sizes and consequently the need for quality of the implementation is obviously higher for the "big" companies. Based on this a grid is developed based on these two dimensions.

Finally, it is worth mentioning that evidence needs to be based on a *"show me"* rather than a *"tell me"* approach. This means that the policy owner mandate for implementation cannot only tell the risk management that a certain policy has been implemented, but also the need to be able to show the supporting evidence.

In this sense the following requirements could define a reasonable minimal level of control. In order to achieve the desired homogeneity the risks have been classified into three categories:

High intrinsic need for controls The minimum standard consists of 5 required controls which need to be implemented. Guidelines relating to asset risk, financial crime, IT etc fall into this category, as such risk can have a major adverse impact on the group's financials. Most of the risk within this category can materialise over a relatively short time horizon, for example during a market crisis. Because of the short term nature of the risk all entities are required to implement these as a first stage.

Average intrinsic need for controls The minimum standard consists of 3 required controls. The risk in this category tends to materialise over a mid to long term and consists of things such as taxation, reserving, legal etc. For this risk category the big operations are expected to implement all of the 3 controls. For the smaller operations 1 control is expected.

Lower intrinsic need for controls The minimum standard consists of 1 required control. These risks tend to be less tangible and materialise typically over an even longer time horizon. Such risks consist of things such as Strategy & Planning, Corporate Social Responsibility etc. For the smaller countries no control is expected in an initial phase.

In order to have a consistent metric to compare the different entities, additional guidance has been given, together with the corresponding weighs. For example in the section derivatives, we have the following requirement: "A derivative strategy" (statement of derivative practise) has been written and implemented. In order to fulfil this requirement, the document needs to be written, a method to determine whether derivatives make sense and a process to handle derivatives needs to be in place. Assume for the moment that there is no method used. In this case one would say that this requirement is 70% implemented. It is clear that considerable judgement is needed in order to come to this point, because there could, for example, be evidence that in this particular case the method is already intrinsic in the process etc. Therefore the checkpoints which are provided aim to give guidance but might not be applicable in any case.

In the following the controls needed for each risk management policy are defined, together with the corresponding measurement criteria.

8.4.1 Strategy & Planning

Nr of Controls checked **1**

	Control Matrix: Requirement	W	Check
1	A local strategy planning process is in place.	0.5	Strategic planning is in place.
		0.5	Evidence is provided that the plan is challenged or is based on different scenarios.

8.4.2 Distribution Management

Nr of Controls checked **3**

	Control Matrix: Requirement	W	Check
1	A local governing body has been established (1x p.a.).	0.3	Meeting minutes.
		0.4	Tasks performed are OK - eg. performance metric has been established.
		0.3	Performance and volume targets are discussed and defined.
2	Due diligence for new Partners.	0.5	Process in place.
		0.5	Show results if any.
3	There are minimal service level agreements in place for distributors.	0.5	List with SLA content and minimal standards.
		0.5	Show an actual SLA and compare with list.

8.4.3 Brand & Marketing Communications

Nr of Controls checked **1**

	Control Matrix: Requirement	W	Check
1	There is a process in place that the brand is not used abusively and that the promises published can be hold.	0.3	All advertisement campaigns are reviewed by senior management.
		0.4	All product offering documents are released according a predefined process.
		0.3	Communication/Training for the relevant employees is in place.

8.4.4 Corporate Social Responsibility

Nr of Controls checked **1**

	Control Matrix: Requirement	W	Check
1	Code of conduct (CoC) is known to everybody in the organisation	0.4	Code of conduct has been given to every employee.
		0.3	The company has processes in place and can demonstrate that the employees have read and understand the CoC.
		0.3	The company shows additional aspects/ processes underlying the embedding within organisation.

8.4.5 *Environment*

Nr of Controls checked **1**

	Control Matrix: Requirement	W	Check
1	There is a local planning and control process in place in respect of climate change and waste management.	0.5	Show process and measurement.
		0.5	Show achievements so far.

8.4.6 *Regulatory*

Nr of Controls checked **3**

	Control Matrix: Requirement	W	Check
1	A responsible person acting as single point of contact for all regulators has been determined, all correspondence from and to the regulator its stored in a structured way in one place.	0.3	Person named.
		0.3	Guideline written.
		0.2	List of contacts established.
		0.2	Repository up and running.
2	There is a process to check the adherence to local regulatory requirement, new requirements are analysed and a process is in place to ensure the timely implementation of the new requirements.	0.6	A check list of deliverables exists - together with responsible persons.
		0.4	Company shows that they are following regulatory developments.
3	All material communication to the regulator - in particular regulatory returns are either signed by the CFO or the CEO.	0.3	There is a 4eye principle in place.
		0.4	Company can show which parts will be signed of by CEO/CFO.
		0.3	Signed documents have been shown.

8.4.7 *Information Technology*

Nr of Controls checked **5**

	Control Matrix: Requirement	W	Check
1	All major projects have a sound governance structure, eg project plan, agreed resources, steering committee and project reporting	0.3	For concrete projects the corresponding structures can be shown.
		0.4	The company can provide a consolidated overview over the global project status.
		0.4	The company provides evidence that there are sufficient project management skills available.
2	There is a well defined change management for programs and data. Access is granted on a need to have bases.	0.4	Process documentation exists.
		0.4	The responsible for this area can explain the process in a credible manner.
		0.2	Access rights can be shown.
3	There is a well defined security and access right process in place, changes of rights are tracked. Access on a need to have basis	0.5	A documentation and a responsible person exists.
		0.2	A list of all admins can be shown for material systems.
		0.3	The admin process for rights is shown.
4	There are robust processes in place for disaster recovery and significant adverse effects. There are processes in place to restore the status ante quo within an agreed time.	0.5	A disaster recovery plan can be shown and explained.
		0.2	The company can show how the backup data is stored; where how often etc.
		0.2	The company has done the process of physical recovery once and can show corresponding results.
5	There is a stringent cost and license management in place.	0.3	The company can show the corresponding begets and deviations from it including explanation.
		0.3	The company has a clear view where it is still inefficient and what could be done against.
		0.4	The company can show its license management process.

8.4.8 Financial Crime

Nr of Controls checked **5**

	Control Matrix: Requirement	W	Check
1	A responsible for this function has been named and enough resources have been allocated.	0.5	Person with commensurate background has been named.
		0.2	Resource allocation.
		0.3	Reporting line.
2	Businesses and Staff are aware of money laundry –> training together with corresponding evaluation processes such as self assessment.	0.5	Training material and training is done on a regular basis.
		0.5	Evaluation process (how and how often).
3	Process in respect of money laundry has been established.	0.3	Regular Process/Reporting.
		0.4	Emergency protocols in place.
		0.3	Show log file in order to underpin evidence that this is working.
4	Process in respect of fraud has been established.	0.3	Regular Process/Reporting.
		0.4	Emergency protocols in place.
		0.3	Show log file in order to underpin evidence that this is working.
5	Whistle blowing procedures have been enacted	0.5	Show process and anonymity requirement
		0.2	Show log - use
		0.3	Show process which is initiated.

8.4.9 Business Protection

Nr of Controls checked **3**

	Control Matrix: Requirement	W	Check
1	IT Security systems in place.	0.3	One key required for internal users in place.
		0.3	2nd key requirement for external users in place.
		0.4	Intrusion detection systems and logging for sensitive IT systems in place.
2	Physical Protection.	0.4	Local Access protocols adhere to guideline.
		0.4	Physical protection of values.
		0.2	Restricted access to IT centres.
3	Regular risk assessment together with mitigation actions.	0.5	Evidence of such a process.
		0.5	Evidence that weaknesses are tackled.

8.4.10 People

Nr of Controls checked **1**

	Control Matrix: Requirement	W	Check
1	A regular performance and talent management process is in place.	0.5	Evidence of a working HR performance process.
		0.3	Evidence of a working talent management process.
		0.2	Show list of key talent together with rationale and corresponding measures.

8.4.11 Outsourcing

Nr of Controls checked **1**

	Control Matrix: Requirement	W	Check
1	A process is in place that all material outsourcing activities are approved by the local CEO which ensures adherence to the relevant guidelines.	0.4	The authority of such processes has been correctly assigned.
		0.4	Process is in place that all activities are channelled to the CEO.
		0.2	Documentation available for exiting outsourcing agreements.

8.4.12 Communications

Nr of Controls checked **1**

	Control Matrix: Requirement	W	Check
1	There is a process in place to avoid unauthorised release of price sensitive information and of unauthorised media contact.	0.2	People receiving price sensitive information have to adhere to corresponding guidelines.
		0.2	There exists a list of persons which deal with price sensitive information.
		0.2	Blackout periods are defined and communicated.
		0.2	Confidentiality agreements for employees are in place for M& A situations etc.
		0.2	Media-contacts process established and in place –> provide evidence (eg conferences etc.)

8.4.13 *Legal*

Nr of Controls checked **3**

	Control Matrix: Requirement	W	Check
1	A senior internal counsel is nominated, overseeing and steering all litigations. External counsels need to be approved by him.	0.4	Formal nomination.
		0.3	List of all current approved external counsels.
		0.3	Guideline defining the corresponding processes.
2	Litigation is centrally managed.	0.4	Governing guidelines.
		0.6	Detailed list of all pending litigations.
3	There is a process in place in order to comply with external listing and disclosure rules.	1.0	Evidence that this process is up and running.

8.4.14 *Financial Reporting*

Nr of Controls checked **3**

	Control Matrix: Requirement	W	Check
1	A system of internal control must be maintained over financial reporting that is consistent with the group's minimum standards, the nature, complexity and risk of the business and is responsive to changes in its environment and conditions.	0.5	BU shows existence of ICS.
		0.5	BU shows efficiency of ICS.
2	A process ensuring that BU adhere with reporting manual is implemented. This process takes in particular care for new and changed requirements	0.4	BU shows existence of process.
		0.4	BU shows how to deal with new requirements (=projects and training).
		0.2	Responsible person has adequate know how.
3	4eye principle, checks and balance.	0.5	BU shows existence of delegated authorities.
		0.5	BU shows 4eye principle for all positions needing estimates.

8.4.15 *External Auditor*

Nr of Controls checked **3**

	Control Matrix: Requirement	W	Check
1	The local Audit Committee evaluates at least yearly the performance and costs/expenses of the Group's auditor and reports this to group; AC oversees and controls rotation requirements of audit partners.	0.5	These tasks are embedded in the bylaws of the AC.
		0.5	Company can provide evidence that this task has been done.
2	A process is in place that the engagement of auditors is according to the policy (Adherence to the Implementation Procedures and Guidance).	0.5	There is a list of external parties (eg auditors) which require special approval.
		0.5	The company shows what has been done in order to adhere to these principles.
3	A process is in place to hire (ex) employees of the primary audit firm according to the corresponding guideline.	0.7	Company shows the corresponding process.
		0.3	Provides a list of candidates which have undergone such a process.

8.4.16 Taxation

Nr of Controls checked **3**

	Control Matrix: Requirement	W	Check
1	A responsible person acting as single point of contact for all tax authorities has been determined, all correspondence from and to the tax authorities is stored in a structured way in one place.	0.3	Person named.
		0.3	Guideline written.
		0.2	List of contacts established.
		0.2	Repository up and running.
2	There is a process to check for adherence to local requirement, new requirements are analysed and a process in place ensuring timely implementation of new requirements.	0.6	A check list of deliverables exists - together with responsible persons.
		0.4	BU shows that they are following tax developments.
3	All material communication to the tax authorities are signed by the CFO. A process to store relevant tax information is up and running	0.3	There is a 4eye principle in place.
		0.4	Company can show what has been signed by CFO.
		0.3	Company can show that the relevant documents are stored.

8.4.17 General Insurance Reserving

Nr of Controls checked **3**

	Control Matrix: Requirement	W	Check
1	A four eyes process is in place for reserving.	0.5	The company can show the process.
		0.5	It can show evidence that it is in pace by for example showing cases of differing opinion.
2	Reserve ranges are produced and analysed based on the tools provided by group at least yearly.	0.5	Describe the tools and methods used.
		0.5	Show results.
3	For every closing there is a reserving report indicating the most relevant facts and possible risks. CEO and CFO read and sign it.	0.6	Show last reserving report.
		0.4	Provide evidence that the CEO/CFO have seen it.

8.4.18 Life Reserving

Nr of Controls checked **3**

	Control Matrix: Requirement	W	Check
1	A four eyes process is in place for reserving.	0.5	The company can show the process.
		0.5	It can show evidence that it is in place by showing cases of differing opinion.
2	Foreseeable changes in mortality et al are analysed and reported	0.5	Describe the tools and methods used.
		0.5	Show results.
3	For every closing there is a reserving report indicating the most relevant facts and possible risks. CEO and CFO read and sign it.	0.6	Show last reserving report.
		0.4	Provide evidence that the CEO/CFO have seen it.

8.4.19 Capital Management

Nr of Controls checked **3**

	Control Matrix: Requirement	W	Check
1	Quarterly calculation of capital positions according to group measures. CEO and CFO read and sign off results.	0.6	Show results.
		0.4	Provide evidence that CEO/CFO have seen it.
2	Process to timely detect a shortfall in capital (eg in case of major market movements etc) is in place. The process covers also the initiation of correcting measures and reporting to group.	0.5	A person looking at this exists
		0.5	The person can explain how often and how the capital is monitored: with which tools.
3	There is a process (at least yearly) in place to monitor and optimise the entities capital structure.	0.5	The process can be described in a convincing manner.
		0.5	Corresponding reports including improvement potential are shown.

8.4.20 Credit

Nr of Controls checked **5**

	Control Matrix: Requirement	W	Check
1	A local governing body for credit risk has been established (4x p.a.).	0.3	Bylaws.
		0.3	Meeting minutes.
		0.4	Tasks performed are OK - eg. limits have been reviewed and agreed ...
2	There is a dedicated, knowledgeable person for running the credit risk management.	0.4	Knowledgeable person nominated.
		0.4	Enough time devoted to this task.
		0.2	Relevant links within organisation established.
3	A limit system is in place in including escalation and reporting processes.	0.6	Show limit system - quality.
		0.4	Show reporting processes.
4	Counter-party risk is measured at least quarterly according to the group's metric.	0.5	Show reports.
		0.5	Review on quality.

8.4.21 Market

Nr of Controls checked **5**

	Control Matrix: Requirement	W	Check
1	An ALM strategy is in place.	0.3	Document.
		0.3	Reasonable method.
		0.4	Process.
2	A governing body meeting at least quarterly (and in case of major market movements) has been established.	0.3	Bylaws.
		0.3	Meeting minutes.
		0.4	Tasks performed are OK - eg. limits have been reviewed and agreed …
3	A limit system is in place in including escalation and reporting processes.	0.6	Show limit system - quality.
		0.4	Show reporting processes.
4	Risk measurement processes and reporting in place according the group's metric.	0.3	Documentation / Tools.
		0.3	Regular reports.
		0.4	Tasks performed are OK - eg. limits have been reviewed and agreed …
5	A Performance measurement and contribution process including reporting is in place (reporting at least quarterly)	0.4	Metrics and Report templates exist.
		0.3	Link to the feeder systems and process up and running.
		0.3	Actual reports.

8.4.22 Derivatives

Nr of Controls checked **5**

	Control Matrix: Requirement	W	Check
1	A derivative strategy ("Statement of derivative practise") has been written and implemented.	0.3	Document.
		0.3	Suitable method used.
		0.4	Process.
2	Persons dealing with derivatives have the required knowledge, commensurate to the complexity.	0.5	Knowledgeable person nominated.
		0.4	Enough time devoted to this task.
		0.1	Relevant links within organisation established.
3	A process ensuring that derivatives are traded only after knowing all consequences is in place.	0.6	Credible description of requirements and implementation.
		0.4	Knowledgeable persons involved knowing the different aspects.
4	There exists a corresponding governing body meeting at least monthly and upon demand.	0.3	Bylaws.
		0.3	Meeting minutes.
		0.4	Tasks performed are OK - eg. limits have been reviewed and agreed ...
5	Risk management, limit systems, independent valuation in place.	0.7	Show limit system (0.2: Independent Valuation, 0.3: Risk Measurement, 0.2: Limits)
		0.3	Show reporting processes.

8.4.23 *Liquidity*

Nr of Controls checked **3**

	Control Matrix: Requirement	W	Check
1	Review and analysis of liquidity exposure risks (cash flow forecasts and scenario analysis).	0.5	Risk measurement in place.
		0.5	Monthly forecast of liquidity.
2	The company maintains an appropriate management structure in place to oversee the daily and long-term management of liquidity risk in line with the principles of this policy.	0.6	Knowledgeable person available.
		0.4	Enough time devoted to this task.
3	A limit system is in place including escalation and reporting processes.	0.6	Show limit system - quality.
		0.4	Show reporting processes.

8.4.24 *Foreign Exchange*

Nr of Controls checked **5**

Control Matrix:

	Requirement	W	Check
1	A local governing body for FX risk has been established (4x p.a.).	0.3	Bylaws.
		0.3	Meeting minutes.
		0.4	Tasks performed are OK - eg. limits have been reviewed and agreed …
2	There is a dedicated, knowledgeable person for running the FX risk mgmt..	0.4	Knowledgeable person nominated.
		0.4	Enough time devoted to this task.
		0.2	Relevant links within organisation established.
3	A process ensuring that FX Exposure is only taken after knowing all consequences is in place.	0.6	Credible description of requirements and implementation.
		0.4	Knowledgeable persons involved knowing the different aspects.
4	A limit system is in place including escalation and reporting processes.	0.6	Show limit system - quality.
		0.4	Show reporting processes.
5	Risk measurement processes and reporting in place according the group's metric.	0.3	Bylaws.
		0.3	Meeting minutes.
		0.4	Tasks performed are OK - eg. limits have been reviewed and agreed …

8.4.25 *Mergers & Acquisitions*

Nr of Controls checked **3**

Control Matrix:

	Requirement	W	Check
1	Ensure that relevant guidelines are known to everyone in the M&A team and that an adequate project organisation is in place.	0.4	Company can show that they have an up to date guideline ready.
		0.4	A senior manager is responsible for adherence to it.
		0.2	Availability of corresponding project mgmt. skills.
2	Ensure that the BU can challenge the external valuation.	1.0	Provide evidence.
3	Ensure adherence to confidentiality agreements and safeguarding of privileged information.	0.5	Existing predefined confidentiality-agreements ready.
		0.5	Company can explain what needs to be done.

8.4.26 Risk Management & Internal Control

Nr of Controls checked **3**

	Control Matrix: Requirement	W	Check
1	The group risk model is applied and the shortcomings at BU level are addressed within the planning.	1.0	Evidence provided.
2	The company regularly self-assesses the quality of its risk management iro quantity and quality.	0.4	Process in place.
		0.4	Review of the people.
		0.2	Review of the Processes and reports.
3	The company applies a three lines of defence approach.	1.0	Show organisational charts.

8.4.27 Life Insurance Risk

Nr of Controls checked **2**

	Control Matrix: Requirement	W	Check
1	Risk measurement processes and reporting in place according the group's metric.	0.3	Bylaws.
		0.3	Meeting minutes.
		0.4	Tasks performed are OK - eg. limits have been reviewed and agreed ...
2	A limit system together with an escalation process is in place.	0.6	Show limit system - quality.
		0.4	Show reporting processes.

8.4.28 Life Insurance Product Development & Pricing

Nr of Controls checked **3**

	Control Matrix: Requirement	W	Check
1	A governing body including performance review process is in place.	0.5	Meeting minutes etc showing body is in place.
		0.5	If actual shortcomings are corrected.
2	There is guideline in place.	0.5	Guideline exists.
		0.5	Limits and delegated authority.
3	There exists a regular performance measurement process.	0.5	Reporting up and running.
		0.5	Relevant Reports.

8.4.29 Unit Pricing

Nr of Controls checked **3**

	Control Matrix: Requirement	W	Check
1	There are robust systems in place for unit pricing.	0.5	Existence of a formal requirement document for the system.
		0.5	Test cases exist underpinning the correctness of the system.
2	Timely calculation of asset values and unit prices.	0.4	Clearly defined valuation methods and processes.
		0.3	Short time lag between market data available and marked data used for unit pricing.
		0.4	Processes and Methods established in case of delayed delivery of asset prices.
3	High quality data and processes in place for asset valuation	0.4	High quality asset valuation data in use.
		0.3	Automated feeding in pricing systems.
		0.4	4eye principle for assets where value is based on models or estimation or entered manually

8.4.30 General Insurance Underwriting

Nr of Controls checked **5**

	Control Matrix: Requirement	W	Check
1	There are yearly reviewed risk statements. Available Capital (Capacity) is allocated to individual product lines. Corresponding limit systems are in place.	0.3	Risk Statements.
		0.3	Capacity allocation.
		0.4	Limit systems.
2	There is a regular performance measurement and management process in place.	0.3	Reasonable metrics.
		0.3	existing reporting.
		0.4	Evidence of management actions.
3	UW Cycle management, Delegated Limits and approval processes.	0.3	The entity can show that a cycle mgmt. is in place.
		0.7	A working delegation and approval process is in place (Note evidence needs to be provided).
4	A feedback process for GI pricing is in place.	0.4	Process.
		0.6	Evidence that is has sufficient feed-back.
5	There is a stringent wording process in place.	0.5	Company shows the governing principles.
		0.5	Review and Feed-back processes take place.

8.4.31 General Insurance Reinsurance

Nr of Controls checked **3**

	Control Matrix: Requirement	W	Check
1	A written, yearly renewable strategy iro reinsurance risk has been established within the delegated authorities from the group.	0.5	Document has been written.
		0.5	Evidence that the content is meaningful and relevant for the company.
2	The required reinsurance has been analysed and documented. The active reinsurance treaties are reviewed in respect of risk mitigation and profitability yearly.	0.5	Analysis of the reinsurance needs.
		0.5	Analysis of the performance of the reinsurance treaties.
3	Processes are in place that no treaties without risk-transfer are written without prior approval of the group.	1.0	Evidence that there is a process which ensures adherence.

8.4.32 General Insurance Claims

Nr of Controls checked **3**

	Control Matrix: Requirement	W	Check
1	Documented claims payment and settlement authorities. Delegated authority for claims handling by external parties is fully documented and conformance overseen.	0.5	Documented levels of authority.
		0.5	Show actual referral process in order to demonstrate that it works.
2	Reserving guidelines, covering early estimation to probable ultimate settlement, are clearly defined, reviewed regularly and followed.	0.5	Existence of guideline.
		0.5	Evidence that guideline has been put in place efficiently.
3	Established Claims handling process.	0.2	for each of iii; ix; xii; xiii; xxii of section C.

Chapter 9
Capital Models and Integrated Risk Management

9.1 Introduction

In this chapter we want to see how the different pieces of the capital models flow together in order to get an integrated capital model, covering all the different risk categories. In a lot of cases the capital models for insurance companies have been designed along the following categories:

M. Koller, *Life Insurance Risk Management Essentials*, EAA Series,
DOI 10.1007/978-3-642-20721-1_9, © Springer-Verlag Berlin Heidelberg 2011

- Financial and ALM risk,
- Life insurance risk,
- General insurance risk,
- Operational risk.

The reason for building these distinct risk modules was that there were people focusing on ALM issues, such as life risk etc. Hence it was a consequence of the relative skill set of the people and of the relative importance of the risks. Sometimes some of the risks were merely modelled as a consequence of a regulatory requirement, such as operational risks. The methods used for the different risk categories are often different and ultimately there is the question on how to link the sub-modules together. This is the same question as linking the individual risk factors within each risk module together.

From a holistic point of view it is important that the company can cover the required capital stemming from all risk types with the available risk capital.

9.2 Bringing the Puzzle Together

From a technical point of view we are in a situation where we have a set of risk categories $(\mathcal{R}_\kappa)_{\kappa \in \mathcal{K}}$ and for each risk category $\kappa \in \mathcal{K}$ we have a corresponding loss function X_κ with a probability density function $F_{X_\kappa}(t)$. So what we actually know per X_κ is its marginal distribution if we consider $(X_\kappa)_{\kappa \in \mathcal{K}}$ as a multidimensional random variable.

One possibility is to link the $(\mathcal{R}_\kappa)_{\kappa \in \mathcal{K}}$ together with *copulas*. In order to understand this concept, we need to look at two random variables X and Y with cumulative distribution functions F_X and F_Y respectively. Furthermore we remark that in this case both of the following random variables $\tilde{X}$ and $\tilde{Y}$ are uniformly $[0, 1]$–distributed:

$$\tilde{X} = F_X(X),$$
$$\tilde{Y} = F_Y(Y).$$

We remark that if (X, Y) are dependent, this holds true also for $(\tilde{X}, \tilde{Y})$. A copula is hence a function which transforms the random variables $(\tilde{X}, \tilde{Y})$. More formally: A copula is a multivariate (n-dimensional) joint distribution on $[0, 1]^n$, such that every marginal distribution is uniform on the interval $[0, 1]$. A function

$$C : [0, 1]^n \to [0, 1], (x_1, \dots x_n) \mapsto C(x_1, \dots x_n)$$

is an n–dimensional copula if the following hold:

1. $C(u) = 0$ for $u = (x_1, \dots x_n) \in [0,1]^n$ if one of the $x_k = 0$,
2. $C(u) = x_j$ for $u = (x_1, \dots x_n) \in [0,1]^n$ if $x_i = 1 \forall i \neq j$,
3. C is increasing for all hyperrectagles $R \subset [0,1]^n$.

Sklar's theorem states the following for the bivariate case: For $H(x, y)$ a bivariate cumulative probability distribution function with marginal cumulative probability functions $F_X(x) := H(x, \infty)$ and $F_Y(y) := H(\infty, y)$, there exists a copula with

$$H(x, y) = C(F_X(x), F_Y(y)).$$

Hence copulas are a means to link individual random variables together and we want to have a look at the most important classes of copulas:

Gaussian copula: For $\rho \in \mathbb{R}$, we denote by

$$\begin{aligned}
C_\rho(x, y) &:= \Phi_\rho(\Phi^{-1}(x), \Phi^{-1}(y)), \text{ with} \\
\Phi(x) &= \int_{-\infty}^{x} \frac{1}{\sqrt{2\pi}} \exp(\frac{\zeta^2}{2}) d\zeta, \\
\Phi_\rho(x, y) &= \frac{1}{2\pi\sqrt{1-\rho^2}} \exp\left(-\frac{1}{2(1-\rho^2)}\left[x^2 + y^2 - 2\rho x y\right]\right).
\end{aligned}$$

Archimedian copula:

$$H(x_1, \dots x_n) = \Psi^{-1}\left(\sum_{j=1}^{n} \Psi(F_{x_j}(x_j))\right).$$

Ψ is known as generator function. If some conditions for Ψ are fulfilled, the resulting C is a copula.

Product copula: The product copula is a special kind of an Archimedian copula with $\Psi(x) = -\ln(x)$. This copula is also known as independent copula, since we get $H(x, y) = F_X(x) \times F_Y(y)$.

After having determined the relationship between the different loss functions $(X_\kappa)_{\kappa \in \mathcal{K}}$ for the different risk categories $(\mathcal{R}_\kappa)_{\kappa \in \mathcal{K}}$, it is possible to calculate the corresponding n-dimensional cumulative probability density function $F_{(X_1, \dots X_n)}$. The distribution function for the total loss $L = \sum_{i=1}^{n} X_i$ can hence be calculated.

9.3 Diversification

Figure 9.1 provides an example for the total capital consumed by an insurance entity. It becomes obvious in this example that consumed capital is not uniform over the risk categories:

	Risk Category		Required Capital
(i)	Financial Risk	X_1	7680
(ii)	Operational Risk	X_2	900
(iii)	Regulatory	X_3	450
	Simple Sum		9030
	Diversification		-3010
(iv)	Required Capital $\sum_{i=1}^{3} X_i$		6020

Assume for the moment that we are using a 99.5 % VaR as a risk measure. In this context the numbers above are determined in the following manner:

$$
\begin{aligned}
7680 &= F_{X_1}^{-1}(0.995), \\
900 &= F_{X_2}^{-1}(0.995), \\
450 &= F_{X_3}^{-1}(0.995), \\
6020 &= F_{\sum_{i=1}^{3} X_i}^{-1}(0.995).
\end{aligned}
$$

This means in particular that the diversification effect $((iv) - (i) - (ii) - (iii))$ is a pure consequence on how we group the different risk factors. It is dependent on the way we construct our risk models and how we link them together. It represents the amount of capital which is released by linking the different risks together. Now looking again at the numbers, one might wonder why the diversification is so high. The solution in the concrete example is actually rather easy, since the numbers such as 7860 already represent simple sums. Hence actually (assuming that we have the 13 risk factors mentioned in the table) we have

$$
3010 = \sum_{i=1}^{13} F_{X_i}^{-1}(0.995) - F_{\sum_{i=1}^{13} X_i}^{-1}(0.995),
$$

which highlights the risk of the concept of diversification. In the table on page 115 you see that diversification can be calculated at different levels of the aggregation.

As with all other relevant risk capital measurements, it makes sense to define a risk appetite with corresponding limits for the total capital consumed by the insurance company.

Financial Risks		Operational Risks		Legal/Regulatory/Compliance Risks	
Risk	**in M €**	**Risk**	**in M €**	**Risk**	**in M €**
Credit Risk	2660	ALM and Investment Processes	c 300	Legislative Changes	c 400
Alternative Investments	520	Reputational Risks	c 150	Quantum / Genesis Regulatory Risk	c 25
Equity and GMDB Risk	2040	Quantum Leap Programme	c 70	Solvency II Risk	c 25
Interest Rate Risk	470	Switch at known prices of Unit Lined Funds	c 250		
Lapse Risk	1990	Underperformance of distribution partners	c 130		
Subtotal	**7680**	**Subtotal**	**c 900**	**Subtotal**	**c 450**
	c 85%		c 10 %		c 5%

Total economic capital	9030
Diversification effect	-3010
Total Required Economic Capital	**6020**

Fig. 9.1 Required Economical Capital Reporting

Chapter 10
Risk Adjusted Performance Metrics

10.1 Introduction

When doing business there are different metrics of great importance, such as

- Statutory profit, since this is the base in order to determine and pay dividends.
- IFRS Profit, since this is one of the most regarded measures, which allows to compare different insurance entities.

M. Koller, *Life Insurance Risk Management Essentials*, EAA Series,
DOI 10.1007/978-3-642-20721-1_10, © Springer-Verlag Berlin Heidelberg 2011

- MCEV earnings and the corresponding value of new business, since this allows to determine an proxy for the economic value of the insurance entity and its ability to write profitable new business.
- Required and available risk capital, allowing to steer and measure the risk the company is running in order to achieve its strategic ambitions.

Most of the above mentioned measures are well known and hence we limit ourselves to remark that in some of the above mentioned performance metrics, there is considerable judgement needed for the set-up and the calibration of the underlying models. Furthermore it is worthwhile to remark that for all the metrics one can distinguish between a value and a performance metric. In case of the IFRS, the absolute profit is the value metric and the profit per shareholder equity (eg return on equity) is the performance metric. In the same sense the value of new business measures the value creation as a consequence of writing new business, and the new business margin (eg the value of new business per present value of new premiums).

10.2 Performance and Value Metrics

Looking at the above metrics it becomes obvious that all of the above metrics are quite complex and some of them such as the market consistent embedded value extremely difficult to understand. Hence the aim of this section is to introduce a value and a performance metric which is easily understandable and which allows to compare different lines of business, channels or companies. We will introduce a cash flow based performance metric and remark that a suitable interpretation of the market consistent embedded value will be consistent with this view.

We have seen in chapter 3 that the market consistent value of the insurance liabilities is based on cash flows and we will show here that this concept is suited to build a fully blown performance management system.

For convenience purposes the formulae already shown in chapter 3, assuming that the cash-flows considered as random variables are stochastically independent of the financial variables:

$$\begin{aligned}
\mathbb{E}[PV] &= \sum_{k=0}^{\infty} \pi_t(\mathcal{Z}_{(k)}) \times \mathbb{E}[CF_k], \\
MVM = CoC &= \sum_{k=0}^{\infty} \beta \times RC_k \times \pi_t(\mathcal{Z}_{(k)}), \\
IRR &= \frac{\mathbb{E}[PV] + \sum_{k=0}^{\infty} i \times RC_k \times \pi_t(\mathcal{Z}_{(k)})}{\sum_{k=0}^{\infty} RC_k \times \pi_t(\mathcal{Z}_{(k)})},
\end{aligned}$$

$$FCC^* = \sum_{k=0}^{\infty} \beta \{\max(0, V_k - \mathbb{E}[PV]_k)\} \times \pi_t(\mathcal{Z}_{(k)}),$$
$$FCC = \max(0, FCC^* - CoC).$$

10.2.1 IBNR Reserves

IBNR is usually calculated using a formula based on actual claim experience for prior years, adjusted for current trends and other factors. IBNR should be calculated for all lines of business in all countries. Companies should determine the IBNR reserve either by relying on past experience modified for current conditions or by determining the actual claims reported up to some point in time, such as 30 days after the balance sheet date, and estimating the claims yet to be reported beyond that date. Because death claims are usually reported quickly, the adequacy of the current year IBNR reserve can generally by determined by developing prior year's IBNR and by comparing this development to the year's reserve and giving consideration to various factors such as premiums in force. IBNR for disability business usually reflects a combination of historical claim experience, reasonable future expectations and the actual waiting periods for the in-force block.

The valuation of IBNR reserves must be according to the corresponding standards. In case of a traditional embedded value there is usually no value attributed to IBNR reserves, which themselves qualify as technical reserves. In case of a market consistent embedded value, IBNR Reserves are the present values of the expected future cash flow streams in relation to the IBNR cash flows using risk free discount rates.

10.2.2 Financial Options

In order to reflect financial options in the value of business in force, such as GMDB's etc, calculations to assess their value need to be performed based on a risk neutral method such as arbitrage free pricing or also Black-Scholes.

This valuation can either be done by explicit formulae (such as in the Black-Scholes context), or can also be based on a general risk neutral valuation method (such as deflators, martingale methods, Monte Carlo simulations, etc.)

The value of options needs to be shown separately and the parameters for its calculation are based on observable data at balance sheet date (for example in relation to the risk free rate, the volatility etc.). These parameters can either be estimated directly by analysing the underlying assets or also by using implicit methods (such as the calculation of the implicit volatility given the price of stock options.)

10.2.3 Frictional Capital Costs

Finally we need to realise that in the real world there are additional constraints, which have an impact on the value of a portfolio or a product sold. The most relevant are listed below:

- Frictional costs and
- Taxes.

Frictional costs stem from the fact, that the company needs to hold at a certain time the corresponding statutory reserves V_t for an underlying block of business. Additional frictional costs are induced by solvency requirements which are higher the economical risk capital. Given the fact that the best estimates liabilities $\mathbb{E}[PV]$ may be inferior, the company needs to hold this additional amount, resulting in the above mentioned (pure) frictional capital costs:

$$FCC^* = \sum_{t=0}^{\infty} \beta \{\max(0, V_t - \mathbb{E}[PV]_t)\} \times \pi(\mathcal{Z}_{(t)}),$$

where $\mathbb{E}[PV]_t$ denotes the expected present value of the future liabilities as seen at time t. Based on the fact that the risk capital also qualifies as capital to fill up missing reserves, the total frictional capital costs amount to:

$$FCC = \sum_{t=0}^{\infty} \beta_2 \{\max(0, V_t - \mathbb{E}[PV]_t - RC_t)\} \times \pi(\mathcal{Z}_{(t)}).$$

In a further step we will distinguish between the sort of frictional capital we need:

- Frictional capital which has been financed by the policyholder: This is the difference between the best estimate liabilities ($\mathbb{E}[PV]_t$) and the carrying amount of reserves in the company's balance sheet. This part of capital is similar to capital provided by letters of credits by bank for a relatively small cost and will therefore be charged less (β_3).
- Frictional capital which needs to be financed by the shareholder: This is the remaining part of the difference as indicated by the formula above and the WACC (weighted average cost of capital) of the company (β_2) will be charged.

Technically the above formula reads now as follows:

$$FCC = \sum_{t=0}^{\infty} [\beta_2 \{\max(0, W_t(1)\} + \beta_3 \{\max(0, W_t(2)\}] \times \pi(\mathcal{Z}_{(t)}),$$

where $W_t(1)$ and $W_t(2)$ denote the above mentioned differences.

As the market consistent valuation is based on a full balance sheet approach the market consistent value of insurance liabilities is to be calculated before tax, as taxes are taken care of in the calculation of the market consistent equity.

10.2.4 Duration of Projection

Projections should be sufficiently long duration to capture all important financial events in the life of a policy. The projections are subject to a minimum projection period of 40 years (or policy duration if less).

10.2.5 Formulae

$$\mathbb{E}[PV] = \sum_{t=0}^{\infty} \pi(\mathcal{Z}_{(t)}) \times \mathbb{E}[CF_t], \tag{10.1}$$

$$\mathbb{E}[PV]_t = \frac{1}{\pi(\mathcal{Z}_{(t)})} \sum_{k=t}^{\infty} \pi(\mathcal{Z}_{(k)}) \times \mathbb{E}[CF_k \mid \mathcal{F}_t], \tag{10.2}$$

$$CoC = \sum_{t=0}^{\infty} \beta_1 \times RC_t \times \pi(\mathcal{Z}_{(t)}), \tag{10.3}$$

$$FCC^* = \sum_{t=0}^{\infty} \beta_2 \{\max(0, V_t - \mathbb{E}[PV]_t)\} \times \pi(\mathcal{Z}_{(t)}), \tag{10.4}$$

$$FCC = \sum_{t=0}^{\infty} \beta_2 \{\max(0, V_t - \mathbb{E}[PV]_t - RC_t)\} \times \pi(\mathcal{Z}_{(t)}), \tag{10.5}$$

$$\text{MV of Ins. Lia.} = \mathbb{E}[PV] - CoC - FCC, \tag{10.6}$$

$$\gamma = (\beta + i) \times (1 - \text{ Tax rate}), \tag{10.7}$$

where i denotes the risk free interest rate for the corresponding period.

10.2.6 Example

In order to show how the different pieces work together we have chosen an annuity portfolio for a valuation as of 31.12.2006. We have the following main characteristics:

Item	Amount in EUR
Balance Sheet Reserve	341723220
of which from in-force	272912264
of which from New Business	25610375
of which IBNR for late reporting	45303338
Annuities to be paid out per Year	16371791

There are principally two effects for which the reserve has to be adjusted. In the concrete example – a reinsurance company – the last settling of annuities payed went back to September 2005. Therefore 16 months of annuity payments are outstanding, leading to the IBNR Reserve of EUR 45.3 M. The data for the projection is as of 30.6.2006. Therefore we need on the one hand roll forward the projection to the valuation date. On the other hand the new production for the 6 missing months needs to be modelled, resulting in an increase of reserve of EUR 25.6 M.

In order to make a market consistent valuation we need to take the following effects into consideration and we have chosen the following parameters:

Item	Parameter
Risk capital	15 % of $\mathbb{E}[PV]$
Unit CoC - riskfree	10 %
FCC	Difference between V_t and $\mathbb{E}[PV]_t$

Figure 10.1 shows the development of the different cash flows. One now needs to calculate the different parts, namely the market value margin (CoC), the change in value due to funds withheld and also the frictional capital costs. The funds withheld can economically be considered as a loan of the reinsurer to the insurer with a valuation corresponding to a cash flow swap; eg reinsurer pays forward interest and receives fixed interest as agreed in the contract. In this particular case this process is in favour of the reinsurer. Figures 10.2 shows the three different pieces.

The above mentioned calculation results in the following results:

Item	Amount in EUR
$+\ \mathbb{E}[FV]$	230158544
− FundsWithheld	-8212193
$-\ MVM$	29581548
FCC^*	20249339
$-\ FCC$	181688
$= \mathbb{E}[FV]+$ FuWi $+CoC + FCC$	**251709588**
Δ Difference to MR	+21202676
+ Gross up for New Business	9.3 %
= Gross Eco Value of Lia.	**275330282**

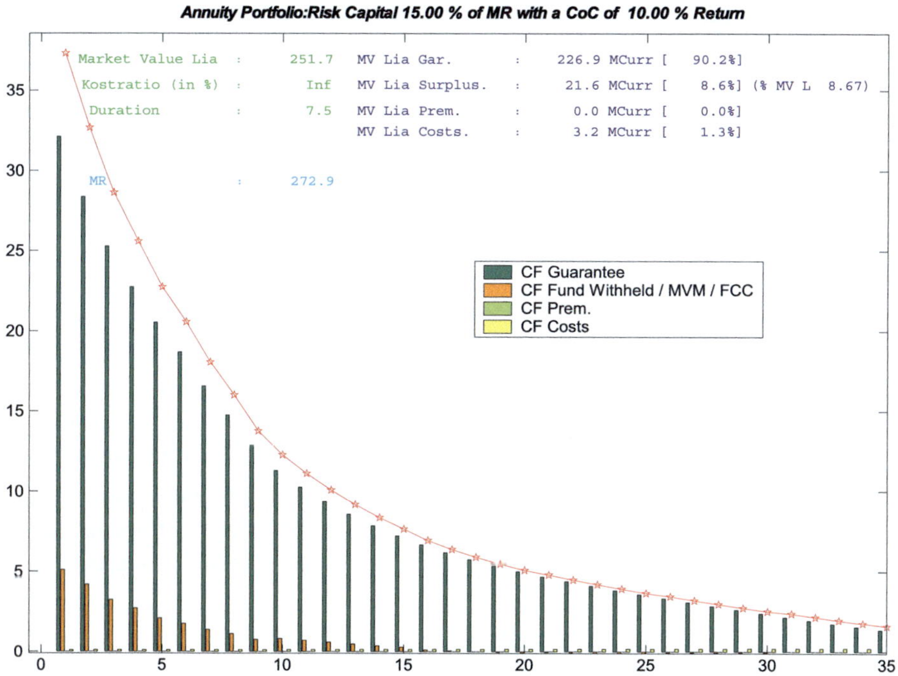

Fig. 10.1 Cash Flows

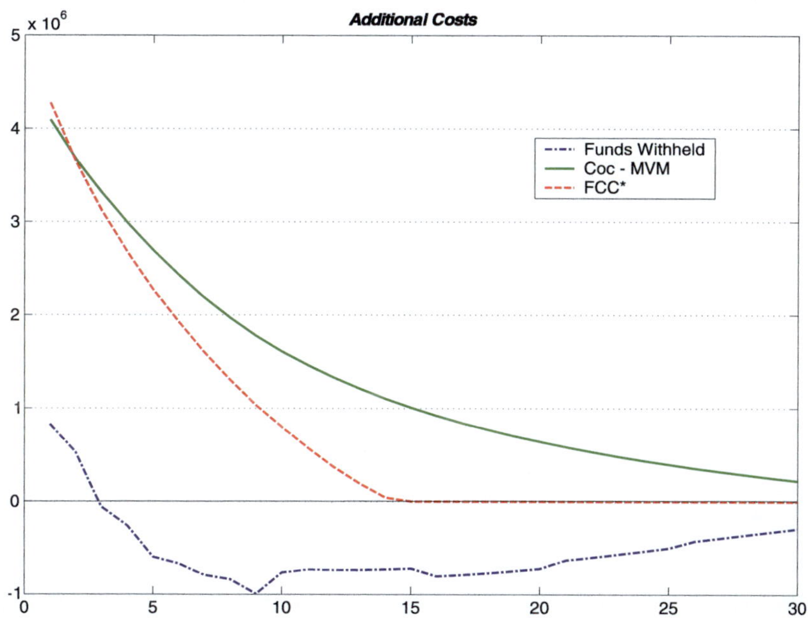

Fig. 10.2 Details

10.3 Examples

This section provides a concrete reporting example in order to see the different parts. The first table shows an overview of all the different product portfolios for an insurance entity, the second one the corresponding details. In order to understand the notation, the table below provides an explanation:

Reserves in B/S: These are the statutory reserves in the balance. In this direct method they serve also as a proxy for the amount of assets covering the liabilities and hence one part of the value attributed to the corresponding product.

PV Premium: The present value of future premiums is the second contributor to value. In the context of market consistent valuation these future premiums are weighted according to persistency and discounted by risk free discount rates.

PV Claims: Here the expected claims are indicated. All different types of claims such as surrender and maturity benefits, but also claims in case of death, etc are subsumed here. It would be possible to be more granular in this position.

PV Exp - Internal: The next three positions relate to expenses. They are split into the different pieces in order to allow a variety of break downs, looking for example at a marginal cost base. Please note that the expenses are important since the company can influence them better than most of the other parts. In this position the present value of future internal costs in relation to the product are displayed.

PV Exp - Overhead: In this position the present value of future overhead costs in relation to the product are displayed.

PV Exp - Commissions: In this position the present value of future commissions in relation to the product are displayed. In case of a value of new business this position carries all the commissions which are paid for the product.

Market Value Margin: The MVM is the amount described above. eg $MVM = \sum_{k=0}^{\infty} \beta \times RC_k \times \pi_t(\mathcal{Z}_{(k)})$.

FCC: Frictional costs of capital as mentioned above.

Funds Withheld: This is the swap arrangement which is often used in a reinsurance treaty, where the cedent pays fixed and receives floating.

Tax: In this position the taxes are deducted.

Total: This line, the sum of the above, represents the total value inherent in the corresponding product from an economic point of view.

PV Profit: Market consistent present value of profits not allowing for risk capital.

PV Capital: Present value of risk capital for the corresponding line of business.

RoRAC (in %): That's the Return on risk adjusted capital.

VIF: This is the expected present value of the profits after tax, using a risk discount rate.

Lock-in: This is the so called lock in effect, eg the opportunity loss as a consequence that shareholder capital is immobilised as a consequence of regulatory capital requirements.

PVFP: Sum of the two above.

Value at B/S Date M EUR	**Prod 1**	**P2**	**P3**	**P4**	**P5**	**P6**	**P7**	**P8**	**P9**	**P10**	**Total**
Reserves in B/S	1.2	0.0	79.0	214.5	8.0	1.6	29.8	448.8	12.1	-41.9	753.4
PV Premium	0.6	0.0	0.5	738.3	0.0	34.3	285.8	0.0	432.9	1563.0	3055.0
PV Claims	-0.0	-0.0	-66.8	-678.6	-3.6	-2.9	-18.4	-387.8	-436.3	-1351.0	-2945.0
PV Exp - Internal	-0.1	0.0	-0.0	-14.7	-0.0	-0.1	-9.7	-2.9	-0.1	-22.3	-50.4
PV Exp - Overhead	-0.2	0.0	-0.0	-29.5	-0.1	-0.2	-14.5	-4.0	-0.2	-36.6	-85.7
PV Exp - Commissions	-0.0	0.0	-0.0	-171.0	0.0	-31.8	-39.9	-33.1	-2.3	-53.8	-332.4
Market Value Margin	-0.0	0.0	-0.0	-14.2	-0.1	-1.1	-17.9	-4.7	-10.0	-48.6	-97.0
FCC	-0.0	-0.0	-0.0	-0.6	-0.0	-2.1	-7.9	-19.7	-0.0	-3.0	-33.5
Funds Withheld	-0.0	-0.0	-0.3	-2.6	0.0	-1.3	0.0	10.6	0.0	0.0	6.2
Tax	-0.4	-0.0	-4.1	-17.5	-1.4	0.0	-79.4	-10.8	-1.3	-12.0	-127.2
Total	**0.9**	**0.0**	**8.0**	**23.7**	**2.4**	**-3.7**	**127.6**	**-3.9**	**-5.4**	**-6.5**	**143.2**
Profitability											
PV Profit	1.0	0.0	8.0	48.7	2.6	-3.7	151.8	-14.0	12.6	77.6	284.7
PV Capital	0.7	-0.0	0.7	300.9	3.4	23.4	378.0	99.3	212.0	1025.0	2044.0
RoRAC (in %)	22.1	15.7	9.2	8.8	14.8	-19.5	27	-25.4	5.5	5.6	8.5
Classical EV											
VIF	1.1	0.0	8.6	37.7	2.6	0.9	131.3	13.0	5.0	33.0	233.5
Lock-in	-0.0	0.0	-0.0	-8.6	-0.0	-0.6	-10.0	-2.8	-6.2	-28.4	-56.9
PVFP	**1.0**	**0.0**	**8.6**	**29.1**	**2.5**	**0.3**	**121.2**	**10.2**	**-1.2**	**4.5**	**176.5**
Sensitivities											
Expenses	-0.0	0.0	-0.0	-3.0	-0.0	-0.0	-1.6	-0.4	-0.0	-4.5	-9.7
Claims	-0.0	-0.0	-4.4	-46.0	-0.2	-0.2	-1.2	-25.4	-34.0	-103.0	-214.7
Capital	-0.0	0.0	-0.0	-1.9	-0.0	-0.1	-2.3	-0.6	-1.5	-7.1	-13.7
Tax	-0.0	-0.0	-0.4	-1.7	-0.1	0.0	-7.9	-1.0	-0.1	-1.2	-12.7
Profits p.a.											
1990 – 2006	0.9	0.0	0.0	1.4	0.0	-0.0	0.7	-0.1	0.0	-0.1	2.9
2007	4.4	0.3	1.1	9.4	0.5	1.8	9.7	-20.6	0.1	1.6	8.6
2008	1.1	0.0	9.4	11.6	0.9	-0.2	20.5	-2.4	0.5	11.3	52.9
2009	0.0	0.0	0.0	10.4	0.8	0.4	19.7	3.0	0.5	9.4	44.6
2010	0.0	0.0	0.0	7.2	0.6	0.3	18.4	2.8	0.5	5.2	35.3
2011	0.0	0.0	0.0	4.7	0.1	-0.4	15.9	2.6	0.5	2.9	26.5
2012 – 2016	0.0	0.0	0.0	2.7	0.1	0.1	13.6	2.0	0.5	1.1	20.5
2017 – 2021	0.0	0.0	0.0	0.4	0.0	0.1	9.5	1.0	0.6	0.3	12.0
2022 – 2031	0.0	0.0	0.0	0.0	0.0	0.0	5.5	0.3	0.2	0.1	6.2
2032 – 2107	0.0	0.0	0.0	0.0	0.0	0.0	0.4	0.0	-0.1	0.0	0.2
Total	1.1	0	8.6	37.7	2.6	0.9	131.3	13	5	33	233.5
Capital p.a.											
1990 – 2006	4.3	0.6	4.5	14.8	0.7	1.2	2.7	2.9	11.2	6.8	50.1
2007	10.9	0.0	14.1	65.7	4.8	3.6	40.9	15.0	33.3	78.9	267.7
2008	0.7	0.0	0.7	71.0	2.3	0.8	33.7	15.1	31.1	105.2	261.1
...											
Total	0	0	0	8.6	0	0.6	10	2.8	6.2	28.4	56.9

Valuation at B/S date

Position	Amount in EUR	Relative Amount
Reserves in B/S	753400000	19.78 %
Present Value Premium	3055000000	80.22 %
Present Value Claims	-2945000000	-77.33 %
PV Exp - Internal	-50400000	-1.32 %
PV Exp - Overhead	-85730000	-2.25 %
PV Exp - Commissions	-332400000	-8.72 %
Subtotal	394800000	10.36 %
Market Value Margin	-97070000	-2.54 %
FCC	-33560000	-0.88 %
Funds Withheld	6281000	0.16 %
Tax	-127200000	-3.33 %
Total	143200000	3.76 %
PV Profit	284700000	
PV Capital	2044000000	
RoRAC		13.93 %

Decomposition of Profit

Time	Φ P/L	Φ Capital	RoRaC w/o FCC
1990 – 2006	2971000	50190000	5.92 %
2007	8633000	267700000	3.22 %
2008	52920000	261100000	20.27 %
2009	44640000	237000000	18.84 %
2010	35370000	221800000	15.95 %
2011	26580000	203100000	13.09 %
2012 – 2016	20540000	166500000	12.34 %
2017 – 2021	12050000	99450000	12.12 %
2022 – 2031	6277000	43480000	14.44 %
2032 – 2107	279600	2526000	11.07 %
PVFP	233500000	56990000	409.80 %

Thc above tables can also be shown in a graphical form, such as in figure 10.3. Here the 100 % mark represents the mathematical reserves plus the present value of future premium. Hence a product is profitable if the sum is below 100 %.

10.4 Capital Allocation Process

Having a performance metric as the one described above and a process such as the one in section 5.3 it is now possible to do a concrete capital allocation process. In order to bring the two things together we have a second look at the process itself. Figure 10.4 provides an example which we want to analyse closer.

In order to do that we need to understand the different parts. The quantifies to be analysed are listed in the table below. It needs to be stressed that the framework

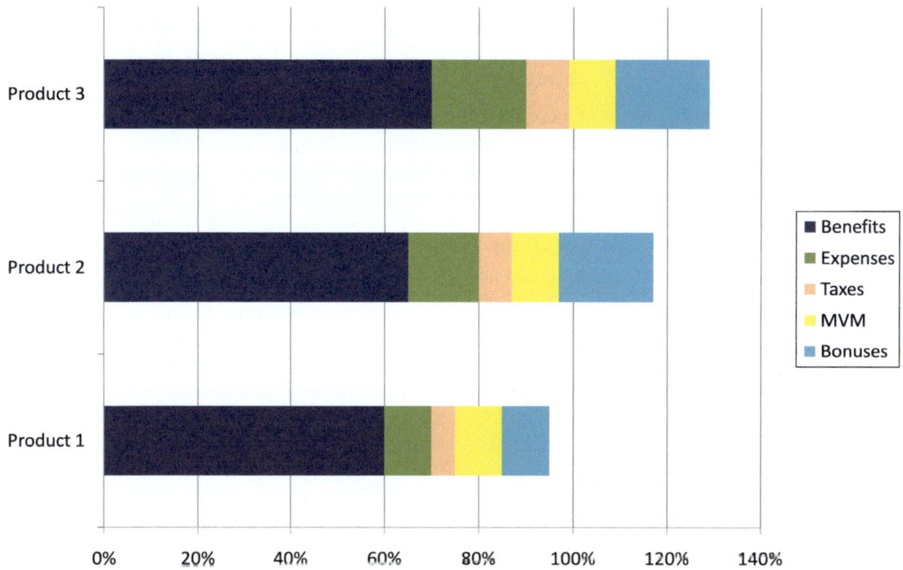

Fig. 10.3 Example Profitability of three products

is very flexible in terms what capital and return actually mean. In a lot of circumstances, capital is a synonym for a VaR (at a one in 200 year event) or for a TVaR (at a one in 100 year event). In some instances it also makes sense to look at two or more capital measures together (say at a VaR in a 1 in 10 and a one in 200 year event). Such considerations are reasonable if one wants to either look at different metrics at the same time or if there are differentiated risk appetite statements for different confidence levels. Also return can be interpreted in different ways and stands in an economic context mainly for the IRR measure introduced before.

Available Capital	This is the maximal capital which can be put at risk for the underlying period. It is normally the available economic capital.
Type of Opportunity	These are the different types of business opportunities, which absorb capital in order to generate shareholder value.
Required Capital	This is capital needed for the corresponding business opportunity.
Break-even return	This is the minimal return required in order to create shareholder value. These numbers are different, since there may also enter strategic and other considerations, which are not captured in the capital model.
Offered return	The expected return offered by the corresponding business opportunity.
Capital allocated	In this next step capital is allocated to each business opportunity.

Hurdle rate for bonus	At the same time the minimal required return is defined in order to incentivice the management accordingly.
Capital limit	Once the capital is allocated, the corresponding numbers become limits and capital is managed in such a way that it is optimally allocated and used, hereby *not* violating capital limits.
Capital used	At every point of time the capital used is compared with its capital limit in order to prevent limit breaches and in order to initiate corrective actions.
Effective Return	At the end of the cycle the realised return is calculated and compared with the agreed hurdle rates in order to determine the value creation for the shareholders and to compensate management accordingly.

In order to offer some alternatives for capital and return, below some possible choices:

Metric	Capital	Return
Economic	VaR or TVar	IRR as mentioned above
Accounting Profit	IFRS Shareholder equity	Return on equity
Risk adj Accounting Profit	VaR of IFRS SHE	Return on equity
Dividends	Free Surplus	Free surplus generated

Finally some remarks to this process:

- The aim of the capital allocation process is twofold. On the one hand one wants to optimise return on capital in order to optimise shareholder returns on a risk adjusted base. On the other hand one wants to limit the risk by using a diversified opportunity portfolio and by agreeing capital limits.
- Normally each business opportunity comes with its capital needs and with its expected returns. In a lot of instances these models are rather crude. Therefore it is essential to robustly challenge the models and their assumptions in order to increase the probability of successful shareholder value creation.
- It is important to fix the management remuneration based on the agreed metric at the time of planning in order to avoid a principal - agent problem. It is essential (while being trivial) to remark that people taking the risk should not be able to determine the exogenous parameters which are used for remuneration purposes in order to avoid self-fulfilling promises. It is key that the yard-stick used for measuring remuneration is reliable.
- In the same sense it is important to remark that a mechanical capital allocation can not replace an expert judgement, in particular in respect to strategic developments and initiates, since such opportunities are very difficult to assess. There are

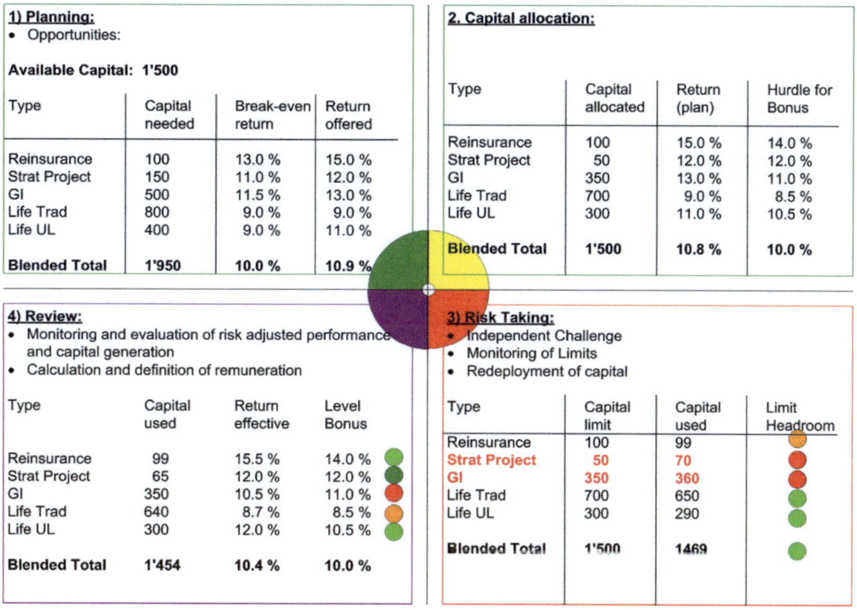

Fig. 10.4 Example Capital Allocation Process

many exogenous factors which also need to be considered outside the traditional models.

Chapter 11

Risk Management in a Group and Intra-group Transactions

The aim of this chapter is to get a deeper insight on the capital of an insurance company. In the past chapters we have seen the shareholder's capital or equity. Besides this capital there are other types of capital which have risk absorbing capabilities and which can serve in consequence as a buffer in case of a financial distress.

11.1 Introduction

Until now the whole world was relatively simple in the sense that we looked only at a single company and did look at the available and the required capital. The re-

M. Koller, *Life Insurance Risk Management Essentials*, EAA Series,
DOI 10.1007/978-3-642-20721-1_11,

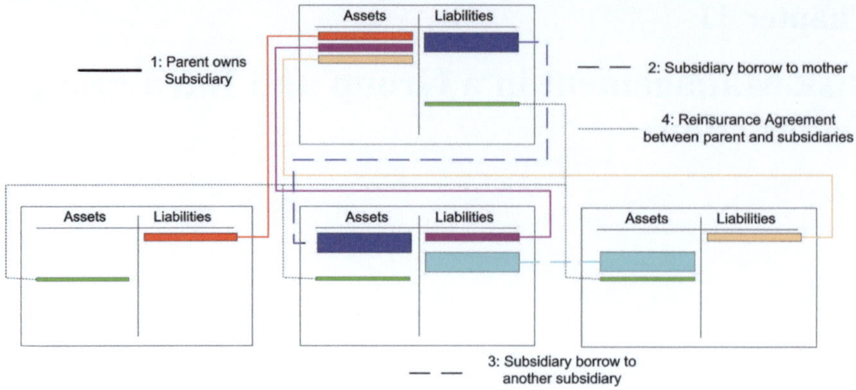

Fig. 11.1 Example of an intra-group relationship

ality is however more complex since insurance groups have several legal entities and more importantly insurance entities may own other insurance entities. In consequence there is a systemic risk in the sense that the different companies can not be considered on a stand alone base but need to be considered all together. Figure 11.1 shows some of the possibilities. In order to better understand these relationships we need to have a look at each of the four:

1. **Parent owns subsidiary:** The owning of a subsidiary is normally done by issuing shares of the subsidiary, which are owned by the parent. In contrast to publicly traded shares there is no market value available for these entities. The shareholder equity of the subsidiary plus intangibles (such as goodwill) represent the asset, which is booked in the balance sheet of the mother. In order to value the shares and the corresponding risk one needs to model the whole balance sheet, of the corresponding subsidiary. If we number the subsidiaries $s_1, s_2, \dots s_n$, the parents owns $\{\mathcal{E}_{s_1}, \mathcal{E}_{s_2}, \dots \mathcal{E}_{s_n}\}$, using the notation of chapter 6. Considering this effect on a stand alone basis is not too difficult if one keeps in mind during the calculation of the required capital that all the risk factors are modelled consistently (!). In addition to the modelling it is important to keep in mind that there might be additional regulatory restrictions, which might lead to a reduction of $\mathcal{E}$ for the parent. It is possible that the subsidiary is not allowed to pay dividends in case some regulatory conditions such as required capital level are not met.
2. **Subsidiary borrow to mother:** In this situation the subsidiary has borrowed money to its mother and in consequence the subsidiary has a credit risk vis-a-vis the mother. This credit risk can have a systemic impact and lead to the default of the child because of financial difficulties of the parent. A well known example is Lehman Europe which would most likely not have defaulted immediately after the default of the parent, wouldn't it have lent money to its parent the weeks before the collapse of the whole group. You can compare such a systemic risk with the failure of an electric network where a fault occurs at a certain knot.

This failure leads to a stress within the rest of the grid and other lines are forced to be shut down, which in turn accelerates the failure of the whole grid. As a consequence of this additional systemic risk it is important to recognise that such intra-group transactions are under regulatory scrutiny and it is important for the risk management to understand these, in particular in a distressed environment. Another important issue in relation to borrowing money is the rank of the money in case of a default. The more junior the debt, the more likely it is that the investor looses the money.

3. **Subsidiary lends to another subsidiary:** This is a variant of the intra-group relationship in the sense that also children can borrow each other money. From a technical viewpoint this might lead to circular relationships which need special care when determining the financial state of the group.
4. **Reinsurance Agreement between parent and subsidiaries:** Here we have a slightly different situation, where there is a risk transfer from the children to the parent. In this case the parent pays a certain amount to the subsidiary. In order to get this cover the subsidiary has paid to the parent a reinsurance premium. In respect to risks two things happen. The risk of the child reduces with respect to the concrete risk. But the child assumes (in the same sense as with borrowing money) a counter-party credit risk towards its mother. This credit risk might be triggered as a consequence of the particular reinsurance treaty or also because of an unrelated event.

11.2 Risk and Capital Transfer Instruments

In a next step we want to have a closer look into the zoo of risk and capital transfer instruments. At this point of time it is important to remark that not all of these RCTI's necessarily reside on the balance sheet and it is important to do the corresponding research. One can split the Risk and Capital Transfer Instruments into the following categories, which we will describe in greater detail

- Internal loans,
- Internal hybrids,
- Cashpools,
- Guarantees,
- GI / Life reinsurance contracts, and
- Derivatives.

In order to analyse the different possibilities in a structured way we use the following grid:

Description of the RCTI: Here we provide a short and concise description of the corresponding RCTI together with some examples.

Intrinsic Risks: Here we describe the corresponding risks.

Possible Modelling approach: What needs to be considered and modelled.

Remarks: Additional remarks.

11.2.1 Internal Loans

Description of the RCTI: The borrower borrows money from the lender which in turn compensates the borrower by paying interest on the money owed. There are different possibilities to structure such internal loans. There are dated and undated (eg perpetual) loans. In case of a perpetual loan there is normally a process how to determine the interest on the outstanding amount. Furthermore there are loans where the coupon is not owed in case of financial difficulties of the borrower. One can for example agree that the interest is deferred in case the solvency level of the borrower is insufficient. In the same spirit the repayment of the principal can be deferred in such cases. Obviously the quality of the capital for the borrower is quite different depending on the structure. If the structure foresees the deferral of the interest payment (and the principal) in certain economic situations or if the loan is perpetual, the ability of the loan to absorb risks is higher. Hence the loan can count towards the regulatory capital for the sake that the lender has a inferior protection.

Intrinsic Risks: The main two risks are the credit counter-party risk and the interest rate risk. Normally counter-party risk is the main risk.

Possible Modelling approach: Modelling the interest rate risk follows the canonical approach for this type of risk. The counter-party risk is more difficult, since the default of the borrower needs to be modelled accurately. This task is equivalent to model $\mathcal{E}$ for the borrower.

Remarks: Internal loans represent the vast majority of all RCTI's used and there are many different possibilities. It needs to be stressed that in some cases there are chains of internal loans for regulatory and tax reasons. Eg entity A borrows B and B borrows C with the same conditions. In such cases it is important to understand whether there are different conditions between $A \rightsquigarrow B$ and $B \rightsquigarrow C$ respectively. Furthermore one might also encounter guarantees in such conditions, which need to be modelled at the same time.

11.2.2 Internal Hybrids

Description of the RCTI: Hybrid capital is a capital layer between debt and equity capital and is therefore able to absorb more risks than debt. The idea behind this type of capital is that it has normally a debt like structure in the sense that the issuer gets regular interest payments in an normal environment. The difference is if a certain trigger is reached, such as an insolvency of the insurance company etc. In this case the capital is converted into pure equity or alike, and hence the capital can then fully absorb risks. From a technical point internal hybrids can be treated similar to internal (subordinated) loans with the additional complexity to model the trigger and the corresponding conversion mechanism.

Intrinsic Risks: In contrast to internal loans where the main risks are default and interest rates, here also the value of the shareholder equity $\pi(\mathcal{E})$ is relevant. To be more precise, whereas the default can be characterised as $\chi_{\pi(\mathcal{E})\leq 0}$, it is for hybrid capital important to not only model the characteristic function in relation to $\pi(\mathcal{E})$, but also the $\pi(\mathcal{E})$ itself.

Possible Modelling approach: The same principal modelling approach as for internal loans can be applied. Particular care is needed for the modelling of the actual conversion, since sometimes option pricing methods are needed.

11.2.3 Cashpools

Description of the RCTI: For an efficient cash management, all cash available to the group is often pooled. Such cash pools can be considered as internal loans and the comments relating to internal loans are applicable, with the only remark that the duration of the cash pooling is much shorter than for an internal loan.

11.2.4 Guarantees

Description of the RCTI: Guarantees are very interesting and also difficult to model, since almost everything can be guaranteed. Often guarantees are used as capital in lieu. A typical situation could be a subsidiary who has difficulties and the regulator intervenes and asks for capital. Instead of providing capital the company provides a guarantee to inject capital into the entity once the regulatory capital falls below a certain threshold. Economically this guarantee can be considered as a call option on $\pi(\mathcal{E})$ and in consequence is quite difficult to model. Another typical example is a chain of internal loans which are at the end financed by an external loan. Since the external lender wants to lend the capital to a well capitalised company – instead of lending to a possibly weakly capitalised company, he normally requests a parental guarantee for the repayment of the loan. Also here the modelling is not trivial.

Intrinsic Risks: Since guarantees can vary so widely all possible risks can occur and the modelling of these guarantees can be very difficult.

Possible Modelling approach: See above.

Remarks: It needs to be stressed that there are guarantees which are overall negligible. Guarantees can be material and can have a systemic impact on the stability of the group. It is of utmost important to understand the economic consequences of these guarantees and to understand what can happen in a distressed environment. Stress tests can help to understand the situation better.

11.2.5 GI / Life Reinsurance Contracts

Description of the RCTI: This is the world of reinsurance with the aim to transfer risks from one legal entity to another. A typical application of a internal reinsurance cover is to pool all risk in one entity. By doing this the company can benefit from a diversification benefit. Assume for example an international reinsurer which writes a lot of earthquake risks in its Japanese subsidiary and a lot of tropical cyclone risk in its American subsidiary. The stand alone capital for both carriers would be higher since these two risks normally diversify. And hence by using two intra-group retrocessions one can bring both risks into one carrier. This approach can not only be done by an international reinsurer but also by primary insurance groups which will normally create their internal reinsurance company which enables the diversification. This internal reinsurance company may consider to buy external reinsurance cover (or to buy insurance linked securities) to reduce its peek risk.

Intrinsic Risks: As with guarantees there are many types of reinsurance and these treaties can cover all types of risks, be it insurance risks, financial and also operational risks. As a consequence the models need to be commensurate to the intrinsic risk. In the narrower sense GI and life reinsurance contract cover GI and life risks, respectively.

Possible Modelling approach: See above. For traditional reinsurance covers the modelling as mentioned for insurance risks can be used. It is important to acknowledge the necessity to model the severity of the different claims, since reinsurance treaties often have option like pay-outs, meaning that the cover starts at a certain attachment point and is limited by a maximal amount.

11.3 Ranking of RCTI's

The quality of the capital in its risk absorbing quality is defined via Tiers. Tier 1 capital (incl using shareholder's equity) has the highest risk absorbing capacity, followed by Tier 2 and Tier 3 capital. Obviously the better the risk absorbing capability of capital, the more risk for the investor. As a consequence of a higher risk, the investor will also want a higher return on this capital.

The whole concept of quality of capital needs to be considered under the view of the company going into default and being wound up. In case of default the different lenders of the company are paid back according to their ranking. A typical ranking of capital could look as follows:

1. Secured Senior Debt,
2. Senior Debt,
3. Subordinated Debt (Tier 2),
4. Convertible Debt (Tier 2),
5. Mandatory Convertible Debt,
6. Equity (Tier 1).

Insurance liabilities classify in the above list normally as Senior Debt or Secured Senior Debt. If an insurance company is wound up the money is paid out first to highest ranking debtors. If there is still remaining money the next lower call of debtors is considered and their claims are paid. In a last stage after paying the money back to the company's debtors the shareholders are considered.

Modelling the level of subordination means that one needs to model $\pi(\mathcal{E})$. Depending on how low this number is after a default of the company, it is possible to repay only the higher ranking debtors or also the lower ranking debtors. Here the concept of the Tiers becomes obvious. Tier 1 represents a very subordinated type of capital. In consequence the corresponding investors expect a higher return, because they assume a higher risk. This capital is more equity like and serves from a regulatory point of view as additional risk bearing capital. Higher ranking types of capital is relatively secure and does not receive the same high risk premium. Furthermore it does not count to the risk bearing capital of the company.

11.4 Modelling

This section will be relatively short since the modelling of a group is from a technical point of view quite similar to the modelling of an individual company. Hence there is a set of risk factors which have to be considered for all companies together and one has to model the shareholder equity for each group company. Until now nothing particular happens.

In a next step the different intra-group instruments have to be modelled just in the same sense as we have seen this in chapter 6 for bonds or equities in order to determine the value of the shareholder equity. Here the task might be complex since some of the instruments are not entirely trivial and hence it is important to model them accurately.

In a next step one has to check whether one of the entities has defaulted. In this case there are two things to consider, both of which are not trivial:

- The model of the subordination of each intra-group instrument and the wind up of them. This task is particularly tricky since there might be circular relationships which can not be handled in a form of a tree. Furthermore one needs to determine the value each company requires to recover post event.
- The other elements which need to be considered are so called management actions. Are there possibilities how the group can save the company which has defaulted. There are two questions which need to be answered: Is there an economic or reputational imperative and an intention to save the company. Secondly: does the group have enough financial resources the save the individual companies.

Chapter 12
Products and Their Risks

The aim of this chapter is to look at some concrete product offerings that went wrong. Looking at the main risk on a large scale for an insurance company we have certainly the relationship between assets and liabilities, the set up of new large scale IT projects (aka insurance administration systems) and products. All of these risks manifest differently. The ALM question is certainly the one which most determines whether a company can survive after a corresponding event. Hence it is characterised by a high impact but also by a continuous evolution and hence one can mitigate it by setting up corresponding processes to govern and limit this risk. The IT risk is a typical project risk which normally has its roots in a too big appetite for systems which can do everything. Also this type of risk can, in principle, be managed in a canonical way by the application of the corresponding change and project

M. Koller, *Life Insurance Risk Management Essentials*, EAA Series,
DOI 10.1007/978-3-642-20721-1_12,

management processes. If we finally look at the product risk we face a very different animal. Product risk is for most products relatively small since there are a lot of similar product designs, which are well known and for which one, in principle, knows very well the corresponding risks. The crystallisation of a product risk is hence a rather rare event. But on the other hand there may be huge impacts. Therefore one needs to be very vigilant when introducing new or adjusting existing products. This chapter aims to show some of the pitfalls to avoid in the form of real case studies. The reader is invited to think whether he would have fallen into the corresponding pit.

12.1 Nuptualite

The first product we want to have a look at is an endowment policy which is sold for young children and mainly provides them with a savings contract, which matures when they reach an age of somewhere between 20 and 25. So far this product is plain vanilla and the risk for the insurance company is modest. The real treat is a small rider which we will see in a second. So the characteristics of the main policy are as follows (for our example):

Entry-age	0
Age at maturity	25
Maturity and Death Benefit	100000
Financed by regular premium payment	
Technical interest rate	3 %

For this rider the following two additional options were sold, respectively given for free:

Premium Holiday: After the first premium payment there is a possibility for a premium holiday. If premium were not paid, the premium holiday starts and the insurance cover is adjusted correspondingly. It is possible to pay the outstanding premiums later if the accrued interest is paid. This option was offered for free.

Nuptualite: The idea was that the child would get the maturity benefit before maturity if he/she got married before this moment. The calculation was based on the most accurate statistical information of the relevant country and premiums were calculated accordingly.

If we have a look at the above cover we would have the following yearly premiums:

Main Policy	2780.50 p.a.
Nuputalite Raider	19.80 p.a.
Total Premium	2800.30 p.a.

This product was sold in the late 80's and early 90's by a mid-sized company with a shareholder's equity of 200+ M. The sales volume for this product was very high - so high that management were worried and stopped selling it. The loss of this product line was higher than the above mentioned shareholder's equity and the company only survived because it had a wealthy caring parent ...

But what had happened? Obviously something did not really work out. In order to analyse the situation a little closer, let's look at the following table:

Question	Answer
Actuarial model correct?	Yes
Statistical basis reliable	For the country and the population it was designed: yes?
Insurability criteria fulfilled?	NO. One of the main criteria is based on the fact that the insured person cannot decide himself whether he is eligible to get a benefit or not and that in consequence the occurrence of paying benefits is random. Obviously the time when getting married can be influenced.
Statistical base relevant for the population?	NO. This type of product was largely sold to an ethical group, which usually get married at the age of 17. As a consequence the statistical basis was not adequate.

Now we know what went wrong: an ill-designed product was sold to a population, where the statistical basis was not adequate.

In a next step we want to look how this product works. To this end we assume the following statistical base, where h_x the probability to get married at age x for the population of the country considered and where $\tilde{h}_x$ denotes the probability to get married for the ethical group mainly buying this sort of policy:

	q_x	p_x	h_x	$\tilde{h}_x$
0	0.00200	0.99800	0.00000	0.00000
5	0.00200	0.99800	0.00000	0.00000
10	0.00200	0.99800	0.00000	0.00000
17	0.00200	0.99800	0.00100	**0.60000**
18	0.00200	0.99800	0.00152	0.00152
19	0.00200	0.99800	0.00374	0.00374
20	0.00200	0.99800	0.00733	0.00733
21	0.00200	0.99800	0.00123	0.00123
22	0.00200	0.99800	0.01742	0.01742
23	0.00200	0.99800	0.02349	0.02349
24	0.00200	0.99800	0.03084	0.03084
25	0.00200	0.99800	0.03953	0.03953

We observe that the "normal" probability to get married is about 1-2 % starting at the age of ca 18. On the other hand we have assumed that for this particular community we have a marriage probability of 60 % at the age of 17. This is naturally a simplification in order to better understand the problematic. Based on the above assumption we get the following prices for the insurance offering:

(A)	(B)	(C) W/o NUP	(D) W NUP	(E) Antisel.	(F) Arbitrage
PV Benefit	Death	0.03410	0.03398	0.02984	0.00000
	Survival	0.45428	0.41613	0.16611	0.00000
	Nupt	0.00000	0.04004	0.35640	0.58739
Total		**0.48839**	**0.49016**	**0.55236**	**0.58739**
PV Prem 1		17.56504	17.50421	15.36879	14.16611
Premium		2780.50	2800.28	3594.07	4146.47
PV	Ben				100000.00
$x = 18$	Prem				79384.28
Delta Prem			19.78	813.56	
Loss			177.20	6396.86	**20615.71**

In the table above column (C) describes the main policy, (D) the one including nuptualite, based on the observed statistics of the country, (E) ditto with the expected behaviour of the specific community. This table tells us for example that the additional premium for the raider costs some 20 per annum. Column (F) is the one to look at. Here the economic effect is indicated if we consider that a lot of policies have used the premium holidays and pay the remaining premium only when they know that the child is going to marry soon. It becomes obvious that in this case the loss per policy amounts to c 20000 per policy. In the concrete set up the in-force portfolio consisted out of ca 12000 policies having therefore in the model the cumulative loss of c 240 M.

12.2 Index Linked Products and Other Contractual Issues

Some three years ago I would have been quite dogmatic in respect of measuring (in terms of capital) operational risk. The following example shows how operational risk can materialise. In the early 21st century structured products for insurance companies were very popular in Italy. In order to better understand this product let's look at the corresponding product characteristics, which I have put in the table below:

Term of contract	10 years
Single Premium	100000 EUR
Benefit at maturity	90 % of Yield of STOXX 50 index, minimally 2 % p.a.

Obviously, one could also offer this product including a mortality cover, but that is not relevant for this example. The concrete numbers for this example are also irrelevant, since for this structured product, the insurer went to an investment bank (Lehmans for example) and gave them say 95000 EUR and the investment bank replicated the guarantee, by issuing a Lehman structured bond, which would pay according to the sometimes complex derivatives structure. We will see in section 12.3 what can go wrong if one tries to replicate these derivatives.

In order to be protected, the insurance contracts were written in a way stating the counter-party risk is explicitly born by the policyholder ...

And now the unexpected happens - the bank defaults. Meanwhile the insurer has issued some 200 M EUR of Lehman structured bonds which trade, for arguments sake, at 4% and have therefore a value of 8 M EUR. So at first glance one could be of the opinion that there is no issue, because people have been advised correctly and are aware of the corresponding risks. What happened in reality in Italy was the following:

1. The lawyers of the company confirm that the insurance company has no legal obligation.
2. It is confirmed that the clients have been correctly advised.
3. The regulator confirms the legal position and mentions that he would be happy if the companies could take some "customer care" action, - eg voluntary payments to help the clients which suffered the loss.
4. Other companies start to compensate the clients for the losses and there is a reputational issue and hence the whole market seeks ways to make good the corresponding loss.
5. The regulator issues a new regulation, which foresees that such products can in the future only be offered if the insurer provides the guarantee.

So at the end, the loss ended up in the balance sheet of the insurer. Most customer care action requests the policyholder to inject additional money, the contract term was prolonged from say 5 to 10 years (because of the interest effect) and the insurer and the distributors (mostly banks in the case of Italy) injected the remaining funds. At the end a considerable part of the loss was taken by the distributor and the insurer, and hence the corresponding loss was an operational loss (reputational category) which was triggered by a credit event.

Now it is necessary to formulate some learnings:

- It is necessary not only to think in legal terms but always to keep the reputational consequences in mind when you believe that a liability has shifted to the customer.
- One must not underestimate the influence of the regulator even though he might not issue legal binding orders.

- The investment strategy chosen by the insurer on behalf of the customer is not acceptable. One would expect that a professional investor (or an adviser) would not put all eggs in one basket. The insurance company would have had concentration limits in place for the funds on its balance sheet.

It is important to recognise that the ex-ante finding of such issues, as the ones mentioned above, is anything but trivial, because one needs always to think what happens if the impossible happens (eg Lehman default). Finally it is worth mentioning that the design of structured products is undergoing considerable change since the issue is at the very end a design issue.

12.3 Variable Annuities

There are several types of performance guarantees for unit linked policies and one may often choose them a la carte, with higher risk charges for guarantees that are riskier for the insurance companies. The first type is comprised of guaranteed minimum death benefits (GMDB), which can be received only if the owner of the contract dies.

GMDBs come in various flavors, in order of increasing risk to the insurance company:

- Return of premium (a guarantee that you will not have a negative return),
- Roll-up of premium at a particular rate (a guarantee that you will achieve a minimum rate of return, greater than 0),
- Maximum anniversary value (looks back at account value on the anniversaries, and guarantees you will get at least as much as the highest values upon death),
- Greater of maximum anniversary value or particular roll-up.

Unlike death benefits, which the contract holder generally can't time, living benefits pose significant risk for insurance companies as contract holders will likely exercise these benefits when they are worth the most. Annuities with guaranteed living benefits (GLBs) tend to have high fees commensurate with the additional risks underwritten by the issuing insurer.

Some GLB examples, in no particular order:

- Guaranteed Minimum Income Benefit (GMIB, a guarantee that one will get a minimum income stream upon annuitisation at a particular point in the future)
- Guaranteed Minimum Accumulation Benefit (GMAB, a guarantee that the account value will be for a certain amount at a certain point in the future)

- Guaranteed Minimum Withdrawal Benefit (GMWB, a guarantee similar to the income benefit, but one that doesn't require annuitising)
- Guaranteed-for-life Income Benefit (a guarantee similar to a withdrawal benefit, where withdrawals begin and continue until cash value becomes zero, withdrawals stop when cash value is zero and then annuitisation occurs on the guaranteed benefit amount for a payment amount that is not determined until annuitisation date.)

In order to value this guarantee, one needs to rely on option pricing techniques such as the Black-Scholes formula. The price for a *put*-option with payout $C(T, P) = \max(K - S; 0)$ at time t and strike price K and equity price S is given by:

$$\begin{aligned}
P &= K \times e^{-r \times T} \times \Phi(-d_2) - S_0 \times \Phi(-d_1), \\
d_1 &= \frac{\log(S_0/K) + (r + \sigma^2/2) \times T}{\sigma \times \sqrt{T}}, \\
d_2 &= d_1 - \sigma \times \sqrt{T}, \\
\Phi(x) &= \int_{-\infty}^{x} \frac{1}{\sqrt{2\pi}} \exp(\frac{-\zeta^2}{2}) d\zeta.
\end{aligned}$$

The reader should be reminded that the formula is based on the efficient market hypothesis which requests that the following holds:

- Deep and friction-less market, and
- Absence of arbitrage.

In order to understand how these options are synthetically "constructed" one needs to understand the concept of a replicating portfolio. Hence one holds at every point in time a portfolio P_t with the aim that this portfolio matches at time T just the payout of the option mentioned above. In order to construct such portfolios one usually uses the "greeks". These greek letters represent the sensitivity of an option in case of a change of the underlying economic parameters such as equity price, interest rate levels, etc. We have the following relationships:

$$\begin{aligned}
\Delta_P &= \frac{\partial P}{\partial S} \\
&= \Phi(d_1), \\
\Gamma &= \frac{\partial^2 P}{\partial S^2} \\
&= \frac{\Phi'(d_1)}{S \times \sigma \times \sqrt{T}}, \\
\Lambda &= \frac{\partial P}{\partial \sigma} \\
&= S \times \Phi'(d_1) \times \sqrt{T - t},
\end{aligned}$$

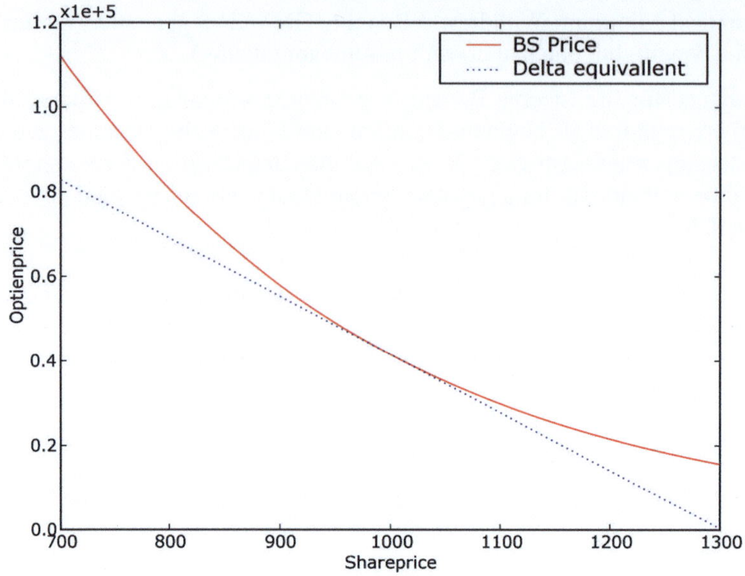

Fig. 12.1 δ–Hedging

$$\begin{aligned} P_P &= \frac{\partial P}{\partial r} \\ &= -(T-t) \times K \times e^{-r\times(T-t)} \times \Phi(-d_2). \end{aligned}$$

Based on the above partial derivatives, it is now possible to define different hedging strategies, one of them being a "delta-hedge". The idea is to define at a point in time t a portfolio P_t consisting of cash and shares, which have the same value and for which the partial derivative with respect to equity price S is the same. Hence we look for a Taylor approximation of order 1 in the variable S. Figure 12.1 shows such a delta hedge. For the concrete example we have the following put option:

Interest	$r = 3.0\%$
Term	10 years
Equity Price	$S_0 = 1000$
Strike	$K = 900$
Volatility	$\sigma = 15\%$
Number of Shares	1000
Value of Put	$P = 41535.7$
Delta	$\Delta_P = -137472$

What becomes obvious is the fact that the hedge is quite good if the stock market does not move too far away during the time between the updating of the replicating portfolio, for example updating the hedge portfolio once a day.

It is interesting to see what starts to happen if there are days with high volatility and market disruption. There are, in principle, two effects of stock market movements which inhibit a perfect hedging of the underlying guarantees. This is the change of stock price overnight. You may observe that the last paid price of a share is 15.5 and that the sentiment overnight has changed and the first paid price is 15.0. The other effect is a high inter-day volatility of the underlying asset. Assume for arguments sake that we consider a company with the following portfolio:

Spot price beginning of day	$K = 1000$
Strike	$K = 950$
Volatility for δ–hedge	$\sigma = 15\%$
Number of Index Baskets	1000000
Value of the portfolio	$P = 1000000000$ USD
Value of Put	$P = 52198940$ USD
Delta	$\Delta_P = -164095996$ USD

Now let's see what happens if we have set up our hedge portfolio at the beginning of the day at an index of 1000 and when the index is at 970 at the end of the day. Obviously this example is fictional. But one could see such extreme market value movements more than once in the autumn of 2008. Before doing the calculation let's see how rare this event actually is with an underlying volatility of $\sigma = 15\%$. Based on the Brownian motion assumption we know that the variance increases linearly in time and hence we know that the log-returns for one day have a $\sigma_{day} = \frac{15}{\sqrt{365}} = 0.79\%$. The probability is that we have a day return of -3 % or less which amounts to 0.000066. The following table shows the corresponding numbers for other moves:

α	$P[X \leq \alpha]$
-3.0 %	0.000066
-2.5 %	0.000725
-2.0 %	0.005427
-1.5 %	0.028034
-1.0 %	0.101391
-0.5 %	0.262116

And we remark again that there were several trading days in the autumn 2008 where we observed daily losses of 2 % or more. Even bigger equity market swings have been observed beginning May 2010, after fears of state bankruptcies in the Euro zone. On Thursday 6.5.2010 the NYSE (Dow Jones Industrial Average) fell temporarily over 9 %, as a consequence of such fears and automated trading. The same day Procter and Gamble lost temporarily more than 35 % of its value. On Monday 10.5.2010 the Euro Stoxx index performed 10.35 % within one day, after the announcement of a EUR 750 bn bail-out plan. Assuming a volatility of 20 % and log-normally distributed equity-market returns, this represents a $9.8 \times \sigma$–event. Such an event has a return period of 5.9×10^{17} years. This number is considerably bigger than the age of the universe of 1.375×10^{10} years and hence it is obvious that the log-normally distributed model is not correct in the tails. From a risk management point of view it becomes obvious that all capital models are prone to *model risk*. Hence it is of utmost importance to test the model in respect to its reliability. This can be done by back-testing and also by the application of statistical tests. In the same sense it is important to understand that all estimates are prone to *parameter risk*, e.g. the risk that a "wrong" parameter is chosen. Concluding, it is important to understand the behaviour of a model with respect to changed parameters.

Now let's have a look at what happens in one single such day. Our hedge portfolio consists of cash of 1.164 bn USD and we are short in the stock market index with an amount of 0.164 bn USD. We also know that the option value amounts to 52.2 M USD. Now let's look at the end of the day.

in M USD	Value at 1000	Value at 970
Bonds	1164	1164
Shares	-164	-159
Options (index impact)	-52	-57
Options (interest - 25bps)		-6
Options (σ to 20 %)		-4
Total	948	938

As you see from the table above we have assumed that the volatility spiked up to 20 % and that the interest rates reduced by 25 bps. Together, all these effects have an adverse impact of 10 M USD which represents 1 % of the funds value. The impact in reality was far bigger than that and some sizable insurers have closed their corresponding portfolios for new business.

But now let's ask ourselves what went wrong. Actually, the underlying model worked in theory but in reality the situation was somewhat different. The whole arbitrage free pricing theory is based on a few relevant assumptions: frictionless, deep and arbitrage free market. The calculation can be interpreted as a market which is not deep and liquid enough, resulting in losses in the corresponding hedge portfolios. The true issue is that there was an over reliance on models and people did not ask what would happen if the model breaks down. Now I do not want to say that models are useless, but rather that we need to be always sure which are the corresponding limitations and what happens if these are not fulfilled.

The other conclusion is that derivatives can be very useful. On the other hand they are dangerous if not managed and analysed carefully.

12.4 Investment Guarantees and Bonus Rates

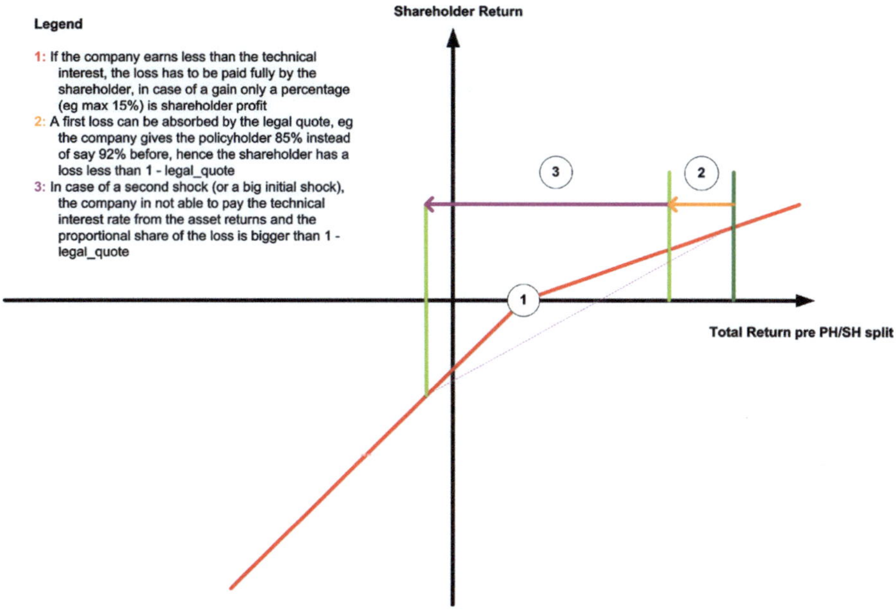

Fig. 12.2 Policyholder – shareholder split

One reason for large issues with insurance products is that sometimes the existence and the level of interest guarantees the absence of a reasonable ALM. In order to understand the corresponding issues we look at Swiss pension schemes. Without going into detail, you can save money during the time you are working and the money saved is converted into an annuity and a corresponding widows pension at a fixed rate. This rate has for a long time been fixed at 7.2 % and is now going to decrease over an extended period. There are also other issues which we will exclude for the moment and we assume the following[1].

Interest during deferral period	$i = 2.0\%$
Current age of the insured	45
Current age of the partner	$\Delta_{XY} = -1$
Conversion rate at 65	7.2 % including widow pension
Contribution rate	15 %
Pensionable salary	100000
Single Premium at age 45	300000
Valuation date	29.12.2006
Profit share mechanism	Reserves for longevity provision can be deducted. Shareholder can claim 10 % of the remaining profit.

[1] We also remark that the example does not fully reflect the Swiss legislation.

We will consider two separate cases. In the first case we assume that the person is a man, in the other we assume that this benefit is offered to a woman. Tables 12.1 and 12.2 clearly shows that there is a sizable difference in future life span between the man and the woman.

Table 12.1 Expected future life span for Swiss men

Age	1881-88	1921-30	1939-44	1958-63	1978-83	1988-93	1998-03
1	51.8	61.3	64.8	69.4	72.1	73.8	76.6
20	39.6	45.2	47.9	51.5	53.8	55.3	58.0
40	25.1	28.3	30.4	32.8	35.1	36.8	39.0
60	12.4	13.8	14.8	16.2	17.9	19.3	21.1
75	5.6	6.2	6.6	7.5	8.5	9.2	10.3

Table 12.2 Expected future life span for Swiss women

Age	1881-88	1921-30	1939-44	1958-63	1978-83	1988-93	1998-03
1	52.8	63.8	68.5	74.5	78.6	80.5	82.2
20	41.0	47.6	51.3	56.2	60.1	61.8	63.4
40	26.7	30.9	33.4	37.0	40.7	42.5	43.8
60	12.7	15.1	16.7	19.2	22.4	24.0	25.2
75	5.7	6.7	7.4	8.6	10.7	11.9	12.8

First we look at the cash flow stream which is induced by the above contact. In order to do that we need to be aware that the conversion of capital in an annuity at age 65 is optional and in particular only the people reaching the age 65 have this option. Because of that we have the following cash flow pattern:

Age	Saving amount	Cash Flow	Discount	MR
45	15000	-15000	1.00000	55609
55	187251	-15000	0.78271	243309
60	290703	-15000	0.68753	358147
64	383169	-15000	0.62467	457903
65	**392749**	28277	0.61004	**484246**
66	0	27990	0.59656	466274
70	0	26626	0.54677	392517
75	0	24441	0.47350	312295
80	0	21605	0.41803	227625
85	0	17953	0.36905	148613
90	0	13183	0.32582	81052
100	0	2621	0.25395	7763

From the above table we see that the savings amount at age 65 equals 392749 CHF and that we need 484246 CHF for paying the liabilities assuming a risk free investment return. Hence c. 23 % of funds are missing at age 65 and the present value of the loss at age 45 equals 55609 CHF which equals 3.7 times the yearly contribution. For women the situation is worse because they live longer. Figure 12.3 shows the development of the available reserves compared with the necessary reserves.

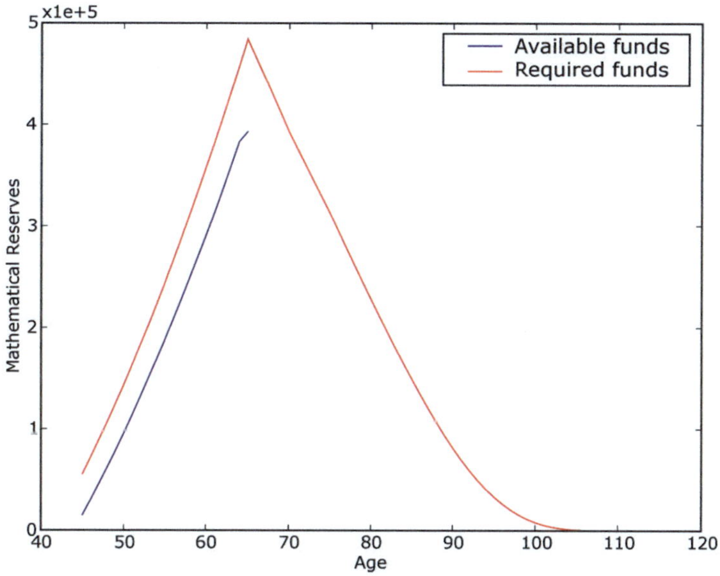

Fig. 12.3 Development of mathematical reserves

But what does this loss now mean for the business. There are three ways how one can look at this:

- Do not do such business,
- Take the loss up front,
- Invest in a asset allocation where one can in average achieve the goal.

It was at this point things started to go wrong. Obviously the big players in the group life scheme market with assets under management above 20bn CHF did not want to cease this business and it was also not acceptable to show such big losses for each new contract and so the companies started to take more investment risk. We need now to know what investment return would be needed to fulfil the given obligations. By backward solving we find that we need to get an investment yield of 3.47% [2]. Hence we have the following situation, assuming that shares yield 400 bps more than risk free:

[2] Please note that the situation at that time was even worse, since the companies needed to credit not only 2 % but rather 4 % to the savings account during the accumulation phase.

Required yield	3.5 %
Risk free yield	2.5 %
Required uplift	1.0 %
Required equity backing ratio	25.0 %

If you now go back to the balance sheets of the Swiss life insurers at the beginning of this century, you will find that they were investing heavily in equities with equity backing ratios of 25 % and more. So this investment strategy worked well, because for most of the latter years of the 20th century, equities had a good return. In order to understand what happened in the year 2001, let's look at the corresponding balance sheets. To this end we look at a mid-sized insurer with assets of 20 bn CHF and insurance liabilities of 18 bn CHF and assume that it has an equity backing ratio of 20 %.

The expected yield of this company and its balance sheet looks as follows

	SAA	in CHF	Yield	Return
Shares	20%	7%	$4000000 \cdot 7\%$ =	280000
Bonds	60%	4%	$12000000 \cdot 4\%$ =	480000
Properties	10%	5%	$2000000 \cdot 5\%$ =	100000
Mortgages	10%	4%	$2000000 \cdot 4\%$ =	80000
			Total	940000
Math. Res.		3.5%	$-18000000 \cdot 3.5\%$ =	-630000
			Total	310000

Hence the insurance has an average return to both shareholders and policyholders of 310 M CHF.

In the year 2001, the equity index fell by 21 %. What has happened to the insurers income statement?

	SAA	in CHF	Yield	Return
Shares	20%	-21%	$4000000 \cdot (-21\%)$ =	-840000
Bonds	60%	4%	$12000000 \cdot 4\%$ =	480000
Properties	10%	5%	$2000000 \cdot 5\%$ =	100000
Mortgages	10%	4%	$2000000 \cdot 4\%$ =	80000
			Total	-180000
Math. Res.		3.5%	$-18000000 \cdot 3.5\%$ =	-630000
			Total	-810000

We can see that while this investment strategy worked for some time, the loss in this one year was so big, that the company lost almost half of its shareholder equity capital. This is also the reason why there were very big insurance companies in Switzerland which had to go to the capital markets in the years 2002 and 2003 to raise capital. There is another pitfall which one needs to be aware of if one takes excessive investment risk. In a lot of countries such as Germany, France but also in the UK there is a policyholder – shareholder split in respect to gross profits. Assume for example, that we have a *legal quote* such as 85 %. Assume for the moment that the gross profit before policyholder – shareholder split and before tax amounts to 1000 M. In this case there is a legal requirement to allocate 850 M to the policyholder and the shareholders get a pre-tax profit of 150 M. Then assume that we have a gross loss before shareholder – policyholder split of -500 M. In this case the shareholder takes the whole loss, since the minimal investment return for the policyholder is guaranteed. In consequence we get a shareholder – policyholder split as indicated in figure 12.2.

The following table shows a comparison between two different investment strategies, assuming a legal quote of 85 % and a tax-rate of 0 %. We assume the following:

Mathematical Reserve	1000000000 EUR
Technical interest	3.0 %
Yield of a bond investment	4.0 %
Expected yield shares	7.0 %
Volatility of shares	18.0 %
Strategy 1	100 % invested in bonds
Strategy 2	25 % invested in shares , 75 % in bonds.

For strategy 1 we know that we have a gross profit of 10 M EUR and hence the shareholder (SH) gets 1.5 M EUR and the policyholder (PH) 8.5 M EUR. For strategy 2, the situation is more complex and we need to look at the corresponding probability distribution:

Return Shares	Probability	Portfolio Return	P/L Gross	P/L $\sum$ SH	P/L PH
-40 %	0.00086	-7.00 %	-100000000	-100000000	0
-35 %	0.00169	-5.75 %	-87500000	-87500000	0
-30 %	0.00426	-4.50 %	-75000000	-75000000	0
-25 %	0.00962	-3.25 %	-62500000	-62500000	0
-20 %	0.01948	-2.00 %	-50000000	-50000000	0
-15 %	0.03530	-0.75 %	-37500000	-37500000	0
-10 %	0.05730	0.50 %	-25000000	-25000000	0
-5 %	0.08331	1.75 %	-12500000	-12500000	0
0 %	0.10851	3.00 %	0	0	0
5 %	0.12659	4.25 %	12500000	1875000	10625000
10 %	0.13229	5.50 %	25000000	3750000	21250000
15 %	0.12383	6.75 %	37500000	5625000	31875000
20 %	0.10383	8.00 %	50000000	7500000	42500000
25 %	0.07799	9.25 %	62500000	9375000	53125000
30 %	0.05247	10.50 %	75000000	11250000	63750000
35 %	0.03162	11.75 %	87500000	13125000	74375000
40 %	0.03097	13.00 %	100000000	15000000	85000000
Expected Value	1.00000	5.34 %	23472093	-1518018	24990112

It becomes obvious that this second investment strategy is much worse for the shareholder since he makes, on average, a loss. As a consequence one needs to be very clear about bonus sharing mechanisms when determining the target asset allocation.

12.5 Longevity and the Ability to Forecast

In this last section I would like to elaborate further on the longevity issue, which I started to introduce in section 12.4. There the conversion rate of 7.2 % was stated, but it was not clear whether it was due to interest guarantees or because people are living longer. Whereas we put the focus on investment guarantees in section 12.4, we want to focus here on the longevity aspect of the issue.

We have seen in tables 12.1 and 12.2 that the future life span of men and women is still increasing at a high pace. The task of the actuary is to develop tables which forecast this (relatively stable) trend in order to avoid future losses. In order to check this need to have a look at the results by comparing the corresponding mortality tables. I use the Swiss tables but I want to stress that I have not encountered yet a single country where I could not observe the same effect:

$\ddot{a}_{65}(i = 3.5)\%$	men	Δ men	women	Δ women
ERM/F 70	12.491	3.958	13.923	3.820
ERM/F 80	13.199	3.250	14.789	2.954
ERM/F 90	14.387	2.062	16.221	1.522
ERM/F 00 @ 2005	16.450		17.744	

The table above compares the single premiums to be paid for an immediate annuity of 1 at age 65. We see that the price for this cover has increased by 3.958 when switching from the table ERM 70 to the table ERM 80.

The table above overstates the situation (because some of the people to whom the older products were sold have already died) but it shows the right direction. So assume that our insurance company has the following portfolio of people (men) who are aged 65 and assume that we expect them to live according to the most recent tables

Tariff generation	MR reserve Original base M CHF	MR reserve ERM/F 2000 M CHF	Difference M CHF
ERM/F 70	400	526	126
ERM/F 80	800	997	197
ERM/F 90	2000	2286	286
ERM/F 00 @ 2005	400	400	0
Total	3600	4210	610

From the above table it becomes obvious that the wrong mortality estimate is quite costly and amounts to CHF 610 M. The question why this happens so consistently cannot be answered easily, but there are some reasons listed below:

- In the past the analysis–tools were not as developed as now.
- There was a disbelief that the existing trend in an increased lifespan would persist in the future.
- Applying lower mortality rates to the in-force book is very expensive and hence one was reluctant to apply tables with stronger trends. Furthermore, there was a fear that the annuity product could not be sold anymore because it would become too expensive.

12.6 Long Term Care

In the previous section we looked at longevity risk and we want now to focus on long term care business. In order to do this, we need to understand the corresponding cover and how to value it. Afterwards we want to have a look at the risks of this cover.

Assume you are a healthy person living at home, able to feed yourself, to wash yourself et cetera. Hence you are able to perform the essential daily living activities (DLA) without help. Once you get older this may not be possible anymore and you are threatened to go into care, which you may not want. You would rather have home help. The long term care (LTC) cover aims to protect you from this, by paying for long term care support. How does this work in practise?

First the insurer defines the main daily living activities which you should be able to perform yourself and an amount which is paid if the person is not able to perform these anymore. Technically speaking we have, for example, 8 DLA which are monitored and you can perform between 0 and 8 of them. One could have a cover where you do not receive anything if your ability is 6 and above and gradually increase for the fewer DLAs you can perform yourself. In the concrete example the respective states are numbered from 1 to 6, where 1 indicates that everything can be autonomously and 6 represents the fact that we need help for all daily living activities. Formally the states are called $S = \{\dagger, 1, 2, 2a, 3, 3a, 4, 5, 6\}$. Assume that the benefits are given by the following table:

Number of DLA	Amount payable pa.
$DLA \geq 7$ (S1, S2)	No benefit, only premium payment
$DLA = 6$ (S2a, S3)	6000
$DLA \leq 5$	12000

In order to price and value this cover we need to use a Markov chain model (see appendix B):

- The Markov model consists of the following states $S = \{\dagger, 1, 2, \dots 8\}$, where $\dagger$ stands for the state of being dead.

- In a next step one needs to define the corresponding transition probabilities $p_{ij}(t, t+1)$, for $(i,j) \in S \times S$.
- For a market consistent valuation the discount rates follow risk free curves as seen before.
- Finally the table above has to be translated in payment functions $a_{ij}^{\text{Post}}(t)$ and $a_i^{\text{Pre}}(t)$. In order not to be overly complicated we assume that the benefits defined above are paid at the beginning of the year. Furthermore we denote with P the premium and we assume that this is only paid in states S1 and S2.

Based on the above we have the following:

$$a_{ij}^{\text{Post}}(t) = 0,$$

$$a_i^{\text{Pre}}(t) = \begin{cases} -P & \text{if} \quad i \in \{S1, S2\}, \\ 6000 & \text{if} \quad i \in \{S2a, S3\}, \\ 12000 & \text{else.} \end{cases}$$

In order to determine the premium and the mathematical reserves we use the recursion (B.1) in appendix B and assume that the person has currently an age $x = 65$. We want to have a closer look at the following questions:

1. What is the price and the mathematical reserves?
2. What happens if we consider an increase in life span and we assume that the remaining probabilities are reduced proportionally?
3. What does it mean if people become older and the time they are healthy remains constant?

Using the elements defined above and Thiele's difference equation, we can calculate the premium P for the two states S1 and S2, where we see an obvious difference in the present value of a premium 1. Figure 12.4 shows this effect. A person buying this cover at age 75 would have to pay about 1000 if he is in state S1 and about 4300 if he is in state S2. This difference shows clearly the risk the company is assuming, as a person who is not able to perform 1 DLA has a materially higher risk. This also explains why the underwriting of this type of policy is of utmost importance. You can imagine what would happen if a person is assumed in state S1 soon becomes unable to perform his daily living activities. Figure 12.4 also shows the relative size of the premiums between states S1 and S2, and we see that there is at age 65 a factor of about 4.5 between the two.

With the same calculation, we can also determine the present value of future cash flows, as shown in Figure 12.5. In order to have a comparison, the mathematical reserves have been scaled relative to state S1. Note that state S4 is the one which is the most expensive and that both states S5 and S6 are cheaper. This is because people in these two states have a higher probability of death and therefore, the time

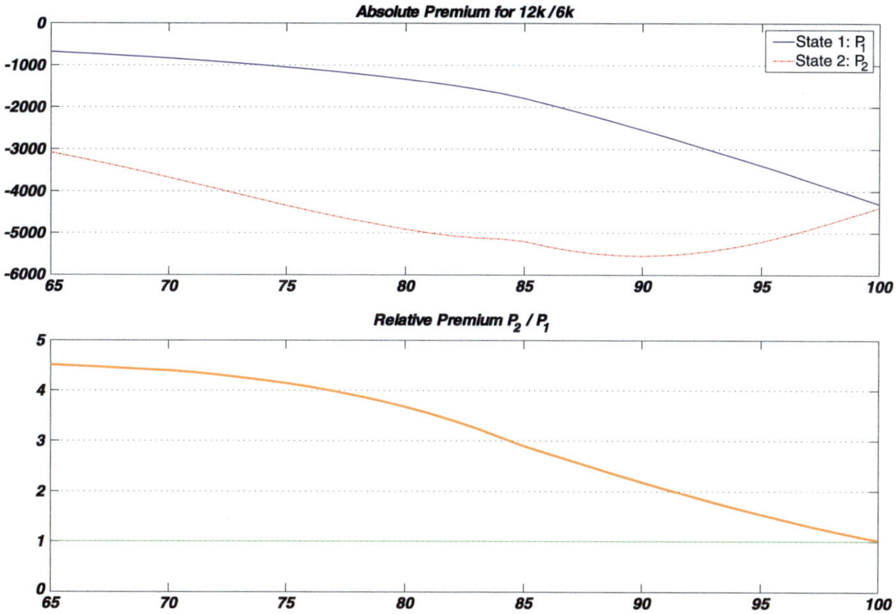

Fig. 12.4 Level Premiums for LTC cover

the insurer has to pay is shorter. We will see, that the future development of mortality could impact this. Finally figure 12.6 shows the distribution of the losses. It also shows the dependency of the corresponding state. Such a calculation can either be performed by recursion, or as in the concrete case by a simulation. From this figure we can for example see that the probability of never receiving a benefit for a person starting in state S1 is about 34%. In the same sense we see that the death probability is higher in state S4 than in state S3.

Next we want to look what happens, if we assume that the mortality reduces faster as shown in Figure 12.7. We see that this improvement has a considerable impact. More concretely two versions have been calculated, one (variant 1) where the people remain healthy and stay longer in state S1. In the other, the reduction in mortality goes in parallel with an increased time where the people are not anymore able to perform the different DLAs. Obviously this has a material impact, which needs to be considered when constructing and pricing this type of product. We finally see in figure 12.8 the way the reduction in mortality leads to higher claims. In respect to variant 1 we see that the main additional cash flows are a consequence of living longer, starting at about age 80. We also see that for variant 2 the higher losses start soon after age 70.

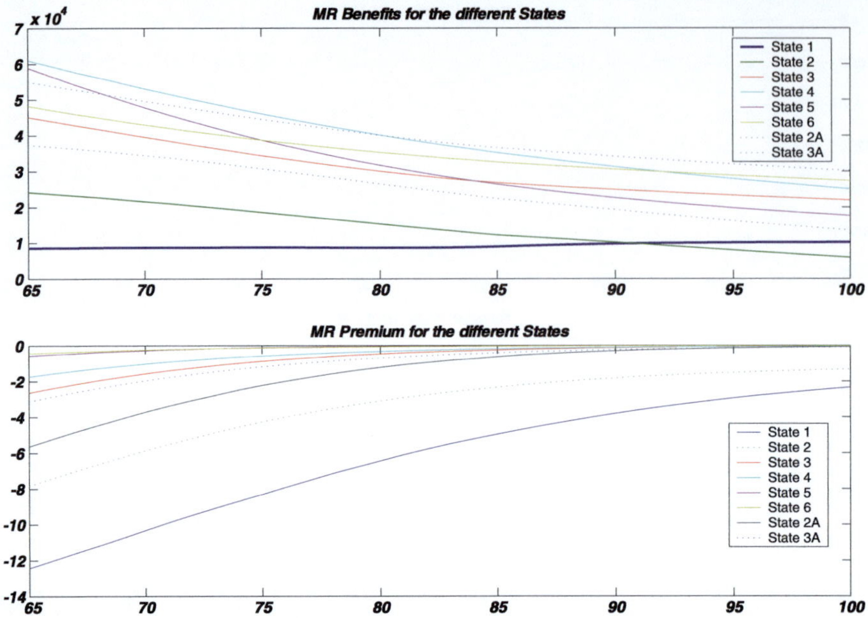

Fig. 12.5 Relative Mathematical Reserves for LTC cover

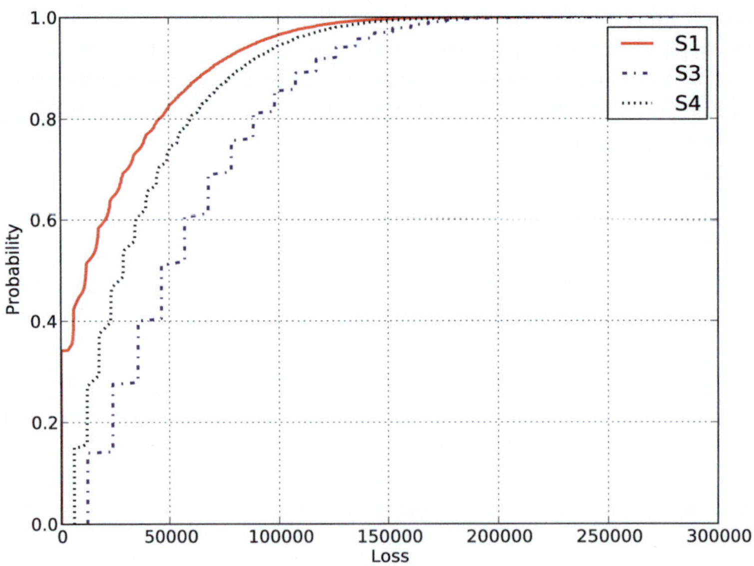

Fig. 12.6 Distribution of Mathematical Reserves for LTC cover

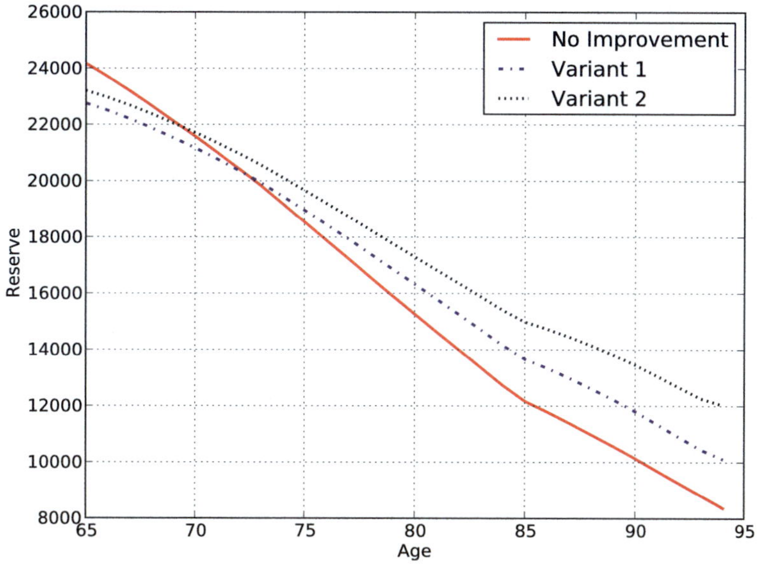

Fig. 12.7 LTC Mathematical reserves when reducing mortality

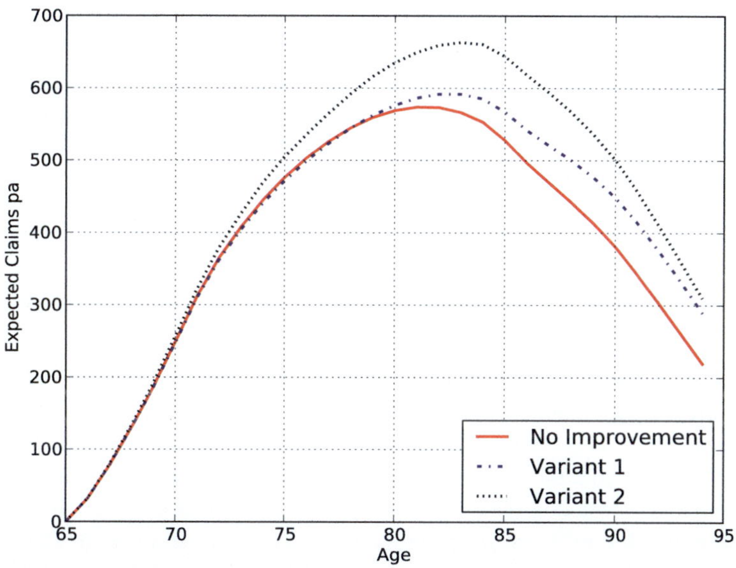

Fig. 12.8 LTC expected losses when reducing mortality

12.7 Imperfect Cash Flows Matching

We have seen in section 2.3 that there are no prices for long dates bonds in some currencies. Moreover even if there are prices for these bonds, there might be only a limited market for long dated bonds, as a consequence of states not wanting to issue long term bonds. A typical example is the CHF, where the market is liquid only up to durations of about 15 years. As a consequence insurance companies and pension funds are not able to match their guaranteed cash flows with corresponding bonds.

In this section we want to have a closer look at this question and the corresponding risks. The best way to understand this risk is to look at concrete examples:

- A portfolio of annuities in payment,
- A portfolio of deferred annuities,
- A portfolio of endowment policies.

In all three cases we assume that the benefits are denominated in CHF and we furthermore assume there is only a liquid market for CHF bonds until year 15 and hence the best thing to do is to use investments according to this. In order to valuate what could happen we look at the following scenarios:

1. Yield curve and investment opportunities as seen today,
2. At time 15 there is a flat yield of 0%, 1%, 2 % and 3 % respectively.

In order to be able to better describe this problem, we denote with $(CF_k)_{k\in\mathbb{N}}$ the vector of expected cash flows and for the moment we neglect the fact that these cash flows are actually random and can depend on the market environment. For the analysis we assume that the company invests as follows in $\sum_{k\in\mathbb{B}} \alpha_k \, \mathcal{Z}_{(k)} \in \mathcal{X}$:

$$\alpha_k = \begin{cases} CF_k & \text{if } k < 14, \\ \sum_{k\geq 15} CF_k & \text{else.} \end{cases}$$

This means that the company actually tries to invest as long as possible. We furthermore assume that the company follows a passive investment strategy and reinvests the excess assets in $\mathcal{Z}_{(15)}$ at time 15 according the investment condition at this time. We need to remark that the chosen investment strategy is obviously a simplification and that reality is more complex. It however exposes the risk the company is facing, when not being able to invest in the corresponding bonds. In order to calculate the corresponding risk we follow a rather easy approach by adjusting the yield curve after year 15. We remember that the prices of zero coupon bonds and corresponding yields have the following relationship:

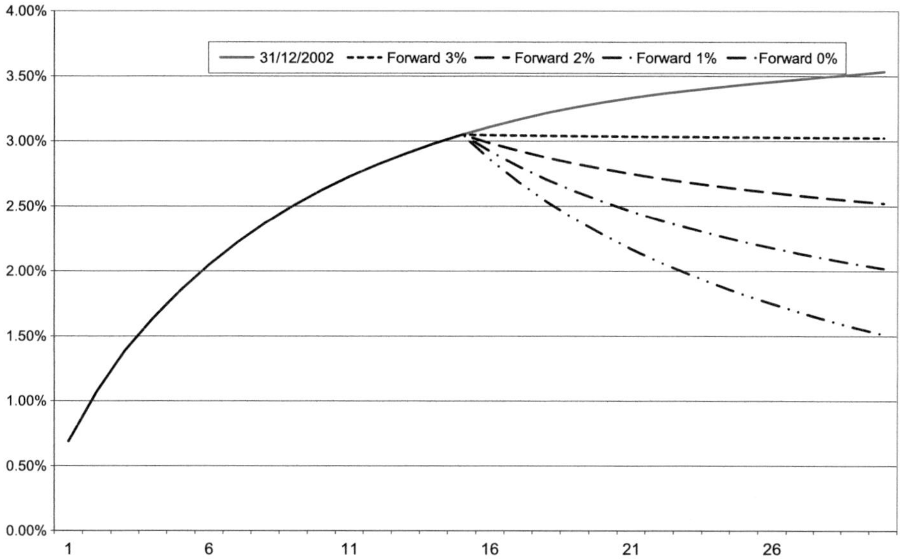

Fig. 12.9 Modified yield curves

$$\begin{aligned}\pi_t(\mathcal{B}) &= \sum_{k=0}^{\infty} CF_k \times \pi_t(\mathcal{Z}_{(k)}) \\ &= \sum_{k=0}^{\infty} CF_k \times (1 + y_t(k))^{-k}.\end{aligned}$$

Moreover the forward rates can in this case be calculated by

$$f_t(n, m) = \left(\frac{\pi_t(\mathcal{Z}_{(n)})}{\pi_t(\mathcal{Z}_{(m)})} \right)^{\frac{1}{m-n}} - 1,$$

and hence the following equation holds:

$$(1 + y_t(n))^n = \prod_{k=0}^{n-1} (1 + f_t(k, k+1)).$$

At this stage it is now easy to "construct" suitable yield curves representing the scenarios above by setting:

$$f_t(n, n+1) = \theta,$$

for all $n \geq 15$, where θ represents the interest rate going forwards, according to the scenario, after year 15. Figure 12.9 shows the corresponding yield curves. It is obvious that the overall yield reduces considerably in particular when using 0% as forward rate. It is worth noting that the case of 0% is the worst case, since we could in this case hold the cash after time 15 until it is used, assuming that both the bonds and the cash is risk-free.

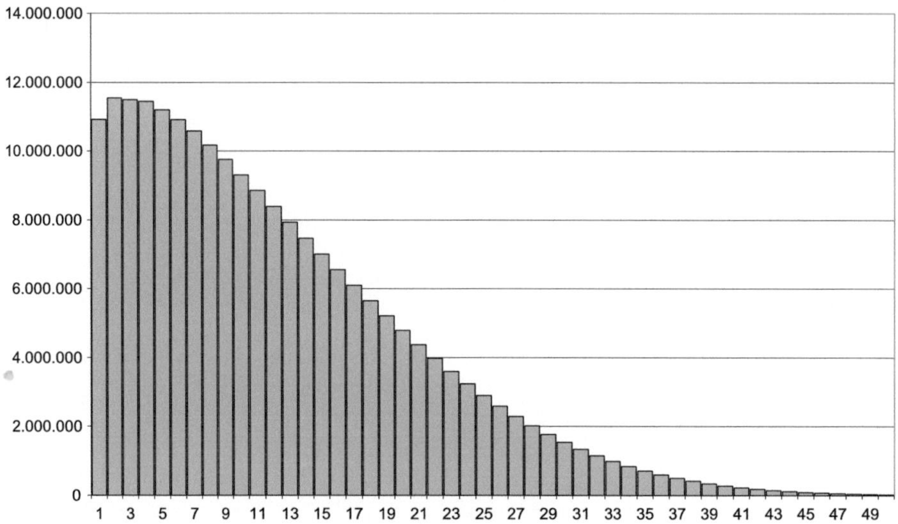

Fig. 12.10 Cash Flow Pattern of Annuities in Payment

Now we need to have a look at the concrete portfolios. We denote the annuity portfolio in payment as (A), the deferred annuity portfolio as (B) and the endowment portfolio as (C)

in M CHF	Portfolio A	Portfolio B	Portfolio C
Benefit	13.6 p.a.	108.1 p.a	4211.1
Statutory Reserves	162.7	841.5	2474.4
Premiums	–	–	20.4
Duration	8.9	25.3	11.0
Figure for Cash Flows	Fig. 12.10	Fig. 12.11	Fig. 12.12

Next we can calculate the corresponding amounts at risk as seen from today by using the different yield curves. The following table summarises this:

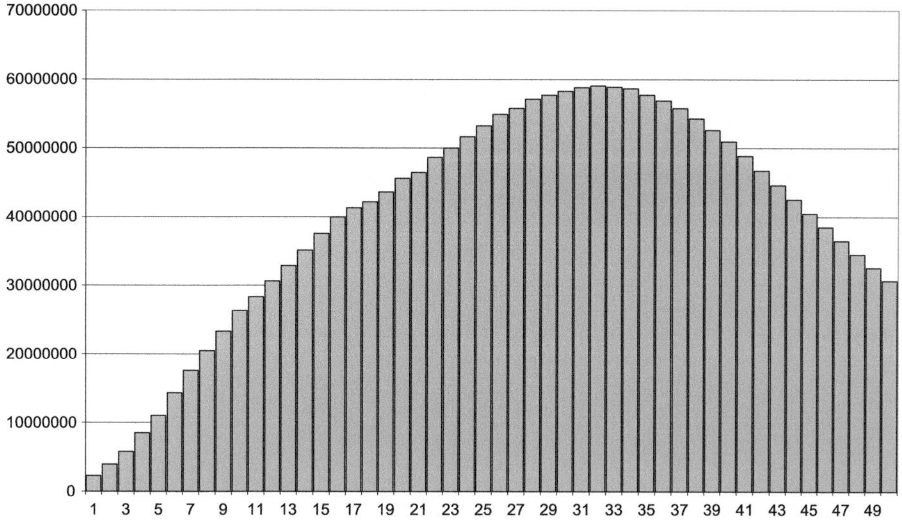

Fig. 12.11 Cash Flow Pattern of Deferred Annuities

in M CHF	Portfolio A	Portfolio B	Portfolio C	Total
Statutory Reserves 3.5 %	162.7	841.5	2474.3	3478.6
Value using yield as at 31.12.2002	159.4	880.0	2617.3	3656.9
Forward 3%	161.5	994.4	2678.9	3834.8
Forward 2%	163.7	1135.6	2746.0	4045.4
Forward 1%	166.3	1322.3	2821.5	4310.2
Forward 0%	169.2	1575.5	2906.7	4651.5
Coverage in %				
Value using yield as at 31.12.2002	102.0 %	95.6 %	94.5 %	95.1 %
Forward 3%	100.7 %	84.6 %	92.3 %	90.7 %
Forward 2%	99.3 %	74.1 %	90.1 %	85.9 %
Forward 1%	97.8 %	63.6 %	87.6 %	80.7 %
Forward 0%	96.1 %	53.4 %	85.1 %	74.7 %
Coverage absolute				
Value using yield as at 31.12.2002	3.2	-38.5	-143.0	-178.3
Forward 3%	1.2	-152.9	-204.5	-356.2
Forward 2%	-1.0	-294.0	-271.7	-566.8
Forward 1%	-3.6	-480.8	-347.2	-831.6
Forward 0%	-6.5	-734.0	-432.3	-1172.9

Looking at the example above one sees that the reinvestment risk, as a consequence of missing long duration assets, can be extremely dangerous for a life insurance company. We also see that deferred annuities in particular are a treat since the have a very long duration. Considering the current yields in CHF we would most likely have to look at a scenario between 1% and 2%. In this case we see that the company

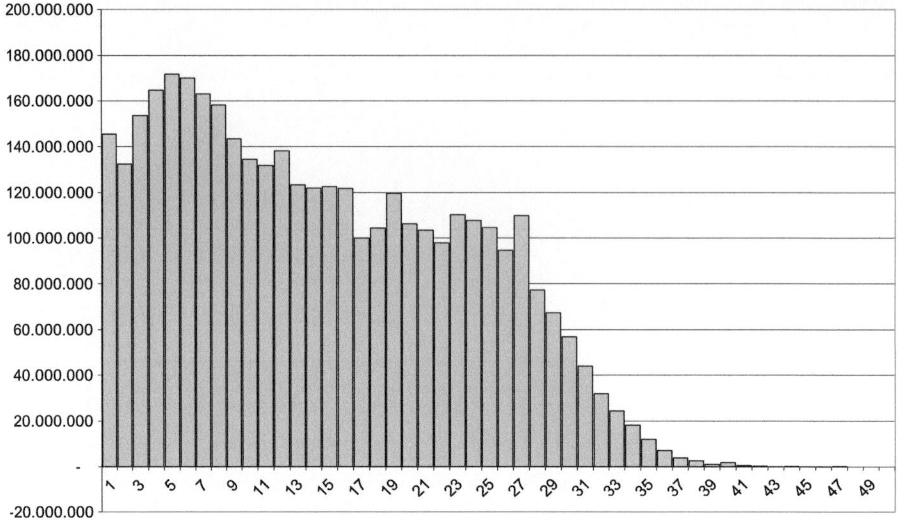

Fig. 12.12 Cash Flow Pattern of Endowment Policies

needs to strengthen their total reserves of c3.5 bn CHF by approximately 17 % (c0.7 bn CHF). This clearly shows the magnitude of this risk and the need for an adequate ALM and risk management.

Another interesting aspect which can be analysed with the above portfolios is the possible impact of the EU gender directive, which postulates the equal treatment of men and women. This would concretely mean that one needs to have to use the same pricing for men and women. The real risk would be a retrospective application of the directive. The table below shows the corresponding impact, which is material for all types of annuities as a consequence of the different future life expectancy.

in M CHF	Portfolio A	Portfolio B	Portfolio C	Total
Men as men	137.2	663.7	1749.0	2549.9
Women as women	25.4	177.8	725.3	928.6
Total	162.7	841.5	2474.3	3478.6
All as men	157.7	804.2	2483.5	3445.5
All as women	197.1	1037.5	2451.1	3685.8
Impact	4.9	37.3	23.2	207.2
Relative Impact	**3.0 %**	**4.4 %**	**0.9 %**	**5.9 %**

Chapter 13
Emerging Risks

The aim of this chapter is to have a look at emerging risks. First this concept needs to be explained and also why it is important. In principle answering this question is at the centre of risk management.

13.1 What Are Emerging Risks and Why Are They Important?

Emerging risk are the ones which are not yet very obvious. Consequently they are not easy to detect. The reason for writing this chapter is based on a person chal-

M. Koller, *Life Insurance Risk Management Essentials*, EAA Series,
DOI 10.1007/978-3-642-20721-1_13, © Springer-Verlag Berlin Heidelberg 2011

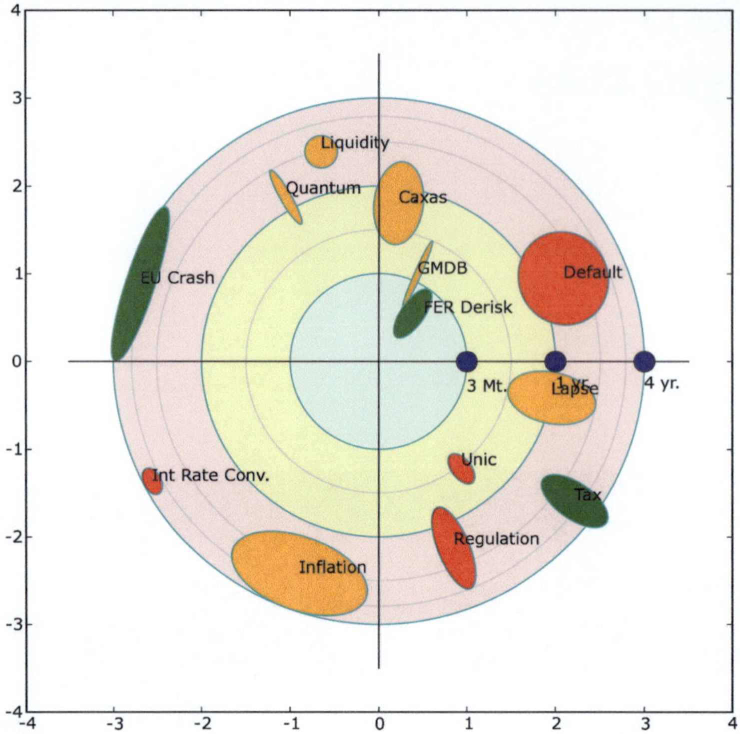

Fig. 13.1 Example of a Risk Radar

lenging me by asking, how risk management can create value. Assume for example some big banks would have seen the consequence of the 2008 financial meltdown early *and* would have reacted accordingly. In this case many losses could have been prevented and the banks would not have lost a lot of their intrinsic value and credibility.

In this sense emerging risks are the ones which are believed to emerge not in the very near future, but rather a little bit later. A typical emerging risk could be global warming. So it is first important to detect these emerging risks and to act accordingly. Both of these tasks are not as easy as they may seem. For the detection of the risks, corresponding methods need to be applied and as a second step it is important to convince the relevant people that these risks are not only issues on the paper, but that there is a likelihood that they materialise over time.

When thinking about emerging risks, I imagined to sit in a valley with high mountains around me. Obviously (if there is no fog), I can see all the mountains and can identify the risks and perils. What I can not do is to look on the other side of the mountain. When imagining to cross over the mountains the perils of the part which I can not see are considered emerging risks. Obviously I would not try to cross a mountain if there is danger (such as a dragon) on the other side and would choose a less risky route, even though the alternative could be steeper in the short run. Based on this trivial idea, I tried to show the risks in the form of a radar (lets call it a *risk radar*). In contrast to a typical radar, where as the radius is the distance, I have chosen time as radius, and in consequence the task is to detect emerging risks, which may occur in say 1 year's time.

This method can not only be applied for regular risk management but this method is particularly useful for projects and change programmes. I have used this method and graphical support in one of my former jobs and it worked really well, because it helped to stimulate a discussion and also to visualise the risks.

13.2 What Process Is Needed for Emerging Risks?

Here I want to explain the above mentioned risk radar in more detail. Figure 13.1 shows a concrete example. The picture is characterised by bubbles of different size and colour and each ellipse is characterised by the following attributes:

Likelihood that risk materialises: This likelihood is expressed as the colour of the bubble. Red means high likelihood and green a relative low likelihood.

Time when the risk is expected to start: This may be in 1 year and that's where the ellipse starts.

Time how long the risk may persist: That's the diameter of the ellipse towards the centre. Hence the ellipse ends when the risk is "over".

Expected monetary impact if risk materialises: The areas of the bubbles have been calibrated in order that the areas represent the relative severity.

Finally it has to be remarked that the figures use a logarithmic scale for the time and that the best way to produce such graphics is to use a small program doing the job for you, since messing around with the ellipses is very time consuming.

Obviously having the figure is not yet sufficient to detect emerging risks. The best way to detect them is an honest and open discussion. For the change programme I mentioned and where I already had a lot of insights, I have taken some time in a silent environment and put together the zoo of emerging risks and produced the graphics, which I then discusses with people working in risk and on the programme at different levels. This helped to improve the content of the risk radar.

The nice thing when using a program is also that you can project in the future and produce the same graphic in a year and look what may hit then. What was really astonishing was the fact that this method provided a good predictor of the upcoming risks and it was possible to avoid some of them in an early stage, hereby helping the change programme considerably.

It is important to note that a risk radar should not be produced once, but it needs to be embedded in a process. In the concrete set up the risk radar is updated every 3 months and the output is discussed, in order to migrate the risks.

Finally I would like to say some words in relation to the required skill set for detecting emerging risks successfully. Obviously models are not really of great help, since this process aims to detect the risk which are somehow hidden and not as easily detectable. Hence the following characteristics are important:

- Good and holistic understanding what is happening for example in a change programme,
- The ability and honesty to analyse what has gone wrong, and what could go wrong also as a consequence of inadequate skills,
- The ability to carefully listen to the programme managers and in particular also to the people working on the project,
- The ability to abstract from the day to day frustrations and fears of the people,
- The creativity to think what could go wrong, and experience,
- The will and ability to these exercises as processes.

Chapter 14
Regulatory View on Risk Management: Solvency II

14.1 Introduction

Solvency regulation in the EU is under reform. The Solvency II ([ECSO2]) project will introduce a new solvency regime which will be characterised by an integrated risk approach better taking into account the risks an insurer is facing than the current solvency regime. Securities for these risks will have to be held in the form of solvency capital. There is however a difference between the risk management for an insurance entity from the company's point of view and from a regulator's point of view. Whereas the assessment of risks and the calculation of the available and re-

M. Koller, *Life Insurance Risk Management Essentials*, EAA Series, DOI 10.1007/978-3-642-20721-1_14, © Springer-Verlag Berlin Heidelberg 2011

quired risk capital *should* follow a strict market consistent approach with no hidden buffers, the regulatory approach focuses mainly on the security for the policyholders. The main aim of the insurance entity is to optimise its risk adjusted returns and it has therefore no incentive to under- or overestimate its capital requirements. On the other hand the regulator puts a bigger emphasis on security and the return on capital is rather a secondary point of view.

Furthermore we need to acknowledge that there is a principal – agent problem from a systemic point of view, since the principal (the policy setters and the general public) aims to have an efficient insurance market with competitive products at reasonable prices. This implies that the capital requirements should not be too onerous. The regulators protects the policyholders' interests and aim a capital requirement at the upper end of the reasonable range, since then they can sleep better. I do not want to state that Solvency II is not reasonable but I just want to say that there is a risk. This issue can be seen when following the discussions between policy setters, regulators and the insurance industry. There is also a principle – agent problem between policy setters on the one hand and the insurance industry, mutatis mutandis.

But what is the ideal outcome from a principal's point of view? In my view the accurate best estimate assessment of available and required capital, for the reason that both a too high capital level with implicit margins and also a too low capital level, is dangerous. For the second case this is obvious. For the first one it is a little bit trickier: since there are in this case implicit margins, one might feel in a secure region, despite the fact that one is not anymore, for the simple reason of not knowing the extent of the implicit margins.

For the insurance companies in Europe, the wave of deregulation in the 1990s brought more freedom - as well as more independent responsibility. This affected the insurance companies, their shareholders and the supervisory bodies. The companies sought to utilise the opportunities offered by deregulation and booming stock markets in order to expand internationally and to enter into more risky investments. At the same time, risk management was often neglected and companies made themselves increasingly dependent on capital gains. This trend became particularly marked among life insurers, who often made huge promises: they promised high surpluses which could only be achieved by assuming a high degree of risk. The turnaround came in 2001, when the climate for the insurance companies deteriorated dramatically - due to the events of 11 September, and to massive stock market losses. As a result, the largest Swiss insurance companies had to contend with very high losses and needed to rebuild their capital base. Outside Switzerland, some insurers even went bankrupt. Another example is Equitable Life, one of the oldest life insurance companies on the European continent. Here the problem were 'quasi' guarantees (say 8% return guarantee including policyholder bonus payments) which have been offered to the policyholders in an interest environment which was then very high (eg GBP interest rate at over 12%). As the company did not valuate these liabilities in a market consistent way, Equitable Life was not able to anticipate its problems in a timely manner and the company was forced to close down their new business (see also section 12.4).

What had happened? It was not the economic principle of diversification that had failed. In fact, what was lacking in the companies was appropriate risk management. On top of this, the instruments of supervision were often not applied with sufficient consistency, nor were they suitable to provide adequate measurements of companies' risks. The yardstick of capital - based on the old solvency regime - was not capable of measuring the asset-liability risk. In other words, the risk that the asset side of the balance sheet (investments) might behave differently than the liabilities side (technical obligations) could not be assessed correctly. The result was that assets were used to enter into risks that were out of any proportion to the insurance portfolio on the liabilities side.

This shows the need for the European regulators to adjust their tools and methods in order to be able to keep up with dramatic increase in complexity in the financial sector. It is however of utmost importance to acknowledge the economic fundamentals of all insurance undertakings: the law of large numbers or the diversification effect. It is from this point of view key that new regulation accepts diversification on all levels: between individual risks, between regions and also between legal entities (group diversification). It is clear that diversification goes hand in hand with capital fungibility and also with mutual trust between the regulators of the different legal entities in a group. Assuming a reduction of the diversification benefit would ultimately lead to either higher costs for the policyholder or to a deterioration of the risk adjusted profitability for the owners of the insurance company. The latter clearly leads to withdrawal of capital from the insurance sector which by itself leads to a reduction in available capacity. Hence it is key for Solvency II to accept the diversification benefit and ensure the corresponding capital fungibility.

14.2 What Is Solvency II?

Currently Solvency II follows a one fits all approach in the sense that one tries to have one big risk based solvency framework which should be applied to all the different insurance entities in the same manner. This however does not reflect the relative importance of the different types of insurers. Figure 14.1 aims to explain this. Looking from a policy holder's standpoint one needs to distinguish at least between life, non-life and reinsurance. From the policyholder's point of view life insurance serves to protect him from the risk of suffering famine in case of ageing - it protects his (or his descendants) individual economic wealth after retirement. Clearly this aspect is one of the most important from the individual's (and also from the social state's) point of view. Therefore the security of the corresponding insurance undertaking needs to be highest. This is even more important under the point of view that these types of contracts consist normally out of very long tailed liabilities involving major financial resources. The pension benefit for a person after retirement is usually the biggest asset of him. In order to protect this wealth one needs to consider

	Life Insurance	P&C Insurance	Re-Insurance
Product	• Savings • Old age - annuities	• B2C Risk	• B2B Risk • Capital Protection
Main Risk	Individual Wealth after Retirement	Claim paying ability in case of event	Counterparty risk
Possible Supervision	• Solvency calculation • Risk governance • Transparency	• Solvency calculation • Risk governance • Transparency	• Solvency calculation • Risk governance • Transparency
	• Capital requirements • Policyholder protection	• Capital requirements • Policyholder protection	
	• ALM requirements • Ring fenced assets		

Fig. 14.1 Supervision Intensity

risk based solvency requirements, ALM requirements and possibly also ring fenced assets to protect the policyholder in case of the default of the insurance company.

P&C insurance has another aim. Here the typical example is the car which had an accident or the house which burned down. The main risk from the policyholder's point of view is the claims paying ability and is correspondingly of less importance for the individual compared with the life covers. Therefore the regulation should be lighter. One could for example only require solvency requirements but no ALM requirements. Moving to reinsurance one has to remark that we are here in a B2B environment where there should not be an explicit retail customer protection. It is however important that there is enough transparency in order that the buyer of such products can access the financial stability of the counter-party (eg reinsurer). Therefore the *counter-party risk* represents the major risk category and correspondingly the regulation should be lighter. Looking to capital markets one can ask the rhetoric question whether junk bonds should be disallowed? The answer is clear: an investor can outperform with respect to his risk return profile by adding junk bonds to his investments. In analogy insurance regulation should also reflect this idea and one could think of a solvency regime which requires only transparency here but does not explicitly require a certain capital level. Even more philosophically one could ask the question whether the state should at all prescribe the level of security in insurance entities - nobody would ever prescribe a private asset portfolio's minimal credit quality.

Now what is the consequence of these ideas? Regulation should not fall in the pit to apply one approach to all possible insurance undertakings. Looking at the proposed *standard formulae* it becomes clear that they were created with the aim to cover all possible risks for every possible insurance company. It is very likely that such an approach neither qualifies as transparent nor as efficient. Furthermore it will be very difficult to develop compatible internal models. It would be much better to start with principles a risk based solvency model should fulfil and in a second step to develop suitable simplifications. This is exactly the way how the *Swiss solvency test* was developed; the first joint working meeting between the industry and the Swiss supervising authorities centred about principal questions. For example should such a model be book-value or market-value based? How should insurance liabilities be modelled - is there a need for the valuation of policyholder options etc. This methodical approach was also recognised by the CEA, the European insurance association and it allows the development of both internal models and standard simplifications in a consistent way. A last general comment: as such models start to become rather complex at a very early stage it is very important to be pragmatic and simple. In the following section we will dig a little bit deeper in some of the relevant areas.

14.3 Economic Balance Sheet and Prudence

The most important additional insight which will be provided by Solvency II are *economic balance sheets* as introduced in chapter 2. The "quasi" guaranteed annuities with a 8% return guarantee indicated above show clearly the need for an economic balance sheet. This means on the asset side that all unrealised capital gains and losses are taken into account in a transparent way. On the liability side the situation is somewhat different, because no tradable instruments exist which can be used to perfectly replicate the liabilities in order to determine their economic price. It is however clear that exactly this information is of utmost importance for managing the risks and one therefore usually uses a model approach to get a reasonable approximation of the market values for the insurance liabilities. In first a step one needs to calculate the expected present value of the future policyholder benefits. On top of this amount one requires a *market value margin* (MVM) or also called *prudence* in a regulatory environment. For details re refer to chapter 2.

In the cost of capital approach (CoC) the required risk capital is projected into the future. In a second step the CoC equals the present value of the corresponding costs for future periods. The parameter corresponds to the unit cost of capital and is usually in the order of between 2% and 6%. Figure 14.2 shows corresponding calculations performed by different Swiss insurers during the field test 2005[1]. These results show that even in the low interest environment for the CHF there are significant margins in the P&C reserves and also to a minor extent in the life reserves. Finally it is im-

[1] The results of the SST field test can be found under http://www.finma.ch.

portant to remark that the CoC approach has two additional benefits: one can quite easily verify the corresponding results and it avoids double counting of capital.

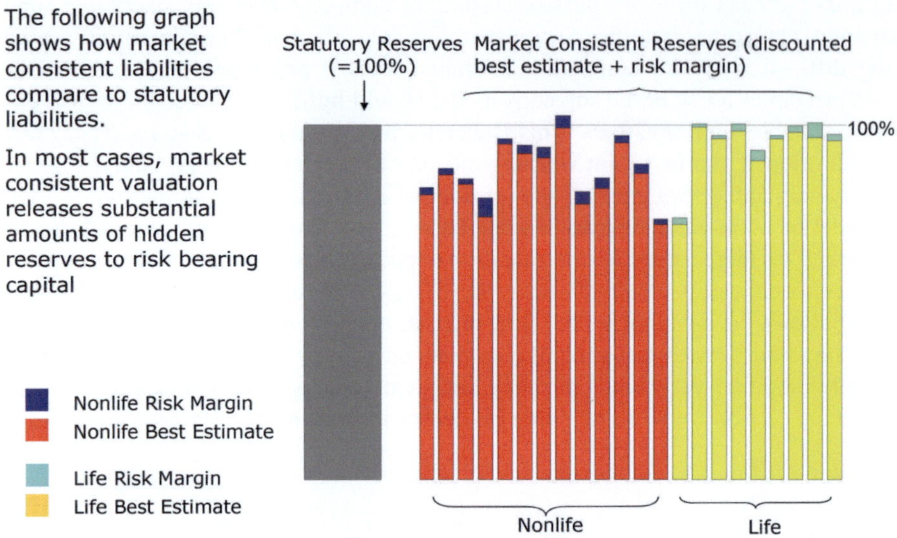

Fig. 14.2 Market Value Margin for the SST

Having stated the importance of basing the Solvency regime on a reliable economic balance sheet, there is another important question relating to the market consistent valuation of liabilities. What is the value of the different policyholder options such as the possibility to surrender a policy or to take capital or annuity in a pension scheme? It is clear that these implicit options can have a considerable value, but there are few reliable methods to value them, which are generally accepted. Therefore a pragmatic approach has to be taken. This means that only the most relevant policyholder options should be quantified. The most prominent example is for the guaranteed unit linked insurance contract. Here the valuation of the corresponding put option on the fund is relatively easy to quantify based for example on the Black-Scholes formula and the corresponding risk management techniques.

14.4 Risk Modelling and Internal Models

After the calculation of the economic balance sheet and the *available capital* it is necessary to define a risk measure and to choose adequate models for the different risks. Within an insurance undertaking, the main risks are ALM risks, including credit risk and liability risk induced by the insurance contracts. The risk model serves to quantify these risks in order to monitor and steer them adequately. Also it is key to keep the model as simple as possible based on predefined principles. Otherwise the model becomes very opaque and model risk increases significantly. Allocating the total risk of an insurance company a typical outcome could be as follows (according to the quantitative impact studies for Solvency II – QIS 4):

Risk	Life Insurer	P&C Insurer
Insurance Risk	44%	75%
ALM Risk	72%	41%
Credit Risk	4%	7%
Diversification	-20%	-23%
Total	100 %	100 %

This table clearly shows that the ALM risk is the most important risk for a life insurer and needs to be modelled very accurately. The insurance risk is very relevant for a typical P&C insurer. Interestingly however is also the relative importance of the ALM risk for the P&C insurer. Looking at the origin of many internal models it can be observed that they have been designed based on this assumption: Replicating portfolio plus a standard ALM model. Only in a second step the pure insurance risk was included. This is also reasonable within Solvency II. Special consideration however needs be taken with respect to bonus reserves and legal quotes. Here it is key to identify the amount of the bonus reserves which can serve as a risk buffer and allow the company to take more risks. Therefore Solvency II is also an opportunity to discuss legal quote regimes in order to make them more efficient resulting in a higher performance for policyholders and shareholders (see also section 12.4). An example is a legal quote where the full bonus reserve can serve as a risk buffer in case the company suffers a loss. Solvency II foresees the application of *internal models* as an alternative to the *standard approach*. Here it is important to recognise the fears around this topic. On the one hand supervisors fear opacity and regulatory arbitrage . On the other hand small undertakings fear to be put into disadvantage with respect of capital requirements. With respect to the first topic it is necessary as mentioned before to base Solvency II on underlying design principles which need to be valid for internal models and also for the standard approach. With respect to the possible capital disadvantage of undertakings using the standard approach there is only one possibility to avoid this: the standard approach must be powerful enough to be close to the reality. Therefore a *standard formula* is likely to be dangerous in contrast to the example the Swiss solvency test approach.

14.5 Good Regulation

In order to understand the need for regulation let's go back in time and think about the roots of insurance. In ancient Rome poor people could not afford their funerals. Therefore they agreed to help each other in the case of death in order to finance the costly funeral ceremonies. This is diversification (raison d'être of insurance).

But did they need regulation and supervisors? No, because the whole was based on trust. Now the insurance industry has become a global play and there is a need for an efficient regulation which does not destroy the underlying principle of diversification. But what does this mean?

- We do not need a lot of regulation but need relevant one.
- Transparency is not the art of producing telephone books full of information, but rather concise and relevant information for transparency.
- Beware of the principal agent problem of regulators.
- It is key that the new regulation is developed in a coordinated effort between regulators and industry:

  Only by this does Regulation becomes relevant and applicable;

  Is accepted by all parties;

  Can enhance the value creation of the sector.

As a mathematician I like Axioms and hence I tried to summarise some relevant Axioms for good regulation:

1. It must be anticipatory – No formulae but principles.
2. It must be nimble – Defined solvency Axioms, pragmatic adaptations.
3. It must cultivate dependable relationships with regulators – Active dialogue between the stakeholders during the design and implementation.
4. It must be capable of implementing strategies to accomplish corporate goals – No monolithic solutions.
5. It must be able to manage a crisis to minimise negative impacts and reputational harm – Try to prevent them by requiring the people to think about risk.

14.6 Swiss Solvency Test

The Swiss solvency test is based on clear design principles, which can also help do design a consistent internal risk model: (nurodyti tiek pt kiek reikia)

1. All assets and liabilities are valued market consistently.
2. Risks considered are market, credit and insurance risks.
3. Risk-bearing capital is defined as the difference of the market consistent value of assets less the market consistent value of liabilities, plus the risk margin (eg. market value margin).

4. Target capital (eg. required capital) is defined as the sum of the expected shortfall of change of risk-bearing capital within one year at the 99% confidence level plus the risk margin.
5. Under the SST, an insurer's capital adequacy is defined if its target capital is less than its risk bearing capital (eg the available capital > required capital).
6. The scope of SST is legal entity and group / conglomerate level domiciled in Switzerland.
7. Scenarios (see also chapter 6) defined by the regulator as well as company specific scenarios have to be evaluated and, if relevant, aggregated within the target capital calculation.
8. All relevant probabilistic states have to be modelled probabilistically.
9. Partial and full internal models can and should be used.
10. The internal model has to be integrated into the core processes within the company.
11. SST report to supervisor such that a knowledgeable 3rd party can understand the results.
12. Disclosure of methodology of internal model such that a knowledgeable 3rd party can get a reasonably good impression on methodology and design decisions.
13. Senior management is responsible for adherence to principles.

Most of the of the design principles above are self explanatory, and so I would like to point only out a few things:

Operational risks have been excluded in the capital calculations in order to focus in a first step on the financial risks. This does not mean that they are not important, but the exclusion allowed the insurance entities to focus on the financial risks and economic balance sheets in order to be able to deliver the required results in a relatively short time.

Risk Margin: The expression risk margin is used as a synonym for the market value margin. The inclusion of the market value margin in the required capital instead of considering it as a part of the market value of insurance liabilities is different to Solvency II. This particular choice has been made, since the concept of a MVM was not yet generally accepted. From a conceptual point of view the inclusion of the MVM in the market value of insurance liabilities is a more sensible choice.

Standard Model for ALM: The standard model for ALM risk for the Swiss solvency test follows the approach shown in chapter 6. In addition to the analytical model there are additional stress scenarios which need to be defined and evaluated. In contrast to the material shown in chapter 6, the Swiss solvency test performs an additional step, in the sense that the total required capital is calculated based on the results of the analytical model and the outcomes of the different stress scenarios. In order to do this a discrete probability is attached to each stress scenario, considering it as a Dirac (point) measure. (Eg the scenario occurs with a certain probability and in all other cases the incremental loss is 0). In a next step one considers the $n+1$ random variables $X_0, X_1, \dots X_n$, where X_0 is the analytical model and the other $(X_k)_{k \in \mathbb{N}_n}$ denote the stress scenarios and forms $X = \sum_{k=0}^{n} X_k$ the total loss. The distribution of the total loss X is calculated by a standard convolution technique[2] for independent random variables.

Own Scenarios: There is a requirement within the Swiss solvency test not only to use the standard stress scenarios, but also the need to define entity specific scenarios, which could occur and threaten the insurance company. The rational behind this is the idea that the SST should not become a pure compliance exercise and that the individual companies should think about their specific risks. Finally it is important to remark that the introduction of scenarios had not only the purpose to use them within the capital models of the different insurance companies, but the outcome of these scenarios allows the regulator to also assess the systemic risk for the whole insurance market within a country, since the standardised scenarios can easily be aggregated.

Core processes: The required use of the SST in the insurance companies' core processes should ensure that the economic capital model is used for making business decisions. In Solvency II this concept is known as *use test.*

The table below summarises the scenarios to be used for the Swiss solvency test:

- Longevity: mortalities fall suddenly and stay low.
- Disability: disability inception rates spike.
- Insufficient P&C reserves.
- Accident: Accident of a tourist group.
- Severe Incident plus panic in a (sport–)stadium.
- Hail Storm.
- Industry Incident (eg Seveso / Bhopal type); besides financial and business interruption loss also casualties.
- Pandemic scenario (1914 Spanish Flu).
- Financial Distress (Run on the bank/insurance company, eg combined distressed asset values and a high liquidity demand).
- Default of the company's biggest reinsurer.
- Terrorist attack (aka 09.11.2001).
- "Own Scenarios" ($4\times$).
- Equity markets drop 60%.

[2] See http://www.finma.ch.

- Real estate crash combined with increase in interest rates.
- Stock market crash (1987).
- Nikkei crash (1990).
- European currency crisis (1992).
- US interest rate crisis (1994).
- LTCM (1998).
- Stock market crash (2000/2001).
- Global deflation.
- Financial crisis 2008.
- Spike in lapse rates.
- Global inflation scenario.

14.7 Solvency II Standard Model

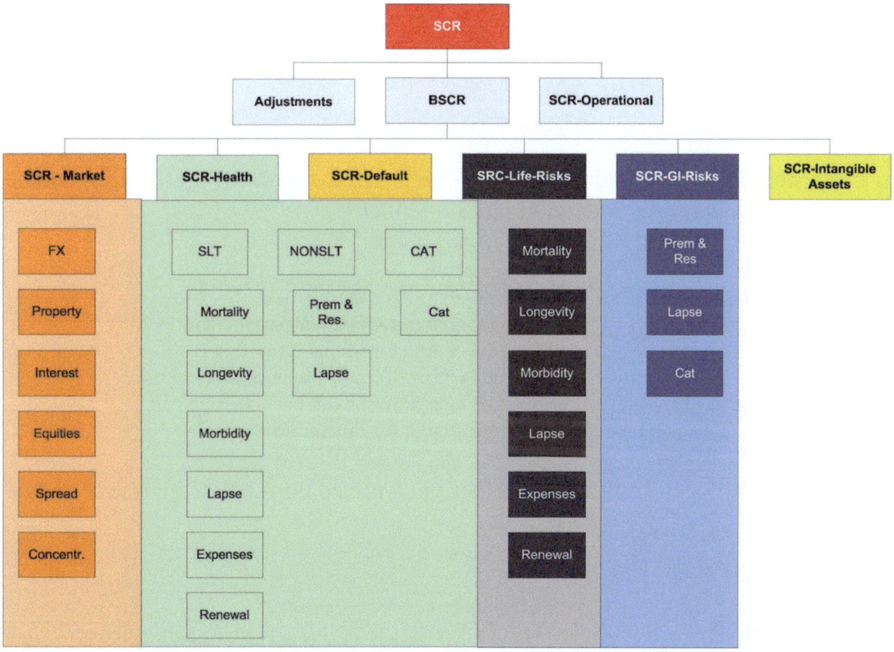

Fig. 14.3 Solvency II Standard Model

It is import to recognise that the Solvency II standard formula is still under development and that this section can not replace the relevant regulation for Solvency II[3]. This section is based on the DRAFT Technical specifications for QIS 5, which were issued in April 2010. Hence it is possible that parts of the framework are still going to change.

In order to define a solvency frame work the following steps need to be performed:

Definition of Capital and Valuation: Define what is capital – eg market value of liabilities minus market value of capital. This question relates to the topics treated in chapters 2 and 3. In technical specification this is treated in section 1.

Time Horizon: The time horizon over which risk capital needs to cover risks. For Solvency II this is one year.

Risks to Quantify: In this step the risk map is defined, see for example chapter 6 for financial risks. Figure 14.3 defines the risk landscape taken for Solvency II.

Definition of Ruin: Solvency II defines ruin, if the market value of assets falls below the market value of liabilities.

Risk Measure: For Solvency II the 99.5 % VaR is taken as risk measure.

Operational Implementation (Standard Model): The standard model is described below.

14.7.1 Structure of the Model

In a first step we need to understand the structure of the model (Figure 14.3): On the highest level we have the following building blocks:

Module	Description
`SCR-mkt`	Capital charge for market risk
`SCR-def`	Capital charge for counter-party default risk
`SCR-life`	Capital charge for life underwriting risk
`SCR-nl`	Capital charge for non-life underwriting risk
`SCR-health`	Capital charge for health underwriting risk
`SCR-Intangibles`	Capital charge for intangible assets risk

These building blocks are linked together via a mixed correlation approach and we have:

$$\text{Basic SCR} = \sqrt{\sum_{i,j \neq \text{Intang.}} \rho_{i,j} \times SCR_i \times SCR_j} + \text{SCR-Intangibles}$$

[3] www.ceiops.eu

All the above building blocks are defined based on more granular risk factors and algorithms to link them together. In the following we will have a closer look at some of the submodules. The correlation coefficients have been chosen as follows:

	Market	Default	Life	Health	Non-Life
Market	1.00	0.25	0.25	0.25	0.25
Default	0.25	1.00	0.25	0.25	0.50
Life	0.25	0.25	1.00	0.25	0.00
Health	0.25	0.25	0.25	1.00	0.00
Non-life	0.25	0.50	0.00	0.00	1.00

14.7.2 Market Submodule

As seen above the market submodule consists itself out of the following submodules, for which the capital requirement SCR is calculated. The respective SCR's are calculated by means of stress scenarios. For each submodule there are two scenarios to be evaluated, an upside movement ($\uparrow$) and a downside movement ($\downarrow$). We denote by $\mathcal{G} = \{\uparrow, \downarrow\}$ and we formally calculate for each $\kappa \in \mathcal{G}$ the following:

- Interest rate risk ($SCR_{irate}(\kappa)$),
- Spread risk ($SCR_{spread}(\kappa)$),
- Concentration risk ($SCR_{Co}(\kappa)$),
- Equity risk ($SCR_{Eq}(\kappa)$),
- Property risk ($SCR_{Prop}(\kappa)$),
- FX risk ($SCR_{FX}(\kappa)$).

All of the above then result into the SCR for the total market risk (SCR_{mkt}) with the following formula:

$$SCR_{mkt} = \max\left(SCR_{mkt}(\uparrow), SCR_{mkt}(\downarrow)\right),$$
$$SCR_{mkt}(\kappa) = \sqrt{\sum_{i,j} \rho_{i,j}(\kappa) \times SCR_i(\kappa) \times SCR_j(\kappa)},$$

with $\rho(\uparrow)$ given as follows:

$\rho(\uparrow)$	Interest	Equity	Property	Spread	Currency	Concentration
Interest	1					
Equity	0	1				
Property	0	0.75	1			
Spread	0	0.75	0.5	1		
Currency	0.25	0.25	0.25	0.25	1	
Concentration	0	0	0	0	0	1

$\rho(\downarrow)$ is slightly different:

$\rho(\downarrow)$	Interest	Equity	Property	Spread	Currency	Concentration
Interest	1					
Equity	0.5	1				
Property	0.5	0.75	1			
Spread	0.5	0.75	0.5	1		
Currency	0.25	0.25	0.25	0.25	1	
Concentration	0	0	0	0	0	1

For the interest rate model a stress scenario has to be applied with an upward and a downward stress of interest rates as follows:

Maturity in years	**Relative change up ($\uparrow$)**	**Relative change down($\downarrow$)**
0.25	70%	-75%
0.50	70%	-75%
1	70%	-75%
2	70%	-65%
3	64%	-56%
4	59%	-50%
5	55%	-46%
6	52%	-42%
7	49%	-39%
8	47%	-36%
9	44%	-33%
10	42%	-31%
15	33%	-27%
20	26%	-29%
25	26%	-30%
30	25%	-30%

Based on the above relative shifts (eg $i_{\mathrm{afterStress}} = i_{\mathrm{beforeStress}} \times (1 + \text{relative amount}))$ all values are recalculated, in the same sense as shown in chapter 6.

For the other risk factors the approach is quite similar and hence we want finally to have a look on how the credit spreads movement looks like. For further details we refer to the relevant literature [ECSO2].

The credit risk factors using the duration approximation mentioned in chapter 6 are defined as follows. Please note that the duration to be taken for the approximation has a floor and a cap.

Rating	↑	↓	Floor	Cap
AAA	1.0 %	-0.4 %	1	∞
AA	1.5 %	-1.0 %	1	∞
A	2.6 %	-1.7 %	1	∞
BBB	4.5 %	-3.0 %	1	7
BB	8.4 %	-6.3 %	1	5
≤ B	16.2 %	-8.6 %	1	3.5
Unrated	5.0 %	-3.3 %	1	7

Finally we want to have a look at the equity sub-model. Also here a stress scenario approach is applied where it is assumed that the equities fall by 30 % for global equity indices and 40 % for all other indices respectively. The corresponding results are aggregated using a correlation matrix. It needs to be stressed that it obviously can not replace the in depth study of the material in [ECSO2].

Please note that Solvency II foresees that insurance companies can use partial or full *internal models*. Figure 14.4 shows how such a modified internal model could look like. Obviously such internal models need to be approved by the relevant regulator.

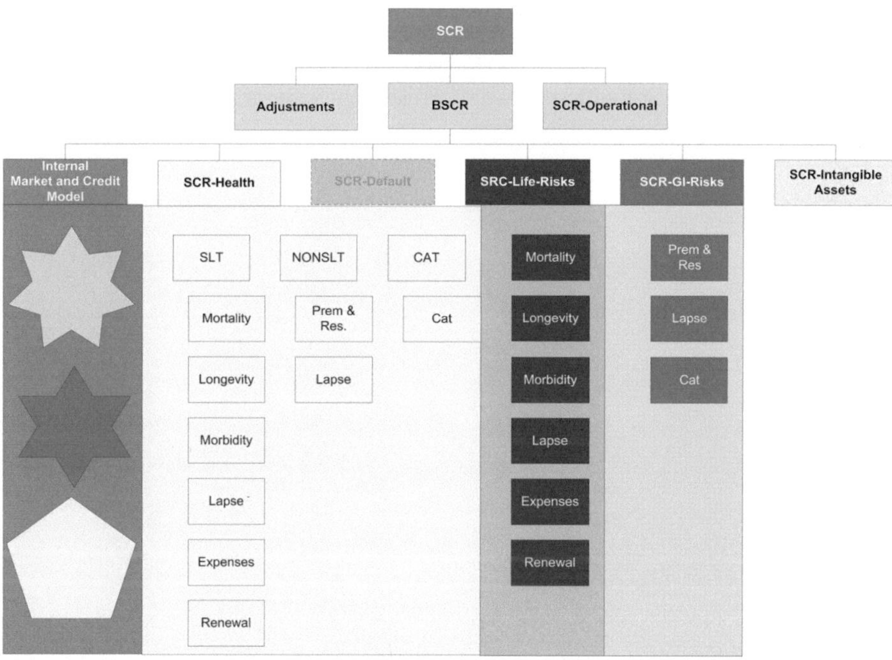

Fig. 14.4 Solvency II Partial Internal Model

14.8 Quo Vadis?

It is clear that Solvency II is a great opportunity for the European insurance industry, for example with respect to product innovation. However there is still a lot to do in order to get Solvency II up and running. It is important to base the whole system on clearly understandable principles. Only by that double counting of capital and capital inefficiencies can be avoided. With respect to valuation it is of utmost importance to base the calculation on an economic view with a transparent method for the market value margin such as the cost of capital approach. The valuation needs to be done on such a level of granularity that companies can calculate their replicating portfolios, in order to do a proper risk management with respect to the ALM risk. Finally the risk management models should be designed in a transparent manner in order that internal models follow logically from the standard principles. This ensures that the main risks are taken into consideration and that the standard approach needs not to be overloaded by additional security measures. Doing that will allow the companies to use their capital more efficiently leading to higher returns for shareholders and for policyholders. Furthermore such a model would also allow the companies to understand their risks better and to do better with respect to ALM. In order to achieve this challenging goal all stakeholders need to engage in an open and constructive discussion. This will ultimately lead to a better result and to a better mutual understanding between the insurance industry and the regulators.

Chapter 15
Governance and Organisation

The aim of this chapter is to show one possible governance form for an effective risk management. It is to be noted that there are many different forms of organisational principles and that the concrete implementation needs to reflect the needs of the company.

15.1 Governance

In this section the aim is to define a segregation of duties needed to provide effective risk management.

M. Koller, *Life Insurance Risk Management Essentials*, EAA Series,
DOI 10.1007/978-3-642-20721-1_15,

15.1.1 Definitions

In order that everybody speaks a common language and to avoid misunderstandings with the aim to anchor an adequate governance within the company it is necessary to define some roles and functions as follows:

Risk owner: Executive committee, through setting limits and appetite for risks and approving risk policies & governance, owns the risks, through the delegation of authority and responsibility for these risks through the company's management processes.

Risk taker / Line Management (1st line of defence:) The business functions (Products / Operation / Distribution), through writing business and implementing the risk policies and governance framework as well as management controls, take risks. In addition, corporate functions take risks, eg. finance through its balance sheet and control management activities.

Risk controlling & reporting (2nd line of defence:) The risk specialist functions, through identification of emerging issues, creation of risk policies, review of the risk taking activities of the business functions, provision of management information and consolidated risk committee / executive committee reporting, perform core controls in the risk management process. The Chief Risk Officer, through periodical review of any part of the risk assurance matrix as he deems appropriate, performs additional controls.

Independent assurance (3rd line of defence): Internal audit, through their audits of process and policy compliance by both business functions and risk specialists, provide independent (from management / risk committee) assurance that framework is complied with.

Risk Policy: The risk policies are governance documents with the aim to ensure that an adequate risk framework is in place for a certain type of risk. These documents are prepared by the risk Management function (second line of defence) and they are adopted by the *Risk Owners*. risk policies are published by the Chief Risk Officer. The company sets the risk appetite for the business. For most of the policies the implementation is the responsibility of the *Line Management*.

Policy Owner: The policy owner is the manager within the first line of defence who is responsible for the corresponding policy.

Policy Coordinator: The policy coordinators act on behalf of the policy owner and ensure the implementation of the corresponding policy in the business.

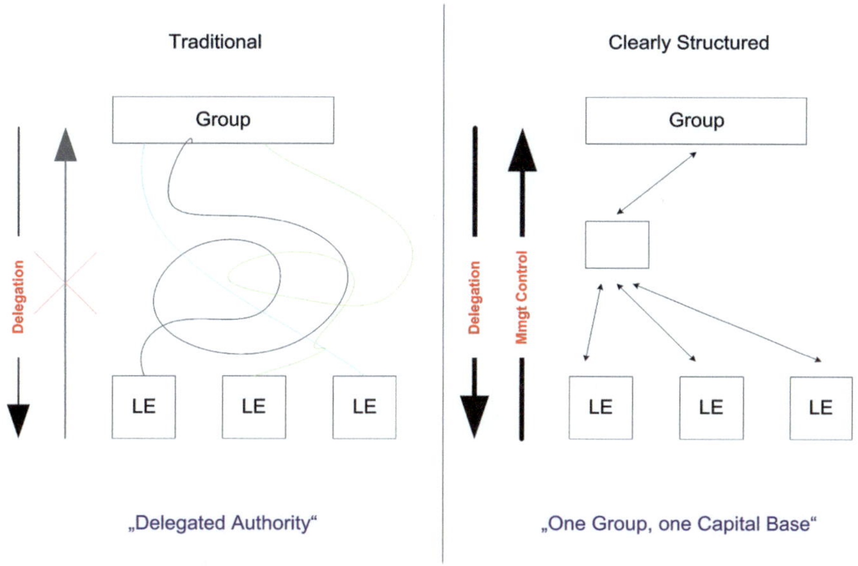

Fig. 15.1 Generic Governance Process

15.1.2 Generic Governance

The risk committee is responsible for the implementation of the group risk policy according to different hierarchical levels. At each level of responsibility within the company, the terms of reference required by the next level, for risk management to function correctly, shall be defined and documented. In particular, responsibilities shall be delegated and the corresponding controlling and reporting responsibilities established. The operational risk management function is responsible for this framework. The generic risk management process is defined in section 1.3.

Top-down processes within this structure are:

- The delegation of limits;
- The issuing of policies and guidelines for identifying, measuring, managing and monitoring/reporting risks.

Bottom-up processes within the company include:

- Applying for limits;
- Reporting of existing risks;
- Reporting any limits that may have been exceeded and any other violations of the risk management directives and guidelines.

Requester for limits:

For all risks (market, credit and insurance, ...), the party who requests to operate outside the risk appetite or requests to adjust the risk appetite carries the responsibility. Approval by a higher instance does *not* release the applicant from his or her responsibility to ensure an adequate risk management.

Separation of powers:

Those persons within the company who are responsible for identifying, assessing and controlling risks must be independent of those responsible for assuming and managing risks.

Figure 15.1 aims to visualise this approach vs a more traditional delegation of authority approach. It is obvious that the structured approach is better suited to manage the capital of the group in order to optimise the risk return profile of the group.

15.1.3 Escalation Processes

In many situations where limits are exceeded or where the organisation is going to operate outside its risk appetite, it is essential that there are clearly defined, efficient and fast processes to get the risk back within acceptable limits. As a consequence there is a need to clearly define the escalation processes which apply. The following list defines the corner stones in respect to escalation processes:

- Material breaches including any identified issues that could lead to a breach, should be notified to the Chief Risk Officer immediately (within 24 hours).
- The Chief Risk Officer will where appropriate advise the relevant oversight committee of these breaches.
- As primary responsibility for risk management lies with *Line Management* it is expected material breaches will also be reported up through functional management. All material breaches should be documented through the quarterly risk reporting cycle.
- CRO escalates disagreements in opinion to the risk committee.

15.2 Effective Governance

Effective governance is dependent on decisions being made and acted upon by the right people, with the right authority, understanding and expertise, using the appropriate information, at the right time.

Committees are a key element of each group's governance and risk management oversight framework different levels to ensure that appropriate governance decisions are made.

The performance management meeting, led by the CEO, is the primary first line of defence mechanism for considering the performance of each region and its businesses.

15.3 Risk Governance and Oversight Framework

Risk management oversight is having full visibility of the issues and risks facing the business and reviewing the appropriateness of the actions being taken to manage them.

The Board has adopted the group risk committee structure as one of the mechanisms that helps to ensure the Board has the appropriate level of risk management oversight and governance across the company.

The structure should enable the operation of effective governance across the group, the regions and business units. Committees reinforce the completeness of risk coverage between the ALCO and the ORC (Operational Risk Committee) and down through their delegated sub-committees.

15.4 Principles of Effective Oversight

Effective risk management oversight relies on:

- Management oversight of risk prior to board oversight;
- Clear distinction and separation between 1st, 2nd and 3rd line risk management responsibilities;
- Clear accountability for coverage of all risk categories and associated group policies;
- Clarity in committee purpose, responsibilities and authority; and
- Effective and appropriate use of senior management time.

15.5 Board Oversight

The Board will look for assurance to demonstrate:

1. Appropriate systems and controls for risk management;
2. Sound operation of governance and internal controls; and
3. The flow of timely management information in order to discharge oversight duties effectively.

To achieve these aims the Board has:

1. Delegated authority of risk oversight from the board to CEO and executive committee; and will
2. Provide challenge to the executive about the operation of risk management; while
3. Maintaining non-executive independence.

15.6 Management Oversight

Management will facilitate risk management oversight by providing:

1. Management information on the risks faced by regions and individual business units;
2. An overall perspective on consolidated group risk;
3. Evidence to show adherence to defined risk appetite and risk tolerances; and
4. Any remediation action required to return to risk appetite by region or business unit.

15.7 Why Risk Management Committees Are Important

The Board has adopted a risk-based approach to establishing a system of internal controls and uses the committee structure to review its overall effectiveness.

Committees provide the board with assurance that all major risks to which the group is exposed are adequately identified, assessed, monitored and controlled. They enable significant movements outside risk appetite to be escalated and allow for the monitoring of the corrective action being taken.

15.8 How Do Risk Management Committees Work

Committees provide independent challenge through oversight by bringing together the different knowledge, perspectives and experience of its members.

The individuals who perform roles within the governance framework are a key element for its success. The correct blend of individuals is essential to ensure that a committee can discharge its duties effectively. Each committee member should be present to provide a valuable contribution to the committee.

Although individual contributions are made, committees operate on the basis of *collective responsibility*; this should allow members to express their views freely in discussion, while maintaining a united front once agreement has been reached. Agreements reached in committee sessions are binding on all members.

15.9 Terms of Reference of a Risk Committee

In this section we will look at a typical terms of reference (ToR) in order to better understand the duties of a risk committee.

The risk committee is established by the executive committee to oversee the company's aggregate risk exposure.

The risk committee will review and monitor the management of financial and operational risks to assess whether the risk profile of the company is within appetite.

The committee will review and monitor risk appetite related to its risks and will compare the regional aggregate risk profile, both current and emerging, against risk appetite.

Further, the committee will review and monitor the implementation and effective adherence to group risk management policies under its oversight.

Scope of committee oversight

1. **Risk policy scope**

   The activities of the risk committee focus on the implementation and management of all group risk management policies as covered by its sub-committees. This includes, but is not limited to the Asset Liability committee, Operational risk committee.

2. **Scope of operations**

   As a governance committee, the activities of the risk committee are applicable within all of its relevant:

- Markets and operations;
- Legal entities; and
- Joint ventures;
- The risk committee will consider risks assumed by entities that the company does not have management control.

3. **Committee authority**

- The risk committee has authority from the executive committee to exercise oversight of all markets and operations within the regions and the corresponding legal entities.
- The committee, in liaison with regional policy owners, has authority to request and receive the type of management information required and any further evidence to support its risk management oversight activities.

4. **Committee accountability**

- The committee is collectively accountable to the executive committee, under the leadership of the chair, for discharging its duties and responsibilities in an appropriate manner as set out in this document.
- The risk committee is also accountable to the company Board.

5. **Committee responsibilities**

The responsibilities of the committee may be delegated, by resolution of the committee unless otherwise reserved to, or determined by the executive committee.

The committee will consider any topics delegated to it by the executive committee.

As a primary responsibility, the committee is collectively responsible for performing oversight over the risk management activities, in respect of the risks inherent in the policies under its oversight, in such a manner that:

- Key risks significant to the achievement of the company's business objectives are identified, assessed and managed; and
- Mitigating actions are taken to bring significant risks within appetite.

In order to execute its oversight duties, the risk committee will make recommendations, in relation to the management of risk within the company for further consideration and review. The committee will track progress against any recommendations.

Specific responsibilities will be to:

Review aggregate risk profile against appetite

- Review and assess the appropriateness of the company's risk profile;
- Review and assess the appropriateness of the drivers and measures for setting risk appetite for key risks;
- Review and monitor significant risk exposures, including risks and issues reported through its sub-committees, and assess whether risks are consistent with the region's appetite.

Recommend aggregate Risk Appetite

- Recommend aggregate risk appetite to the executive committee annually or more frequently if required.

Assess whether mitigating actions are in place in the event of a policy breach or an activity taking a market or operation significantly out of appetite

- Undertake a quarterly review of the company's emerging financial and operational risks and changes to the region's aggregate risk profile against appetite;
- Oversee aggregate financial and operational risk exposures within appetite and:

    1. Ensure the operation of minimum standards of controls that are proportionate to managing financial and operational risks associated with the region's operations;
    2. Ensure appropriate review of waiver and exception applications to group risk policies; and
    3. Consult with the relevant regional policy owner on matters related to the content, applicability, implementation, adherence and enforcement of a group risk policy;

- Assess and implement strategies that improve the region's risk profile, at an appropriate cost, and review how this is monitored in practise to ensure the operational and financial risk position remains within the region's aggregate risk appetite.

Notify the executive committee of any policy breaches or events which have taken the region significantly out of risk appetite and the actions in place to return within appetite

- Notify and escalate to the executive committee, and relevant group risk committees as appropriate;
- Any actual movement outside of risk appetite or control deficiencies, as outlined in the risk policies under the remit of risk committee, occurring which may require changes in local plans or regional intervention;

- Corrective actions taken by a market or operation in the event of any movement outside risk appetite which is material at regional level.

Review the effectiveness governance

- Review the committee's effectiveness and terms of reference:

  1. Receive a half-yearly attestation of compliance of markets and operations with, and the embeddedness of the group policies applicable to the company;
  2. Undertake a self assessment of the committee's own performance during the year by evaluating its activities against the terms of reference; and
  3. Receive an annual report from the committee secretary on the appropriateness of the terms of reference, including any appropriate changes; and
  4. Recommend any changes to the terms of reference of the risk committee to the executive committee.

Once approved, the chair of the committee is responsible for the committee terms of reference.

Oversight relationships

- The risk committee has delegated its responsibility (but not accountability) for oversight by forming the following risk sub-committees:

  1. Asset liability committee;
  2. Operational risk committee; and
  3. Regulatory & compliance committee;

Committee members and invitees

In carrying out their responsibilities and fulfilling their duties, all members of the risk committee shall adhere to the external regulatory and internal control and policy restrictions pertaining to the business of the company that are in force from time to time.

Individual members of the committee and all papers relating to its business shall be subject to the terms of controls relating to price sensitive information.

The risk committee shall comprise the following executive committee members:

- Chief Executive Officer (chair)
- Chief Financial Officer
- Chief Risk Officer
- ...

Risk Category	Risk Ownership	Risk Taking	Controllership	Risk Management	Independent Assurance
Overall responsibility	**Group CEO / Group GEC**	**CEO Europe / EEC**	**CEO Europe / EEC**	**CRO**	**GIA**
Non-Life (re)insurance risk	GEC	EEC	EEC	Insurance Risk Management	GIA
Life (re)insurance risk	GEC	EEC	EEC	Insurance Risk Management	GIA
Financial markets & credit risk	GEC	EEC	EEC	Credit & Financial Risk Management	GIA
Funding & liquidity risk	GEC	EEC	EEC	Economic Capital	GIA
Operational risk	GEC	EEC	EEC	Operational Risk Management	GIA

Fig. 15.2 Allocation of Responsibilities

15.10 Roles and Responsibilities

15.10.1 Duties of the Line Management

Head of business:

The business head, together with the local executive and operational management team, is responsible for achieving the agreed strategic and operational objectives of the business. As such they have a responsibility to:

- Ensure their business operates in accordance with the risk strategy of the company.
- The business head is the local policy owner for the risk management and internal control policy.
- Identify and manage risk, including emerging risk, throughout the business based on the minimum standards.
- Set up an appropriate control structure and culture to ensure effective internal controls and to manage exposures within risk appetite.

- Ensure delegated authorities are clear and fully documented with appropriate segregation of duties and that people are not assigned conflicting responsibilities.
- Establish a local risk committee or ensure this is a clearly defined role of the business executive committee.
- Ensure adequacy of internal financial, operational and compliance information and review external market information about emerging events and conditions.
- Meet Group governance requirements including the need to provide information within agreed thresholds to respective committees with ultimate oversight by either the ORC or the ALCO.
- Establish effective channels of communication to ensure staff are fully aware of policies and procedures affecting their duties.
- Promote an environment where management and staff can report without fear, control breaches, suspicions of fraud, theft, malpractice and any near misses, while guaranteeing anonymity when requested.
- Meet the risk management requirements of local and group regulators.
- Reinforce the "three lines of defence" model by encouraging close working relationships between line management and the local risk function whilst facilitating independent assurance by internal audit.

Staff:

It is the responsibility of all staff to understand and manage the risks faced in relation to the core activities and processes under their control and stewardship. The duties of the line management in respect of risk management are summarised in section 15.10.4.

15.10.2 Duties of the Chief Risk Officer

Governance:

- Ensure statutory and regulatory requirements are met on a timely basis.
- Ensure the ongoing compliance culture is reinforced in area of own influence.
- To ensure policies and committee terms of reference are appropriate, effectively communicated, continuously reviewed and updated to reflect internal/external change.

Context:

- The role brings together enterprise and operational risk, and risk frameworks and financial management in a single place, with a remit to ensure these key control and business functions are operating effectively and in congruence within the company.
- Significant input into the finance strategy which supports the overall strategy of the company.
- To provide consultancy support to the company executives and markets/business units.
- Input into implementation, management and communication of the financial framework for the company.
- Build effective relationships with the senior management of the operating entities which comprise the company's business.

Duties:

- Lead role in financial and risk management and business controls.
- Lead an operationally independent second line of defence, facilitate implementation of effective first line of defence risk management practises.
- To lead risk management interfaces between risk, functions, products and distribution.
- Ensure the company meets its control standards for risk management.
- Advise the company's executive on improving risk adjusted returns from the company's portfolio maximising returns through superior risk management techniques.
- To lead the education and development of the risk management community.
- To drive improvements in product development and life cycle management and management of life reinsurance across the company.
- Responsible for the company's business and corporate governance controls.
- Responsible for the ongoing compliance with the company's risk governance framework.
- To provide assurance on the effectiveness of risk management across the company through the implementation of group policy on risk management.
- To report on and provide assurance to the executive on the effectiveness of financial and risk management across the company.
- Undertake/co-ordinate financial risk management at a regional level i.e. in respect of the major risks on the company balance sheet.

Accountabilities:

- Create and manage appropriate risk frameworks and policies that will ensure the company's operational risk across the business is kept within appetite and with effective oversight.
- Ownership (assessment, management, monitoring and reporting) of functional risk area - accountability to give assurance to operational risk committee as to appropriateness of risk rating.

15.10.3 Duties of the Group Internal Audit

Context:

Internal audit's mission is to provide reliable independent assurance to the group audit and group risk and regulatory committees, local audit committees, board members and executive management the company and its subsidiaries on the adequacy, effectiveness of the control frameworks which include governance and risk management.

Role and Responsibilities:

- Internal audit is the "third line of defence" in the group's risk governance structure. Internal audit provides independent and objective assurance over the design and effectiveness of controls in place to manage the key risks impacting the group's business performance. Internal audit has a key role in supporting the accomplishment of objectives of the group.
- Internal audit is accountable for developing and delivering a programme of assurance aimed at validating the effective management of key business risks. Internal audit is accountable for reporting its findings, conclusions, and recommendations to the audited parties, regional and local management, oversight and local audit committees, group management, the group audit committee and, where appropriate, the risk and regulatory committee. Management is responsible for the effective identification of risk and the maintenance of adequate systems of controls.
- Internal audit is responsible for ensuring that issues that could impact on the achievements of the group's objectives are brought to the attention of the local, regional and group management, oversight and local audit committees, the group audit committee and, where appropriate, the group risk and regulatory committee and that timely follow-up on management actions occurs. Management is

responsible for corrective actions on reported weaknesses. Internal audit, as part of delivering its assurance programme, validates that risks are identified and addressed.

Scope of Work:

The primary scope of internal audit's activities is the examination and evaluation of the adequacy and effectiveness of the group's systems of risk management, internal control and governance processes, operating in the first and second lines of defence. This includes all legal entities, joint-ventures and other businesses partnerships, outsourcing and reinsurance arrangements. In addition to this, the scope of Internal audit includes financial crime investigation activities by the Investigation team.

15.10.4 Risk Relevant Parts Within Job Descriptions

In the following section relevant parts in respect to risk management job descriptions have been put together.

For Line Management Functions

- To understand and manage risk in their area of operation in accordance with the company risk strategy and applicable group policies.
- Process design and implementation in order to meet the group's risk management standards (guidelines).
- Primary responsibility to risk identification and management. Initiation of mitigation strategies once operating outside of risk appetite or in case of limit breach.
- Requests limits and risk tolerances and ensures adherence to limits granted.
- Management assurance that all required standards are met and that it operates within risk limits.

For Risk Management Function

- Facilitate implementation of effective first line of defence risk management practises.
- Ensure the company meets its control standards for financial management.
- To provide assurance on the effectiveness of risk management across the company through the implementation of group policy on risk management.

- To report on and provide assurance to the group and the company executive on the effectiveness of financial and risk management across the company.
- Create and manage appropriate risk frameworks and policies that will ensure the company's operational risk across the business is kept within appetite and with effective oversight.
- Ensure statutory and regulatory requirements are met on a timely basis.
- To ensure policies and committee terms of reference (for which this role is responsible) are appropriate, effectively communicated, continuously reviewed and updated to reflect internal/external change.

15.11 Decomposition of the Risk Management Process

In order to make the process in figure 1.3 more concrete the following sub-steps can be applied.

Task	Definition	Owner	Assurance
1 Limit & appetite setting, approval of risk policies & governance framework	The *Risk Owners* are responsible for the approval of risk limits and Risk Appetite. Moreover they are responsible for the over all risk governance and hence approve risk policies.	Executive Committee	–
2 Business process design	The design of the business processes in order to adhere to internal (risk policies et al.) and external (legal, regulatory, ...) requirements.	Line Mgmt	Risk Experts/ ORM
3 Business process implementation	Implementation of the above.	Line Mgmt	GIA
4 Risk assessment, measurement, mitigation & control strategies	Regular risk assessment is done by the first and the second line of defence for relevant risks. Measurement for financial risks is performed by the second line of defence in order to check for the limits and for risk adjusted performance. First line of defence is responsible that business is done not violating any limits. Furthermore the *Line Management* develops risk control and mitigation strategies.	Line Mgmt/ Risk Experts	Risk Experts/ ORM
5 Business assurance (management assurance)	*Line Management* gives assurance that its business is adhering to defined limits and tolerances.	Line Mgmt	GIA
6 Identification of emerging issues & risk policy creation	Risk Experts prepare and review regularly the risk policies. Furthermore they are responsible for identifying emerging risks and for addressing them.	Risk Experts	GIA

	Task	Definition	Owner	Assurance
7	Review of process design	Cf. 2.	Risk Experts/ ORM	–
8	Exposure monitoring & risk concentration analysis	Exposure and adherence to defined risk limits is regularly monitored by the second line of defence. In case of limit breaches the corresponding processes are initialised quo reporting and mitigation.	Risk Experts	–
9	Review of risk assessment, mitigation & control strategies	Cf. 4.	Risk Experts	–
10	Management information	Risk Management is being prepared.	Risk Experts	–
11	Independent challenge of 1st line	Independent challenge of the first line of defence in order to improve the over all risk management process.	Risk Experts/ ORM	–
12	Executive Committee & Board reporting	The collection, interpretation and aggregation or risk management information for the relevant internal (risk committees, BoD, ...) and external stakeholders is prepared.	ORM	–
13	Auditing of policy and process compliance	GIA performs its third line of defence duty by auditing the policies and the process compliance.	GIA	–
14	Review of adequacy of risk controlling	GIA reviews the adequacy of the risk controlling and the second line of defence.	GIA	–

15.11.1 Data and Systems

One of the strategically important factors for risk management is the data, processes and systems involved. The aim of this section is to define the corresponding processes and governance. In principle: the risk management function defines data requirements, which are either entered into a system or which is collected via ad hoc methods such as EXCEL. It is key that the following governance principles apply:

1. Risk management defines the data needs and the methods to calculate results.
2. Markets and functions provide data, which is signed off for quality and accountability purposes. It needs to be kept in mind that some of the data will be used for external purposes such as for regulators, rating agencies and shareholders.
3. Risk management will provide analyses based on this data.

It is the aim to use systems for data which is collected in a granular form on a regular basis such as asset and liability information. Therefore the risk function will be responsible for running some of the systems which must adhere to professional standards.

Data collected for the aim of risk analysis, limit and concentration checks and for risk adjusted performance analysis, the risk function is the data owner. Once the input is checked and approved by the markets and functions, it serves as master data on which the various calculations are performed. The risk function will provide the markets and functions with standardised reports.

Appendix A
Stochastic Processes

A.1 Definitions

In this section we will introduce the definitions which we will use throughout this book and we assume that the reader is familiar with elementary calculus, measure theory and probability theory.

Definition 1 (Sets) *In the following we denote with*

$$\begin{aligned}
\mathbb{N} &= \text{the set of all natural numbers including } 0,\\
\mathbb{N}_+ &= \{x \in \mathbb{N} : x > 0\},\\
\mathbb{R} &= \text{the set of real numbers},\\
\mathbb{R}_+ &= \{x \in \mathbb{R} : x \geq 0\}.
\end{aligned}$$

Furthermore we use the following notation for intervals. For $a, b \in \mathbb{R}, a < b$ we denote

$$\begin{aligned}
[a,b] &:= \{t \in \mathbb{R} : a \leq t \leq b\},\\
]a,b] &:= \{t \in \mathbb{R} : a < t \leq b\},\\
]a,b[&:= \{t \in \mathbb{R} : a < t < b\},\\
[a,b[\ k &:= \{t \in \mathbb{R} : a \leq t < b\}.
\end{aligned}$$

Definition 2 (Characteristic Function) *For a set $A \subset \Omega$ we denote $\chi_A : \Omega \to \mathbb{R}, \omega \mapsto \chi_A(\omega)$ the characteristic function, where*

$$\chi_A(\omega) := \begin{cases} 1, & \text{if } \omega \in A,\\ 0, & \text{if } \omega \notin A. \end{cases}$$

M. Koller, *Life Insurance Risk Management Essentials*, EAA Series,
DOI 10.1007/978-3-642-20721-1_16, © Springer-Verlag Berlin Heidelberg 2011

With δ_{ij} *we denote the Kronecker-Delta, which is equal to* 1*, if* $i = j$ *and* 0 *otherwise.*

Definition 3 *For a function*

$$f : \mathbb{R} \to \mathbb{R}, x \mapsto f(x)$$

we define the limit from below and from above, if they exist, as follows:

$$f(x^-) := \lim_{\xi \uparrow x} f(\xi),$$
$$f(x^+) := \lim_{\xi \downarrow x} f(\xi).$$

Definition 4 *A real valued function* $f : \mathbb{R} \to \mathbb{R}$ *is of order* $o(t)$*, if*

$$\lim_{t \to 0} \frac{f(t)}{t} = 0.$$

In this case we write $f(t) = o(t)$*.*

Definition 5 (Functions with Bounded Variation) *Let* $I \subset \mathbb{R}$ *be a finite interval. The total variation for a function* f

$$f : I \to \mathbb{C}, t \mapsto f(t)$$

with respect to the interval I *is given by*

$$V(f, I) = \sup \sum_{i=1}^{n} |f(b_i) - f(a_i)|,$$

where the supremum is taken over all decompositions of the interval I*, with*

$$a_1 \leq b_1 \leq a_2 \leq b_2 \leq \ldots \leq a_n \leq b_n.$$

The function f *has a* bounded variation *on* I*, if* $V(f, I)$ *is finite.*

Properties of functions with bounded variation can be found in [DS57].

It is important to know that functions with bounded variation are both an algebra and a lattice. Hence for f, g functions with bounded variation and $\alpha \in \mathbb{R}$ the following functions have also a bounded variation: $\alpha f + g$, $f \times g$, $\min(0, f)$ and $\max(0, f)$.

Definition 6 (Probability spaces, stochastic processes) *By* $(\Omega, \mathcal{A}, P)$ *we denote always a probability space which fulfils the axioms of Kolmogorov.*

Let $(S, \mathcal{S})$ be a measurable space (eg S a set and $\mathcal{S}$ a σ-algebra over S) and T a set. We denote by $\mathcal{R} = \sigma(\mathbb{R})$ the σ-algebra over the Borel set of the real numbers.

A family $\{X_t : t \in T\}$ of random variables

$$X_t : (\Omega, \mathcal{A}, P) \to (S, \mathcal{S}), \omega \mapsto X_t(\omega)$$

is a stochastic process *over $(\Omega, \mathcal{A}, P)$ with* State space S.

For every $\omega \in \Omega$ we define by

$$X_.(\omega) : T \to S, t \mapsto X_t(\omega)$$

the corresponding trajectory. *We assume that these trajectories are right continuous and that the left side limit exists.*

Definition 7 (Expected Value) *For a random variable X on $(\Omega, \mathcal{A}, P)$ and $\mathcal{B} \subset \mathcal{A}$ a σ-algebra we denote:*

- $\mathbb{E}[X]$ *the* expected value *of the random variable X,*
- $Var[X]$ *the* variance *of the random variable X,*
- $\mathbb{E}[X|\mathcal{B}]$ *the* conditional expected value *of X with respect to $\mathcal{B}$.*

Definition 8 *For a stochastic process $(X_t)_{t \in T}$ on $(\Omega, \mathcal{A}, P)$ with values in a countable set S and $i \in S$ we define*

$$I_j(t)(\omega) = \begin{cases} 1, & falls \;\; X_t(\omega) = j, \\ 0, & falls \;\; X_t(\omega) \neq j \end{cases}$$

the indicator functions with respect to the stochastic process $(X_t)_{t \in T}$ at time t.

Analogously we define for $j, k \in S$ with

$$N_{jk}(t)(\omega) = \#\{\tau \in]0, t[: X_{\tau^-} = j \text{ and } X_\tau = k\}$$

the number of jumps from j to k during the time interval $]0, t[$.

Definition 9 (Normal distribution) *A random variable X on $(\mathbb{R}, \sigma(\mathbb{R}))$ with probability density function*

$$f_{\mu,\sigma^2}(x) = \frac{1}{\sqrt{2\pi\sigma^2}} \exp\left(-\frac{(x-\mu)^2}{2\sigma^2}\right), \qquad x \in \mathbb{R}$$

is called normally distributed *with expected value μ and variance σ^2. We denote $X \sim \mathcal{N}(\mu, \sigma^2)$.*

Examples of stochastic processes are:

Example 10 (Brownian motion) *An example of a non-trivial stochastic process is the Brownian motion. This process* $X = (X_t)_{t \geq 0}$ *in continuous time* ($T = \mathbb{R}_+$) *with state space* $S = \mathbb{R}$ *is used for describing many natural phenomena.*

The Brownian motion can be characterised by the following properties:

1. $X_0 = 0$ *almost surely.*
2. X *has independent increments: For all* $0 \leq t_1 < t_2 < \ldots < t_n$ *and all* $n \in \mathbb{N}$ *we know that the random variables:* $B_{t_1} - B_{t_0}, B_{t_2} - B_{t_1}, \ldots, B_{t_n} - B_{t_{n-1}}$ *are independent.*
3. X *has stationary increments.*
4. $X_t \sim \mathcal{N}(0, t)$.

One can show that X *has almost surely continuous paths, which are nowhere differentiable.*

Example 11 (Poisson process) *The Poisson process* $N = (N_t)_{t \geq 0}$ *is a* count process *with state space* $\mathbb{N}$, *which is used for example for the modelling of number of claims within an insurance company. The time homogeneous poison process can be characterised by the following properties:*

1. $N_0 = 0$ *almost surely.*
2. N *has independent, stationary increments.*
3. *For all* $t > 0$ *and all* $k \in \mathbb{N}$ *the following property holds:* $P[N_t = k] = \exp(-\lambda t) \frac{(\lambda t)^k}{k!}$.

A.2 Markov Chains with Countable State Space

In the following we denote by S a countable set.

Definition 12 *Let* $(X_t)_{t \in T}$ *be a stochastic process over* $(\Omega, \mathcal{A}, P)$ *with state space* S *and* $T \subset \mathbb{R}$. *The process* X *is called a Markov chain, if for all*

$$n \geq 1, t_1 < t_2 < \ldots < t_{n+1} \in T, i_1, i_2, \ldots, i_{n+1} \in S$$

with

$$P[X_{t_1} = i_1, X_{t_2} = i_2, \ldots, X_{t_n} = i_n] > 0$$

the following equation holds:

$$P[X_{t_{n+1}} = i_{n+1} | X_{t_k} = i_k \forall k \leq n] = P[X_{t_{n+1}} = i_{n+1} | X_{t_n} = i_n]. \quad \text{(A.1)}$$

Definition 13 *Let $(X_t)_{t \in T}$ be a stochastic process over $(\Omega, \mathcal{A}, P)$. In this case we denote*

$$p_{ij}(s,t) := P[X_t = j \,|\, X_s = i], \qquad \text{where } s \le t \text{ and } i,j \in S,$$

the conditional probability, to change from time s to t from state i to state j.

The Chapman-Kolmogorov-theorem states the relationship of $P(s,t)$, $P(t,u)$ and $P(s,u)$ for $s \le t \le u$:

Theorem 14 (Chapman-Kolmogorov-equation) *Let $(X_t)_{t \in T}$ be a Markov chain and let $s \le t \le u \in T$, $i,k \in S$ with $P[X_s = i] > 0$. Then we have the following equations:*

$$p_{ik}(s,u) = \sum_{j \in S} p_{ij}(s,t)\, p_{jk}(t,u), \tag{A.2}$$

$$P(s,u) = P(s,t) \times P(t,u). \tag{A.3}$$

Hence we can calculate $P(s,u)$ for $s \le t \le u \in T$ by matrix multiplication of $P(s,t)$ and $P(t,u)$.

Proof. For $t = s$ or $t = u$ the equation is obviously true and hence we can assume that $s < t < u$. We denote by:

$$\begin{aligned} S^* &= \{j \in S : P[X_t = j \,|\, X_s = i] \neq 0\} \\ &= \{j \in S : P[X_t = j\,,\, X_s = i] \neq 0\}\,. \end{aligned}$$

The Chapman-Kolmogorov-equation can be proved by the use of the following equations:

$$\begin{aligned} p_{ik}(s,u) &= P[X_u = k \,|\, X_s = i] \\ &= \sum_{j \in S^*} P[X_u = k,\, X_t = j \,|\, X_s = i] \\ &= \sum_{j \in S^*} P[X_t = j \,|\, X_s = i] \times P[X_u = k \,|\, X_s = i,\, X_t = j] \\ &= \sum_{j \in S^*} p_{ij}(s,t) \times p_{jk}(t,u) \\ &= \sum_{j \in S} p_{ij}(s,t) \times p_{jk}(t,u), \end{aligned}$$

where we have used the Markov property.

Definition 15 (Transition matrix) *A family $p_{ij}(s,t)$ is called transition matrix if the following four conditions are fulfilled:*

1. $p_{ij}(s,t) \geq 0$.
2. $\sum\limits_{j \in S} p_{ij}(s,t) = 1$.
3. $p_{ij}(s,s) = \begin{cases} 1, \text{ if } i = j, \\ 0, \text{ if } i \neq j, \end{cases} \quad \text{if } P[X_s = i] > 0.$
4. $p_{ik}(s,u) = \sum_{j \in S} p_{ij}(s,t)\, p_{jk}(t,u)$ *for* $s \leq t \leq u$ *and* $P\,[X_s = i] > 0$.

Proposition 16 *For a Markov chain* $(X_t)_{t \in T}$ *is* $p_{ij}(s,t)$ *is a transition matrix.*

Proof. This is a direct consequence of the Chapman-Kolmogorov-Theorem (Thm. 14).

Definition 17 *A Markov chain* $(X_t)_{t \in T}$ *is called* time homogeneous, *if for all* $s, t \in \mathbb{R}, h > 0$ *and* $i, j \in S$ *with* $P[X_s = i] > 0$ *and* $P[X_t = i] > 0$ *the following homogeneity condition is fulfilled:*

$$P[X_{s+h} = j \,|\, X_s = i] = P[X_{t+h} = j \,|\, X_t = i].$$

In this case we write:

$$\begin{aligned} p_{ij}(h) &:= p_{ij}(s, s+h), \\ P(h) &:= P(s, s+h). \end{aligned}$$

A.3 Mean Excess Function

The mean excess function for a random variable X is given by

$$e_X(x) = \mathbb{E}[X - x \,|\, X > x] = \frac{\int_x^\infty \{1 - F_X(\xi)\} d\xi}{1 - F_X(x)},$$

and that we have $e^\alpha(X) = e_X(VaR^\alpha(X))$.

For the convenience of the reader we have listed below some mean excess functions for different probability distributions:

Normal Distribution: 1. Probability density function:

$$f_X(x) = \frac{1}{\sqrt{2\pi}\sigma} \exp(-\frac{(x-\mu)^2}{2\sigma^2}).$$

2. Cumulative probability density function for standard normal distribution:

$$\Phi(x) = \frac{1}{\sqrt{2\pi}} \int_{-\infty}^{x} \exp(\frac{-\xi^2}{2}) d\xi.$$

3. Mean: $\mathbb{E}[X] = \mu$.
4. Variance: $Var[X] = \sigma^2$.
5. Mean excess function: For the standard normal distribution we have the following:

$$e(x) = \frac{\int_x^{\infty} (\xi - x) \frac{1}{\sqrt{2\pi}} \exp(-\frac{\xi^2}{2}) d\xi}{1 - \Phi(x)} = \frac{1}{\sqrt{2\pi}} \frac{\exp(-\frac{x^2}{2})}{1 - \Phi(x)} - x.$$

We remark that based on the $e(x)$ for the standard normal distribution we can easily calculate $e(x)$ for every normal distribution. In particular we can calculate the difference between VaR and TVaR for a normal distribution with standard deviation σ by $\sigma \times (\frac{1}{\sqrt{2\pi}} \frac{\exp(-\frac{x^2}{2})}{1-\Phi(x)} - x)$.

Log–Normal Distribution: 1. Probability density function:

$$f_X(x) = \frac{1}{\sqrt{2\pi}\sigma\, x} \exp(-\frac{(\log(x) - \mu)^2}{2\sigma^2}), \text{ for } x > 0.$$

2. Mean: $\mathbb{E}[X] = \exp(\mu + \frac{\sigma^2}{2})$.
3. Variance: $Var[X] = (\exp(\sigma^2) - 1)\exp(2\mu + \sigma^2)$.
4. Mean excess function:

$$e_X(x) = \frac{\exp(\mu + \frac{\sigma^2}{2}(1 - \Phi(x)(\frac{\log(x) - \mu - \sigma^2}{\sigma}))}{1 - \Phi\left(\frac{\log(x) - \mu}{\sigma}\right)} - x.$$

Exponential Distribution: 1. Probability density function: $f_X(x) = \beta \exp(-\beta\, x)$, for $x > 0$.

2. Mean: $\mathbb{E}[X] = \frac{1}{\beta}$.
3. Variance: $Var[X] = \frac{1}{\beta^2}$.
4. Mean excess function: $e_X(x) = \frac{1}{\beta}$.

Pareto Distribution: 1. Probability density function: $f_X(x) = \frac{\alpha\, \lambda^\alpha}{(\lambda + x)^{\alpha+1}}$, for $x > 0$.

2. Mean: $\mathbb{E}[X] = \frac{\lambda}{\alpha-1}$. Note exists only if $\alpha > 1$.
3. Variance: $Var[X] = \frac{\alpha\,\lambda^2}{(\alpha-1)^2(\alpha-2)}$. Note exists only if $\alpha > 2$.
4. Mean excess function: $e_X(x) = \frac{\lambda+x}{\alpha-1}$.

Gamma Distribution: 1. Probability density function: $F_X(x) = \beta(\beta\,x)^{\alpha-1}\frac{e^{-\beta\,x}}{\Gamma(\alpha)}$, for $x > 0$.

2. Mean: $\mathbb{E}[X] = \frac{\alpha}{\beta}$.
3. Variance: $Var[X] = \frac{\alpha}{\beta^2}$.
4. Mean excess function:

$$\begin{aligned} e_X(x) &= \frac{\alpha}{\beta} \times \frac{1 - F_X(x, \alpha+1, \beta)}{1 - F_X(x, \alpha+1, \beta)} \\ &= \frac{1}{\beta}(1 + o(1)), \end{aligned}$$

where $F_X(x, \alpha, \beta)$ denotes the cumulative probability density function of the gamma distribution with parameters α and β.

A.4 Deterministic Cash Flow Streams

Definition 18 (Payout function) *A deterministic payout function A is a function*

$$A : T \to \mathbb{R}, t \mapsto A(t),$$

with the following properties, with $T \subset \mathbb{R}$:

1. *A is right continuous,*
2. *A has bounded variation.*

We interpret $A(t)$ as the amount of money which has been paid until time t. One can show the following properties for functions with bounded variation [DS57]:

1. A function of bounded variation A can be extended to a measure on $\sigma(\mathbb{R})$, which we denote also with A. This measure is called *Stieltjes measure*.
2. For a function A on $\mathbb{R}$ with bounded variation there exist two positive, increasing functions with bounded variation and disjoint support such that $A = B - C$. We interpret B inflow of money and C as outflow of money.

Definition 19 (Decomposition of measures) *Let f be a function of bounded variation and denote with A the corresponding Stieltjes measure. In this case we define:*

$$\mu_f := A.$$

Since this decomposition into $A = B - C$ is unique (with B and C positive and with disjoint support), we define

$$A^+ := B,$$
$$A^- := C.$$

In order to define the mathematical reserves we need to introduce the discounting functions.

$$v(t) = \exp(-\int_0^t \delta(\tau)\, d\tau)$$

Now we can define the present value of a cash flow as follows

Definition 20 (Value of a cash flow) *Let A be a deterministic cash flows and $t \in \mathbb{R}$. In this case we define:*

1. *The value of a cash flow A at time t is defined by:*

$$V(t, A) := \frac{1}{v(t)} \int_0^\infty v(\tau)\, dA(\tau).$$

2. *The value of the future cash flow is given by*

$$V^+(t, A) := V(t, A \times \chi_{]t,\infty]}).$$

A.5 Random Cash Flows

Definition 21 (Random cash flow) *A* random cash flow *is a stochastic process $(X_t)_{t \in T}$, for which almost all paths are of finite variation.*

Consider now a stochastic process A with bounded variation and $\omega \in \Omega$ with $t \mapsto A_t(\omega)$ right continuous and increasing. For a bounded Borel function f we can now define the integral $\int f(\tau) d\mu_{A.(\omega)}(\tau)$ define. Analogously it is possible to define the integral $\int f(\tau, \omega) d\mu_{A.(\omega)}(\tau)$ P-a.e. for a bounded function $F_t = f(t, \omega)$, which is measurable with respect to the product sigma algebra. This construction can be extended to stochastic processes with bounded variation by using the decomposition of functions with bounded variation into its positive and negative parts.

Definition 22 *For a stochastic process $(A_t)_{t \in T}$ on $(\Omega, \mathcal{A}, P)$ with bounded variation and a product measurable, bounded function $F : \mathbb{R} \times \Omega \to \mathbb{R}$ we define as described above:*

$$(F \cdot A)_t(\omega) = \int_0^t F(\tau, \omega) dA_\tau(\omega) = \int_0^t F \, dA,$$

and we write

$$d(F \cdot A) = F \, dA.$$

Now we can define the present value of a cash flow as follows

Definition 23 (Value of a random cash flow) *Let A be a stochastic cash flows and $t \in \mathbb{R}$. In this case we define:*

1. *The expected value of a cash flow A at time t is defined by:*

$$V(t, A) := \mathbb{E}[\frac{1}{v(t)} \int_0^\infty v(\tau) \, dA(\tau)].$$

2. *The value of the future cash flow is given by*

$$V^+(t, A) := \mathbb{E}[V(t, A \times \chi_{]t,\infty]})].$$

Appendix B
Application of the Markov Model to Life Insurance

B.1 Traditional Rating of Life Contracts

Before starting with the Markov model, I would like to summarise how traditional calculations using commutation functions are performed. Usually one starts with the probabilities of death and then calculates a decrement table starting with, say, 100000 persons at age 20.

After that one, has to calculate the different commutation functions, which I assume everybody knows by heart. These numbers depend on the persons alive and on the technical interest rate i. Only when you have done this it is (in the classical framework) possible to calculate the necessary premiums. In the following we will look a little bit closer at the calculation of a single premium for an annuity. To do this we need the following commutation functions:

$$D_x = v \times l_x \text{where } l_x \text{ denotes the number of persons alive at age } x.$$
$$C_x = v \times (l_{x+1} - l_x)$$

Having this formalism it is well known that

$$\ddot{a}_x = \frac{N_x}{D_x}$$

From this example is easily seen that almost all premiums can be calculated by summation and multiplication of commutation functions. Such an approach has its advantages in an environment where calculations have to be performed by hand, or where computers are expensive. Calculation becomes messy if benefits are considered with guarantees or with refunds.

M. Koller, *Life Insurance Risk Management Essentials*, EAA Series, 271
DOI 10.1007/978-3-642-20721-1_17, © Springer-Verlag Berlin Heidelberg 2011

The Markov model here presented offers rating of life contracts without using commutation functions. It starts with calculation of the reserves and uses the involved probabilities directly. In order to see such a calculation let's review the abovementioned example: We will use ${}_np_x$ to denote the probability of a person aged exactly x surviving for n years.

$$\begin{aligned}\ddot{a}_x &= \sum_{j=0}^{\infty} {}_jp_x \times v^j \\ &= 1 + p_x \times \ddot{a}_{x+1}\end{aligned}$$

The above formula gives us a recursion for the mathematical reserves of the contract. Hence one can calculate the necessary single premiums just by recursion. In order to do this, we need an initial condition, which is in our case $V_\omega = 0$.

The interpretation of the formula is easy: The necessary reserve at age x consists of two parts:

1. The annuity payment, and
2. The necessary reserve at age x+1. (These reserves must naturally be discounted.)

It should be pointed out that the calculation does not need any of the commutation functions; only p_x and the discount factor v are used. As a consequence this method does not produce the overheads of traditional methods.

In the following paragraphs the discrete time, discrete state Markov model is introduced and solutions of some concrete problems are offered.

At this point, it is necessary to stress the fact that the following frame work can be used, with some modifications, in an environment with stochastic interest. But as we are limited in space and time we have to restrict ourselves to deterministic constant discount rates.

B.2 Life Insurance Considered as Random Cash Flows

The starting point of the Markov model is a set of states, which correspond to the different possible conditions of the insured persons. In life insurance the set of states usually consists of alive, dead. The set of states will be denoted by S.

The second point which originates from the life contract has to do with the so-called contractual functions which depend on the states and the time. Hence the structure of a generalised life contract can be thought of:

Contractual situation between time t and time $t + 1$

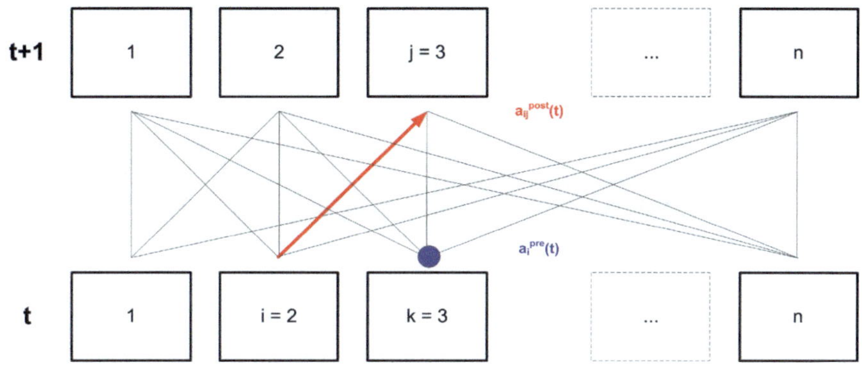

From the above diagram it can be seen that a finite number of states is considered, and that for

each transition $i \rightarrow j$ two different sums are paid, namely $a_{ij}^{\text{Post}}(t)$ at the end of the considered time interval and $a_i^{\text{Pre}}(t)$ at the beginning of it. It is clear that the value of the payment stream $a_{ij}^{\text{Post}}(t)$ has to be discounted by v in order to be compatible with $a_i^{\text{Pre}}(t)$. Probably it is worth remarking that the use of the two payment streams $a_i^{\text{Pre}}(t)$ and $a_{ij}^{\text{Post}}(t)$ eases the solution of things like payments during the year and the distinction between lump

sums (generally payable at the end of the period) and annuities (at the beginning). Finally it must be said that premiums payable to the insurer can (not must (!)) be considered as benefits with the opposite sign.

Until now we have defined the sums which are payable if a certain insured event occurs. Now there has to be a probability law in order to rate the different transitions. In the following we denote by $p_{ij}(t, t+1)$ the probability of transition at time t from state $i \rightarrow j$. Hence in the language of the above diagram there is one transition probability assigned to each line between two states.

So summarising a Markov life insurance model consists of the following:

S	A finite state space (set).
$((p_{ij}(t))_{(i,j)\in S^2})_{t\in(1,2,\dots\omega)}$	The transition probabilities describing the Markov chain X_t on S.
$((a_i^{\text{Pre}}(t))_{i\in S})_{t\in(1,2,\dots\omega)}$	The prenumerando benefits relating, paying at the beginning of the corresponding period.
$((a_{ij}^{\text{Post}}(t))_{(i,j)\in S^2})_{t\in(1,2,\dots\omega)}$	The postnumerando benefits relating, paying at the end of the corresponding period, if a transition $i \rightarrow j$ happens.
$((v_i(t))_{i\in S})_{t\in(1,2,\dots\omega)}$	The yearly discount rate from $[t, t+1[$. We have $v_t = \sum_{j\in S} I_j(t)\, v_i(t)$.

B.3 Reserves, Recursion and Premiums

One of the most important quantities in actuarial science is the prospective reserve, as the insurer must have this amount of money for each policy. Therefore the concept of the prospective reserve is known to all actuaries. It is defined to be the present value of the future cash flow A given the information at present. Formally we write

$$V_j^+(t,A) := \mathbb{E}[V(t, A \times \chi_{]t,\infty]}) \,|\, X_t = j],$$

(where j denotes the state at time t). This notation tells us, that the reserve depends heavily on the state of the policy.

In the context of the above we have

$$\begin{aligned}
\Delta A(t) &= \sum_{j\in S} I_j(t) \times a_i^{\mathrm{Pre}}(t) + \sum_{(i,j)\in S\times S} \Delta N_{ij}(t) \times a_{ij}^{\mathrm{Pre}}(t),\\
A(t) &= \sum_{k\leq t} \Delta A(k),\\
\Delta V(t,A) &= v(t)\Delta A(t),\\
&= v(t)\left[\sum_{j\in S} I_j(t) \times a_i^{\mathrm{Pre}}(t) + \sum_{(i,j)\in S\times S} \Delta N_{ij}(t) \times a_{ij}^{\mathrm{Pre}}(t)\right],\\
v(t) &= \prod_{\tau\leq t}\left[\sum_{j\in S} I_j(\tau) \times v_j(\tau)\right].
\end{aligned}$$

The direct calculation of the necessary reserves for the different states is not too easy if you consider a general time continuous Markov model. An advantage of this model is the existence of a powerful backwards recursion. The following formula (Thiele difference equation) allows the recursive calculation of the necessary reserves and hence of the necessary single premiums:

$$V_i^+(t) = a_i^{Pre}(t) + \sum_{j\in S} v_i(t)\, p_{ij}(t) \{a_{ij}^{Post}(t) + V_j^+(t+1)\}. \tag{B.1}$$

The interpretation of the formula is almost the same as in the trivial example at the beginning. In principle the present reserve consists of payments due to the different possible transitions and the discounted values of the future necessary reserves. It can be seen that the above recursion uses only the different benefits, the probabilities and the discount factor. In order to calculate the reserve for a certain age one has to do a backwards recursion starting at the expiration date of the policy. For annuities this is usually the age ω when everybody has died. Starting the recursion it is necessary to have boundary conditions, which depend on the payment stream at the expiration date. Usually the boundary conditions are taken to be zero for all reserves. It should be pointed out that one has to do this recursion for the reserves of all states simultaneously.

After the calculation of the different reserves one can naturally determine the corresponding necessary single premiums by the principle of equivalence.

We want to end this section with a short proof of the above mentioned Thiele recursion:

We know that $A(t) = \sum_{k \le t} \Delta A(k)$ and also that

$$\Delta V(t, A) = v(t) \left[\sum_{j \in S} I_j(t) \times a_i^{\mathrm{Pre}}(t) + \sum_{(i,j) \in S \times S} \Delta N_{ij}(t) \times a_{ij}^{\mathrm{Pre}}(t) \right].$$

Hence we have

$$\begin{aligned} V_i^+(t) &= \frac{1}{v(t)} \mathbb{E} \left[\sum_{\tau=t}^{\infty} v(\tau) \times \Delta A(\tau) \,|\, X_t = i \right] \\ &= \frac{1}{v(t)} \mathbb{E} \left[\sum_{j \in S} I_j(t+1) \times \sum_{\tau=t}^{\infty} v(\tau) \times \Delta A(\tau) \,|\, X_t = i \right], \end{aligned}$$

remarking that $\sum_{j \in S} I_j(t+1) = 1$. If we now consider all the terms in $\Delta A(t)$ for a given $I_j(t+1)$ for $j \in S$, it becomes obvious that the Markov chain changes from $i \to j$ and in consequence only $N_{ik}(t)$ increases by one for $k = j$. If we furthermore use the projection property and the linearity of the conditional expected value and the fact that $\mathbb{E}[I_j(t+1) \,|\, X_t = i] = p_{ij}(t, t+1)$, together with the Markov property, we get the formula if we split $V_i^+(t)$ as follows:

$$\begin{aligned} V_i^+(t) &= \frac{1}{v(t)} \mathbb{E} \left[\sum_{\tau=t}^{\infty} v(\tau) \times \Delta A(\tau) \,|\, X_t = i \right] \\ &= \frac{1}{v(t)} \mathbb{E} \left[\left\{ \sum_{\tau=t}^{t} + \sum_{\tau=t+1}^{\infty} \right\} v(\tau) \times \Delta A(\tau) \,|\, X_t = i \right]. \end{aligned}$$

Doing this decomposition we get for the first part:

$$\text{Part}_1 = a_i^{Pre}(t) + \sum_{j \in S} v_i(t)\, p_{ij}(t) a_{ij}^{Post}(t),$$

and for the second:

$$\text{Part}_2 = \sum_{j \in S} v_i(t)\, p_{ij}(t) V_j^+(t+1).$$

Adding the two parts together we get the desired result:

$$V_i^+(t) = a_i^{Pre}(t) + \sum_{j \in S} v_i(t)\, p_{ij}(t) \{a_{ij}^{Post}(t) + V_j^+(t+1)\}.$$

More concretely we have

$$\begin{aligned}
V_i^+(t) &= \frac{1}{v(t)} \mathbb{E}\left[\sum_{j \in S} I_j(t+1) \times \sum_{\tau=t}^{\infty} v(\tau) \times \Delta A(\tau) \,|\, X_t = i\right] \\
&= a_i^{Pre}(t) + \sum_{j \in S} \mathbb{E}\left[I_j(t+1) \times \sum_{\tau=t}^{\infty} \frac{v(\tau)}{v(t)} \times \Delta A(\tau) \,|\, X_t = i\right] \\
&= a_i^{Pre}(t) + \sum_{j \in S} \mathbb{E}\Bigg[I_j(t+1) v_i(t) \Bigg\{ a_{ij}^{Post} + \\
&\quad + \mathbb{E}\left[\sum_{\tau=t+1}^{\infty} \frac{v(\tau)}{v(t+1)} \times \Delta A(\tau) \,\middle|\, X_t = i, X_{t+1} = j\right]\Bigg\}\Bigg| X_t = i\Bigg] \\
&= a_i^{Pre}(t) + \sum_{j \in S} v_i(t)\, p_{ij}(t) \{a_{ij}^{Post}(t) + V_j^+(t+1)\}.
\end{aligned}$$

We remark that this section can only be a short introduction to this topic and we refer to [Kol10] for a more extensive discussion.

Appendix C
Abstract Valuation

This appendix follows closely [Kol10] and creates a link between the Markov chain model for life insurance on the one hand and abstract valuation and the concepts used in this book. Furthermore it aims to explain the concept of replicating portfolios and the cost of capital approach in a more general and abstract manner. We assume that the reader is familiar with elementary functional analysis such as Hilbert spaces and we refer to [DS57], [Con91] or [Ped89] for the corresponding mathematical proofs. Finally it is worth mentioning that [Duf92] covers the theoretical approach in some greater detail and we would encourage everybody to deepen its know-how with respect to this topic.

C.1 Framework

Definition 24 (Stochastic Cash Flows) *A* stochastic cash flow *is a sequence* $x = (x_k)_{k\in\mathbb{N}} \in L^2(\Omega, \mathcal{A}, P)^{\mathbb{N}}$, *which is* $\mathbb{F} = (\mathcal{F}_t)_{t\geq 0}$ *adapted.*

Definition 25 (Regular Stochastic Cash Flows) *A* regular stochastic cash flow x with respect to $(\alpha_k)_{k\in\mathbb{N}}$, *with* $\alpha_k > 0 \forall k$ *is a stochastic cash flow such that*

$$Y := \sum_{k\in\mathbb{N}} \alpha_k X_k \in L^2(\Omega, \mathcal{A}, P).$$

We denote the vector space of all regular cash flows by $\mathcal{X}$.

Remark 26 *1. We note that for all* $n \in \mathbb{N}$ *the image of* $\psi : L^2(\Omega, \mathcal{A}, P)^n \to \mathcal{X}, (x_k)_{k=0,\ldots n} \mapsto (x_0, x_1, \ldots, x_n, 0, 0 \ldots)$ *is a sub-space of* $\mathcal{X}$.

2. $\mathcal{X}$ *has been defined this way in order to capture cash flow streams where the sum of the cash flows is infinite with a finite present value. In this set up* α_k *can be interpreted as a majorant of the price of the payment* 1 *at time* k.

M. Koller, *Life Insurance Risk Management Essentials*, EAA Series,
DOI 10.1007/978-3-642-20721-1_18, © Springer-Verlag Berlin Heidelberg 2011

Proposition 27 *1. For $x, y \in \mathcal{X}$, we define the scalar product as follows:*

$$< x, y > = \sum_{k \in \mathbb{N}} < \alpha_k\, x_k, \alpha_k\, y_k > = \mathbb{E}[\sum_{k \in \mathbb{N}} \alpha_k^2\, x_k\, y_k],$$

and remark that the scalar product exists as a consequence of the Cauchy-Schwarz inequality.

2. $\mathcal{X}$ equipped with the above defined scalar product is a Hilbert space with norm $||x|| = \sqrt{< x, x >}$.

Proof. We leave the proof of this proposition to the reader.

In a next step we introduce the concept of a positive valuation functional and we closely follow [Büh95].

Definition 28 (Positivity) *1. $x = (x_k)_{k \in \mathbb{N}} \in \mathcal{X}$ is called* positive *if $x_k > 0$ P-a.e. for all $k \in \mathbb{N}$. In this case we write $x \geq 0$.*

2. $x = (x_k)_{k \in \mathbb{N}} \in \mathcal{X}$ is called strictly positive *if $x_k > 0$ P-a.e. for all $k \in \mathbb{N}$ and there exists a $k \in \mathbb{N}$, such that $x_k > 0$ with a positive probability. In this case we write $x > 0$.*

Definition 29 (Positive Functionals) $Q : \mathcal{X} \to \mathbb{R}$ *is called a* positive, continuous and linear functional *if the following hold true:*

1. *If $x > 0$, we have $Q[x] > 0$.*
2. *If $x = \lim_{n \to \infty} x_n$, for $x_n \in \mathcal{X}$ we have $Q[x] = \lim_{n \to \infty} Q[x_n]$.*
3. *For $x, y \in \mathcal{X}$ and $\alpha, \beta \in \mathbb{R}$ we have $Q[\alpha x + \beta y] = \alpha Q[x] + \beta Q[y]$.*

Remark 30 *1. We note that $Q \in \mathcal{X}'$ the dual space of $\mathcal{X}$ equipped with its canonical norm.*

2. Instead of L^2 we can also use the Hilbert space L^p, remarking that the dual of L^p can be identified with L^q with $\frac{1}{p} + \frac{1}{q} = 1$.

Theorem 31 (Riesz representation theorem) *For Q a positive, linear functional as defined before, there exists $\phi \in \mathcal{X}$, such that*

$$Q[y] = < \phi, y > \forall y \in \mathcal{X}.$$

Proof. This is a direct consequence of Riesz representation theorem of continuous linear functionals of Hilbert spaces.

Definition 32 (Deflator) *The $\phi \in \mathcal{X}$ generating $Q[\bullet]$ is called* deflator.

Proposition 33 *For a positive functional $Q : \mathcal{X} \to \mathbb{R}$, with deflator $\psi \in \mathcal{X}$ we have the following:*

1. *$\phi_k > 0$ for all $k \in \mathbb{N}$.*
2. *ϕ is unique.*

Proof. 1. Assume $\phi_k = 0$ for some $k \in \mathbb{N}$. In this case we have $Q[(\delta_{kn})_{n\in\mathbb{N}}] = 0$ which is a contradiction.

2. Assume $Q[y] =< \phi, y >=< \phi^*, y >$ for all $y \in \mathcal{X}$. In this case we have $< \phi - \phi^*, y >= 0$, in particular for $y = \phi - \phi^*$. Hence we have $||\phi - \phi^*|| = 0$.

Definition 34 (Projections) *For $k \in \mathbb{N}$ we define the following projections:*

1. *$p_k : \mathcal{X} \to L^2(\Omega, \mathcal{A}, P), x = (x_n)_{n\in\mathbb{N}} \mapsto (\delta_{kn}\, x_n)_{n\in\mathbb{N}}$, the projection on the k-th coordinate.*
2. *$p_k^+ : \mathcal{X} \to L^2(\Omega, \mathcal{A}, P), x = (x_n)_{n\in\mathbb{N}} \mapsto (\chi_{k\leq n}\, x_n)_{n\in\mathbb{N}}$, the projection starting on the k-th coordinate.*

Remark 35 • *We remark that the both above defined projections are linear operators with norm ≤ 1.*

- *As a consequence of that we have for $x \geq 0$ with $x \in \mathcal{X}$ the following two relations:*

$$Q[p_k(x)] \leq Q[x],$$
$$Q[p_k^+(x)] \leq Q[x].$$

Definition 36 (Valuation at time t) *For $t \in \mathbb{N}$ we define the valuation of $x \in \mathcal{X}$ at time t by*

$$Q_t[x] = Q[x|\mathcal{F}_t] = \frac{1}{\phi_t}\, \mathbb{E}[\sum_{k=0}^{\infty} \phi_k\, x_k|\mathcal{F}_t].$$

In the same sense as for mathematical reserves we define the value of the future cash flows at time t by

$$Q_t^+[x] = Q[p_t^+(x)].$$

Definition 37 (Zero Coupon Bonds) *The zero coupon bond $\mathcal{Z}_{(k)} = (\delta_{kn})_{n\in\mathbb{N}}$ is an element of $\mathcal{X}$. We remark that*

$$\pi_0\left(\mathcal{Z}_{(t)}\right) = Q[\mathcal{Z}_{(t)}] = \mathbb{E}[\phi_t].$$

Definition 38 *The cash flow $x = (x_k)_{k \in \mathbb{N}}$ in the discrete Markov model (cf. appendix B) is given by:*

$$x_k = \sum_{(i,j) \in S^2} \Delta N_{ij}(k-1) a_{ij}^{Post}(k-1) + \sum_{i \in S} I_i(k) a_i^{Pre}(k),$$

where we assume that $\Delta N_{ij}(-1) = 0$.

Proposition 39 *For $x \in \mathcal{X}$, as defined above we have the following:*

1. $\mathbb{E}[\Delta N_{ij}(s) | X_t = k] = p_{ki}(t,s) p_{ij}(s, s+1)$,
2. $\mathbb{E}[I_i(s) | X_t = k] = p_{ki}(t,s)$,
3. $\mathbb{E}[x_s | X_t = k] =$

$$\sum_{(i,j) \in S^2} p_{ki}(t, s-1) p_{ij}(s-1, s) a_{ij}^{Post}(s-1) + \sum_{i \in S} p_{ki}(t,s) a_i^{Pre}(s),$$

where we assume that $p_{ki}(t, s-1) = 0$ if $t \geq s$.

Proof. We leave the proof of this proposition to the reader as an exercise.

Definition 40 *The abstract vector space of financial instruments we denote by $\mathcal{Y}$. Elements of this vector space are for example all zero coupon bonds, shares, options on shares etc.*

Remark 41
- *We remark we can canonically embed $y \in \mathcal{Y}$ in $\mathcal{X}$, by means of its corresponding cash flows $(\xi(y)_k)_{k \in \mathbb{N}}$. Hence applying $Q[\bullet]$ to $y \in \mathcal{Y}$, is a shortcut for $Q[\xi(y)]$.*
- *Link to the arbitrage free pricing theory: If we assume that Q does not allow arbitrage, see appendix D. In proposition 83 we can see that $\pi(X) = \mathbb{E}^Q[\beta_T X]$, where β_T denotes the risk free discount rate. In the context of the above, we would have $\pi_0(x) = Q[x] = \mathbb{E}^P[\phi_T x]$. Hence we can identify $\phi_T = \frac{dQ}{dP} \beta_T$. In consequence we can interpret a deflator as a discounted Radon-Nikodym derivative with respect to the two measures P and Q.*

Proposition 42 *Let Q be a positive, continuous functional $Q : \mathcal{X} \to \mathbb{R}$, and assume $Q[\bullet] =< \phi, \bullet >$, with $\phi = (\phi_t)_{t \in \mathbb{N}}$ $\mathbb{F}$ – adapted. In this case $(\phi_t\, Q_t[x])_{t \in \mathbb{N}}$ is an $\mathbb{F}$-martingale over P.*

Proof. Since $\mathcal{F}_t \subset \mathcal{F}_{t+1}$ and the projection property of the conditional expectation we have

$$\mathbb{E}^P[\phi_{t+1}\, Q_{t+1}[x] | \mathcal{F}_t] = \mathbb{E}^P[\mathbb{E}^P[\sum_{k \in \mathbb{N}} \phi_k x_k\, Q_{t+1}[x] | \mathcal{F}_{t+1}] | \mathcal{F}_t]$$

$$= \mathbb{E}^P[\sum_{k \in \mathbb{N}} \phi_k x_k \, Q_{t+1}[x] | \mathcal{F}_t]$$
$$= \phi_t \, Q_t[x].$$

Example 43 (Replicating Portfolio Mortality) *In this first example we consider a term insurance, for a 50 year old man with a term of 10 years, and we assume that this policy is financed with a regular premium payment. Hence there are actually two different payment streams, namely the premium payment stream and the benefits payment stream. For sake of simplicity we assume that the yearly mortality is* $(1 + \frac{x-50}{10} \times 0.1)\%$. *We assume that the death benefit amounts to 100000 EUR and we assume that the premium has been determined with an interest rate* $i = 2\%$. *In this case the premium amounts to* $P = 1394.29$ *EUR. The replicating portfolio in the sense of expected cash flows at inception is therefore given as follows (cf proposition 39). We remark that the units have been valued with two (flat) yield curves with interest rates of 2% and 4% respectively, and that the use of arbitrary yield curves does not imply additional complexity.*

Age	*Unit*	*Units for Mortality*	*Units for Premium*	*Total Units*	*Value* $i = 2\%$	*Value* $i = 4\%$
50	$\mathcal{Z}_{(0)}$	–	-1394.28	-1394.28	-1394.28	-1394.28
51	$\mathcal{Z}_{(1)}$	1000.00	-1380.34	-380.34	-372.88	-365.71
52	$\mathcal{Z}_{(2)}$	1089.00	-1365.16	-276.16	-265.43	-255.32
53	$\mathcal{Z}_{(3)}$	1174.93	-1348.77	-173.84	-163.81	-154.54
54	$\mathcal{Z}_{(4)}$	1257.56	-1331.24	-73.67	-68.06	-62.97
55	$\mathcal{Z}_{(5)}$	1336.69	-1312.60	24.09	21.82	19.80
56	$\mathcal{Z}_{(6)}$	1412.12	-1292.91	119.20	105.85	94.21
57	$\mathcal{Z}_{(7)}$	1483.67	-1272.23	211.44	184.07	160.67
58	$\mathcal{Z}_{(8)}$	1551.18	-1250.60	300.57	256.54	219.62
59	$\mathcal{Z}_{(9)}$	1614.50	-1228.09	386.41	323.33	271.48
60	$\mathcal{Z}_{(10)}$	1673.52	–	1673.52	1372.87	1130.57
Total					**0.00**	**-336.47**

Exercise 44 (Replicating Portfolio Disability) *Consider a disability cover and calculate the replicating portfolios for a deferred disability annuity and a disability in payment.*

C.2 Cost of Capital

In section C.1 we have seen how to abstractly valuate $x \in \mathcal{X}$ by means of a pricing functional Q. For some financial instruments $y \in \mathcal{Y}^\star$ we can directly observe $Q[y]$ such as for a lot of zero coupons bonds $\mathcal{Z}_{(\bullet)}$. On the other hand this is not always possible.

Definition 45 *We denote by $\mathcal{Y}^\star$ the set of all financial instruments in $x \in \mathcal{Y}$ such that $Q[x]$ is observable. With $\tilde{\mathcal{Y}} = span < \mathcal{Y}^\star >$ we denote the vector space generated by $\mathcal{Y}^\star$ and we define:*

1. *$x \in \mathcal{Y}^\star$ is called of level 1.*
2. *$x \in \tilde{\mathcal{Y}}$ is called of level 2.*
3. *$x \in \mathcal{Y} \setminus \tilde{\mathcal{Y}}$ is called of level 3.*

Remark 46 *It is clear that the model uncertainty and the difficulties to value assets or liabilities increases from level 1 to level 3. Since we are interested in market values only the valuation of level 1 assets and liabilities is really reliable. For level 2 assets and liabilities on has to find a sequence of $x_n = \sum_{k=1}^n \alpha_k\, e_k$ with $e_k \in \mathcal{Y}^\star$ such that $x = \lim_{n\to\infty} x_n$. Since we assume that Q is linear and continuous we can calculate*

$$\begin{aligned} Q[x] &= \lim_{n\to\infty} Q[x_n] \\ &= \lim_{n\to\infty} \sum_{k=1}^{n} Q[\alpha_k\, e_k] \\ &= \lim_{n\to\infty} \sum_{k=1}^{n} \alpha_k\, Q[e_k]. \end{aligned}$$

For level 3 assets and liabilities the situation is even more difficult, since there is no obvious way to do it. The best, which we can be done is to define $\tilde{Q}[x]$ such that $\tilde{Q}[x] = Q[x] \forall x \in \mathcal{Y}^\star$ and hope that $\tilde{Q}[x] \approx Q[x]$ for the $x \in \mathcal{Y}$ we want to valuate. In most cases such $\tilde{Q}[\bullet]$ are based on first economic principles. In the following we want to see how the Cost of Capital *concept works for insurance liabilities and how we can concretely implement it.*

Definition 47 (Utility Assumption) *If we have $x, y \in L^2(\Omega, \mathcal{A}, P)^+$, with $x = \mathbb{E}[y]$. A rational investor would normally prefer x, since there is less uncertainty. The way to understand this, is by using utility functions. For $x \in L^2(\Omega, \mathcal{A}, P)^+$ and u a concave function, the utility of x is defined as $\mathbb{E}[u(x)]$. The idea behind utilities is that the first 10000 EUR are higher valued than the one 10000 EUR from 100000 EUR to 110000 EUR. Hence the increase of utility per fixed amount decreases if amounts increase. As a consequence of the Jensen-inequality, we see that the utility of a constant amount is higher than the utility of a random payout with the same expected value.*

Definition 48 *Let $x = (x_k)_{k\in\mathbb{N}} \in \mathcal{X}$ be an insurance cash flow, for example generated by a Markov model.*

1. *In this case we define the expected cash flows by*

$$CF(x) = (\mathbb{E}[x_k])_{k \in \mathbb{N}}.$$

2. *The corresponding portfolio of financial instruments in the vector space $\mathcal{Y}$ we define by*

$$VaPo^{CF}(x) = \sum_{k \in \mathbb{N}} CF(x)_k \mathcal{Z}_{(k)} \in \mathcal{Y}$$

3. *By $R(x)$ we denote the residual risk portfolio given by*

$$\begin{aligned} R(x) &= x - VaPo^{CF}(x) \\ &= \sum_{k \in \mathbb{N}} (x_k - CF(x)_k \, \mathcal{Z}_{(k)}) \in \mathcal{Y} \end{aligned}$$

4. *For a given $x \in \mathcal{X}$ we denote by $VaPo^*(x)$ an approximation $y \in \tilde{\mathcal{Y}}$ of x, such that $||x - VaPo^*(x)|| \leq ||x - VaPo^{CF}(x)||$. In analogy to $R(x)$, we can define $R^\star(x)$ with respect to $VaPo^*(x)$.*

Since we are sometimes interested in conditional expectations, we will also use the following notations for $A \in \mathcal{A}$:

$$\begin{aligned} CF(x \,|\, A) &= (\mathbb{E}[x_k \,|\, A])_{k \in \mathbb{N}}, \\ VaPo^{CF}(x \,|\, A) &= \sum_{k \in \mathbb{N}} CF(x \,|\, A)_k \mathcal{Z}_{(k)} \in \mathcal{Y}, \end{aligned}$$

Proposition 49 *The value of $x \in \mathcal{Y}$ can be decomposed in*

$$Q[x] = Q[VaPo^{CF}(x)] + Q[R(x)],$$

and we have

$$Q[VaPo^{CF}(x)] \geq Q[x]$$

if we use the utility assumption.

Remark 50 *1. We will denote $x \in \mathcal{X}$ with $x \leq 0$ as a liability. Proposition 49 hence tells us that we need to reserve more than $Q[VaPo^{CF}(x)]$ for this liability as a consequence of the corresponding uncertainty.*

2. *A risk measure is a functional (not necessarily linear) $\psi : \mathcal{X} \to \mathbb{R}$ which aims to measure the capital needs in an adverse scenario. There are two risk measures, which are commonly used the* Value at Risk *and the* Expected Shortfall *to a given quantile $\alpha \in \mathbb{R}$. The value at risk (VaR) is defined as the corresponding quantile minus the expected value. The expected shortfall is the conditional expectation of the random variable given a loss bigger than the corresponding loss, again minus the expected value. We can hence speak about a 99.5% VaR or a 99% expected*

shortfall. It is worthwhile to remark that these two concepts are normally applied to losses. Hence in the context introduced above one would strictly speaking calculating the VaR$(-x)$*, when considering* $x \in \mathcal{X}$*. Furthermore in a lot of applications, such as Solvency II, we assume that there is a Dirac measure (aka stress scenario), which just represents the corresponding VaR-level for example. So concretely the stress scenarios, which are used under Solvency II should in principle represent the corresponding point (Dirac) measures at to the confidence level 99.5 %. In the concrete set up, one would for example assume that* $q_x(\omega) \in L^2(\Omega, \mathcal{A}, P)$ *is a stochastic mortality and one would define the* $A, B \in \mathcal{A}$*, as the corresponding probabilities in the average and in the tail. In consequence for a policy* $x \in \mathcal{X}$*, we would have two replicating portfolios, namely* $VaPo^{CF}(x \mid A)$ *for the average and* $VaPo^{CF}(x \mid B)$ *for the stressed event according to the risk measure chosen. The corresponding required risk capital is then given (in present value terms) by* $Q[VaPo^{CF}(x \mid B) - VaPo^{CF}(x \mid A)]$*.*

3. *It is important to remark that the concept of cash flow representation of insurance policies* $x \in \mathcal{X}$ *makes particularly sense when the corresponding insurance cash flows are independent from the capital market induced stochastic variables. This is the case for non-profit products and also for annuities in payment without discretionary benefits. For other insurance products, such as classical with profits products or also GMDB types of covers, the cash flow representation* $VaPo^{CF}(x)$ *is not suited to represent* x*. In this case a replicating portfolio needs to take also into consideration the corresponding effects, as we will see in the following. In this set up one has to determine a suitable* $VaPo^{*}(x)$*–representation and in consequence use* $R^{\star}(x)$*. Also the cost of capital approach has in this case to be performed with respect to* $VaPo^{*}(x)$*. Having remarked this we will always the notation* $VaPo^{CF}(x)$ *even though that in certain of the above mentioned cased we actually mean a suitable* $VaPo^{*}(x)$*–representation also taking into consideration dependencies on the capital markets, mutatis mutandis.*

Definition 51 (Required Risk Capital) *For a risk measure* ψ_α *such as VaR or expected shortfall to a security level* α *we define the required risk capital at time* $t \in \mathbb{N}$ *by*

$$RC_t(x) = \psi_\alpha(p_k(x - VaPo^{CF}(x))).$$

Remark 52 1. *If we use* $VaR_{99.5\%}$ *the required risk capital at time* t *corresponds to the capital needed to withstand a 1 in 200 year event.*

2. *The definition above could apply to individual insurance policies, but is normally applied to insurance portfolios* $\tilde{x} = \sum_{k=1}^{n} x_k$*, where* $(x_k)_{k=1,\dots n}$ *are the individual insurance policies. As we can see in section 10.3 of [Kol10] the pure diversifiable risk disappears for* $n \to \infty$*.*

3. *What is more material than the diversifiable risk is the risk, which affects all of the individual insurance policies at the same time, such as a pandemic event,*

where the overall mortality could increase by 1 % in a certain year such as 1918 (see for example figure 3.3).

Definition 53 (Cost of Capital) *For a unit cost of capital $\beta \in \mathbb{R}^+$ and an insurance portfolio $\tilde{x} \in \mathcal{X}$, we define:*

1. *The present value of the required risk capital by*

$$PVC(\tilde{x}) = Q[\sum_{k \in \mathbb{N}} RC_t(\tilde{x}) \mathcal{Z}_{(k)}].$$

2. *The cost of capital $CoC(\tilde{x})$ is given by:*

$$CoC(\tilde{x}) = \beta \times PVC(\tilde{x}),$$

and $\tilde{Q}$ is defined by $\tilde{Q}[x] = Q[VaPo^{CF}(\tilde{x})] + \beta PVC(\tilde{x})$.

Remark 54 1. *The concept as defined before is somewhat simplified, since one normally assumes that the required capital C from the shareholder is $\alpha \times C$ after tax and investment income on capital. Assume a tax-rate κ and a risk-free yield of i. In this case we have*

$$\alpha \times C = i \times (1 - \kappa) \times C + \beta \times C,$$

and hence $\beta = \alpha - i \times (1 - \kappa)$. In reality the calculation can still become more complex since we discount future capital requirements risk-free and because of the fact that the interest rate i is not constant. In order to avoid these technicalities, we will assume for this book that i is constant.

2. *We remark $\tilde{Q}[\tilde{x}]$ is not uniquely determined, but depends on a lot of assumptions such as ψ_α, α, β, ...*
3. *For the moment we do not yet see how to actually model $\tilde{x}$ and we remark that one is normally focusing on the non-diversifiable part of the risks within $\tilde{x}$.*

Example 55 *We continue with example 43 and we assume that the risk capital is given by a pandemic event where $\Delta q_x = 1\%$ for all ages. This roughly corresponds to the increase in mortality of 1918 as a consequence of the Spanish flu pandemic. The aim of this example is to calculate the required risk capital and the market value of this policy based on the cost of capital method using $\beta = 6\%$. The required risk capital in this context can be calculated as $\Delta q_x \times 100000$ and we get the following results:*

Age	Unit	Units for Risk Capital	Units for Benefits	Total Units	$-\tilde{Q}[x]$ $i=2\%$	$-\tilde{Q}[x]$ $i=4\%$
50	$\mathcal{Z}_{(0)}$	1000.00	-1394.28	-1334.28	-1334.28	-1334.28
51	$\mathcal{Z}_{(1)}$	990.00	-380.34	-320.94	-314.65	-308.60
52	$\mathcal{Z}_{(2)}$	979.11	-276.16	-217.41	-208.97	-201.01
53	$\mathcal{Z}_{(3)}$	967.36	-173.84	-115.80	-109.12	-102.95
54	$\mathcal{Z}_{(4)}$	954.78	-73.67	-16.38	-15.14	-14.00
55	$\mathcal{Z}_{(5)}$	941.41	24.09	80.57	72.98	66.22
56	$\mathcal{Z}_{(6)}$	927.29	119.20	174.84	155.25	138.18
57	$\mathcal{Z}_{(7)}$	912.45	211.44	266.19	231.73	202.28
58	$\mathcal{Z}_{(8)}$	896.94	300.57	354.39	302.47	258.95
59	$\mathcal{Z}_{(9)}$	880.80	386.41	439.26	367.55	308.61
60	$\mathcal{Z}_{(10)}$	–	1673.52	1673.52	1372.87	1130.57
Total					**520.69**	**143.98**

We remark that the value of the policy at inception becomes positive, which means nothing else, that the insurance company does need equity capital to cover the economic loss. It is obvious that this is the case for $i = 2\%$, since the premium principle did not allow for a compensation of the risk capital. More interestingly even at the higher interest rate the compensating effect is not big enough to turn this policy into profitability.

Exercise 56 *In the same sense as for the mortality example calculate the respective risk capitals and $\tilde{Q}$ for a disability cover.*

Example 57 (GMDB mortality cover) *In this example we want to have a look at a unit linked insurance policy ($x \in \mathcal{X}$) where the death benefit amounts to the maximum of the fund value and of a fixed (minimal) death benefit (see also section D.2.2).*

We assume that the policyholder aged 40 invests 100000 EUR single premium in a fund $\mathcal{U}$ with a volatility $\sigma = 20\%$. The term of the policy is 10 years, the guaranteed mortality benefit 150000 EUR and we assume a flat yield curve of 2 %. The mortality follows example 43. We want to calculate the following things:

- *Calculate $VaPo^*(x)$ for this GMDB death benefit, where we assume that the policyholders die according to the given mortality law.*
- *Calculate the single premium of this GMDB cover (w/o CoC).*
- *Calculate some sensitivities for $Q[VaPo^*(x)]$, namely for a volatility of 30 % and for an interest rate of 1 %.*

In a first step we need to calculate the expected number of death people by ${}_tp_x\, q_{x+t}$. For this people we need to be able to sell the funds value S_t at a value of 150000 EUR. This represents a put option $\mathcal{P}_t \in \mathcal{Y}$ with a current funds value $S_0 = 100000$, maturity t and strike 150000. The price for a put-option *with payout $C(T,P) = \max(K - S; 0)$ at time t and strike price K and equity price S is given by:*

$$
\begin{aligned}
P &= K \times e^{-r \times T} \times \Phi(-d_2) - S_0 \times \Phi(-d_1), \\
d_1 &= \frac{\log(S_0/K) + (r + \sigma^2/2) \times T}{\sigma \times \sqrt{T}}, \\
d_2 &= d_1 - \sigma \times \sqrt{T}, \\
\Phi(x) &= \int_{-\infty}^{x} \frac{1}{\sqrt{2\pi}} \exp(\frac{-\zeta^2}{2}) d\zeta,
\end{aligned}
$$

and we get the following for $VaPo^*(x)$:

Age	*Unit*	*Expected Deaths*	*Units Death*	*Value Units*	*Value* $\sigma = 30\%$	*Value* $i = 1\%$
50	$\mathcal{P}_0$	*0.01000*	*0.01000*	–	–	–
51	$\mathcal{P}_1$	*0.01089*	*0.01089*	*472.78*	*487.26*	*487.26*
52	$\mathcal{P}_2$	*0.01174*	*0.01174*	*497.45*	*539.39*	*526.90*
53	$\mathcal{P}_3$	*0.01257*	*0.01257*	*524.96*	*592.78*	*569.82*
54	$\mathcal{P}_4$	*0.01336*	*0.01336*	*552.39*	*644.22*	*613.24*
55	$\mathcal{P}_5$	*0.01412*	*0.01412*	*578.51*	*692.85*	*655.91*
56	$\mathcal{P}_6$	*0.01483*	*0.01483*	*602.75*	*738.31*	*697.20*
57	$\mathcal{P}_7$	*0.01551*	*0.01551*	*624.83*	*780.41*	*736.70*
58	$\mathcal{P}_8$	*0.01614*	*0.01614*	*644.58*	*819.02*	*774.14*
59	$\mathcal{P}_9$	*0.01673*	*0.01673*	*661.93*	*854.08*	*809.32*
60	$\mathcal{P}_{10}$	–	–	*676.85*	*885.55*	*842.09*
Total				**5837.06**	**7033.92**	**6712.62**

We see here very clearly the dependency of the value of the replicating portfolio on the different parameters.

C.3 Inclusion in the Markov Model

In this section we want to have a look how we could concretely use the recursion technique for the calculation of the cost of capital in a Markov chain similar environment. In order to do that we look at an insurance policy with a term of one year.

We assume that we have a mortality of q_x in case of a "normal" year with a probability of $(1 - \alpha)$ and an excess mortality of Δq_x in an extreme year with probability α. We denote with $\Gamma = \frac{q_x + \Delta q_x}{q_x}$. Furthermore we assume a mortality benefit of 100000. In this case we get the following by some simple calculations:

$$
VaPo^{CF}(x) = (\delta_{1k}(q_x + \alpha(\Gamma - 1)q_x \times 100000))_{k \in \mathbb{N}},
$$

$$\begin{aligned} RC_1(x) &= (\delta_{1k}(1-\alpha)(\Gamma-1)q_x \times 100000)_{k\in\mathbb{N}}, \\ \tilde{Q}[x] &= Q[(\delta_{1k}(q_x + \alpha(\Gamma-1)q_x \times 100000 + \\ &\quad + \beta(1-\alpha)(\Gamma-1) \times 100000))_{k\in\mathbb{N}}]. \end{aligned}$$

We see that the price of this insurance policy with only payments at time 1 can be decomposed into a part representing best estimate mortality:

$$\delta_{1k}\{q_x\,(1+\alpha\,(\Gamma-1))\},$$

where we can arguably say that this $\tilde{q}_x = q_x\,(1+\alpha\,(\Gamma-1))$ is our actual best-estimate mortality. On top of that we get a charge for the excess mortality Δq_x with an additional cost of β. Hence we get the following:

1. There is a contribution to the reserve from the people surviving the year with a probability p_x.
2. There is a contribution to the reserve from the people dying in normal years with probability q_x and the defined benefit $a_{*\dagger}^{\text{post}}$, and
3. There is finally a contribution of the people dying in extreme years with probability Δq_x and the additional cost of defined benefit of $\beta \times a_{*\dagger}^{\text{post}}$.

The interesting fact is that we can actually use the same recursion of the reserves for the Markov chain model as in formula B.1 with 3 states and the exception that now the "transition probabilities" do not fulfil anymore the requirement that their sum equals 1. However this method provides a pragmatic way to implement the cost of capital in legacy admin systems.

The main problem for the determining of the corresponding Markov chain model is the underlying stochastic mortality model. For the QIS 5 longevity model a similar calculation can be used. In this model it is assumed that the mortality drops by 25 % in an extreme scenario. Hence the calculation goes along the following process:

1. Determine $x_1 = VaPo^{CF}(\tilde{x}|A)$ for standard mortality A.
2. Determine $x_2 = VaPo^{CF}(\tilde{x}|B)$ for stressed mortality B.
3. $\tilde{Q}[x] = Q[x_1] + \beta\, Q[x_2 - x_1]$

Example 58 *In this example we want to revisit the exercise 43 and we want again to calculate the market value of the insurance liability using the cost of capital approach, but this time with the recursion. We get the following results:*

Age	*Benefit Normal*	*Benefit Premium*	*Excess Risk*	*Math Res.* $i = 2\%$	*Value* $i = 2\%$	*Value* $i = 4\%$
50	*100000*	*-1394.28*	*6000*	**0.00**	**520.69**	**143.98**
51	*100000*	*-1394.28*	*6000*	*426.43*	*901.09*	*542.82*
52	*100000*	*-1394.28*	*6000*	*765.56*	*1193.21*	*861.67*
53	*100000*	*-1394.28*	*6000*	*1015.22*	*1394.79*	*1096.96*
54	*100000*	*-1394.28*	*6000*	*1172.95*	*1503.20*	*1244.68*
55	*100000*	*-1394.28*	*6000*	*1235.88*	*1515.45*	*1300.33*
56	*100000*	*-1394.28*	*6000*	*1200.79*	*1428.16*	*1258.89*
57	*100000*	*-1394.28*	*6000*	*1064.00*	*1237.49*	*1114.74*
58	*100000*	*-1394.28*	*6000*	*821.42*	*939.18*	*861.64*
59	*100000*	*-1394.28*	*6000*	*468.45*	*528.45*	*492.63*
60				*0*	*0*	*0*

We remark that this calculation is much faster to calculate since it is based on Thiele's difference equation for the mathematical reserves, and we get at the same time the corresponding results for the classical case and also for the case using the cost of capital approach.

As seen in the calculation above there is a small second order effect, which we can detect, when looking more closely. The results below correspond to the 2% *valuation:*

```
Direct Method 520.698380872792
Recursion 520.698380872793
```

Exercise 59 *Perform the corresponding calculation for the disability example.*

C.4 Asset Liability Management

Until now we have looked only at insurance liabilities as an $x \in \mathcal{Y}$. An insurance company needs to cover its insurance liabilities $l = \sum x_\iota \in \mathcal{X}$ with corresponding assets, which are also elements in $\mathcal{Y} \subset \mathcal{X}$.

Definition 60 (Assets and Liabilities) *An* $x \in \mathcal{X}$ *with a valuation functional* Q *is called*

1. *an asset if* $Q[x] \geq 0$ *and*
2. *a liability if* $Q[x] \leq 0$.

Remark 61 *At this point is important to have a closer look at the convention what is an asset and what is a liability and the corresponding signs of the cash flows. In* *a actuarial context payments from the insurance company to the policyholder have a "+" sign and premium payments a "−". Hence the whole appendix up to here is based on this convention. Now we are looking at an entire balance sheet and we will in consequence apply the respective accounting conventions, where assets have a "+" sign and liabilities a "−". Hence the reserves and stochastic cash flows as defined and calculated until now represent actually the* negative *liability in term of a balance sheet.*

Definition 62 (Insurance balance sheet) *An insurance balance sheet consists of a set of assets $(a_i)_{i \in I}$ and a set of liabilities $(l_j)_{j \in J}$. The equity of an insurance balance sheet is defined as*

$$e = \sum_{i \in I} a_i + \sum_{j \in J} l_j.$$

The insurance entity is called bankrupt if $Q[e] < 0$.

Definition 63 *In a regulated insurance market, each insurance company is required to hold an adequate amount of risk capital in order to absorb shocks. In order to do that, the regulator defines a risk measure ψ_α to a security level α. In this context an insurance company is called solvent if:*

$$Q[e] \geq \psi_\alpha(e).$$

Remark 64 *Note that an insurance regulator may not want to use a market consistent approach. Never the less the above definition can be used, be suitably adjust ψ.*

Definition 65 (Asset Liability Management) *Under asset liability management we understand the process of analysing $(l_j)_{j \in J}$ and the (dynamic) management of $(a_i)_{i \in I}$ in order to achieve certain target , such as remaining solvent.*

Definition 66 *For an insurance liability $l \in \mathcal{X}$ an asset portfolio $(a_i)_{i \in I}$ is called:*

1. Matching *if $\sum_{i \in I} a_i + l = 0$, and*
2. Cash flow matching *if $\sum_{i \in I} a_i + VaPo^{CF}(l) = 0$.*

Remark 67 *We remark that is normally not feasible to do a perfect matching, and hence one normally uses a cash flow matching to a achieve a proxy for a perfect match. We also remark that in this case the shareholder equity needs still be able to absorb the basis risk $l - VaPo^{CF}(l)$.*

Definition 68 (Macaulay Duration) *The duration for an $x \in \mathcal{Y}$ with*

$$x = \sum_{k \in \mathbb{N}} \alpha_k \, \mathcal{Z}_{(k)} \text{ and } \alpha_k \geq 0$$

is defined by

$$d(x) = \frac{Q[\sum_{k\in\mathbb{N}} \alpha_k \times k \times \mathcal{Z}_{(k)}]}{Q[\sum_{k\in\mathbb{N}} \alpha_k \times \mathcal{Z}_{(k)}]}$$

We say that an asset portfolio $(a_i)_{i\in I}$ is duration matching a liability l if the following two conditions are fulfilled:

1. $Q[\sum_{i\in I} a_i + l] = 0$, *and*
2. $d(\sum_{i\in I} a_i) = d(-VaPo^{CF}(l))$.

Example 69 *In this example we want to further elaborate on the example 43 and we want to see how the replicating scenario changes in case a pandemic occurs in year three, with an excess mortality of 1 %. We want also to have a look on what risk is implied in this, assuming that the pandemic at the same time leads to a reduction of interest rates down from 2% to 0 %. Finally we want to see an example how we could do a perfect cash flow matching portfolio.*

Definitions *We assume that $A \in \mathcal{A}$ represents the information that we have going to have average mortality after year 3 and three and that the person survived until then (year 2). In the same sense we assume that $B \in \mathcal{A}$ represents the same as A but with the exception that we assume a pandemic event in the year 3 with an average excess mortality of 1%. For simplicity reasons (to avoid notation) we use $x, y \in \mathcal{X}$ as abbreviations for the corresponding conditional random variables.*

Calculation of the Replicating Portfolios *In a first step we will calculate the replicating portfolios (starting at time 2) with respect to both A and B. Doing this we get the following results for case A:*

Age	*Unit*	*Units for Mortality*	*Units for Premium*	*Total Units*	*Value* $i = 2\%$	*Value* $i = 4\%$
52	$\mathcal{Z}_{(0)}$	–	-1394.28	-1394.28	-1394.28	-1394.28
53	$\mathcal{Z}_{(1)}$	1200.00	-1377.55	-177.55	-174.07	-170.72
54	$\mathcal{Z}_{(2)}$	1284.40	-1359.64	-75.24	-72.32	-69.57
55	$\mathcal{Z}_{(3)}$	1365.21	-1340.61	24.60	23.18	21.87
56	$\mathcal{Z}_{(4)}$	1442.25	-1320.50	121.75	112.48	104.07
57	$\mathcal{Z}_{(5)}$	1515.33	-1299.37	215.95	195.59	177.49
58	$\mathcal{Z}_{(6)}$	1584.27	-1277.28	306.99	272.59	242.61
59	$\mathcal{Z}_{(7)}$	1648.95	-1254.29	394.65	343.57	299.90
60	$\mathcal{Z}_{(8)}$	1709.23	–	1709.23	1458.81	1248.91
Total					**765.56**	**460.30**

For case B we get:

Age	*Unit*	*Units for Mortality*	*Units for Premium*	*Total Units*	*Value* $i = 2\%$	*Value* $i = 4\%$
52	$\mathcal{Z}_{(0)}$	–	*-1394.28*	*-1394.28*	*-1394.28*	*-1394.28*
53	$\mathcal{Z}_{(1)}$	*1200.00*	*-1377.55*	*-177.55*	*-174.07*	*-170.72*
54	$\mathcal{Z}_{(2)}$	**2272.40**	*-1345.87*	**926.52**	*890.54*	*856.62*
55	$\mathcal{Z}_{(3)}$	*1351.38*	*-1327.03*	*24.35*	*22.95*	*21.65*
56	$\mathcal{Z}_{(4)}$	*1427.64*	*-1307.12*	*120.51*	*111.34*	*103.01*
57	$\mathcal{Z}_{(5)}$	*1499.97*	*-1286.21*	*213.76*	*193.61*	*175.70*
58	$\mathcal{Z}_{(6)}$	*1568.22*	*-1264.34*	*303.88*	*269.83*	*240.16*
59	$\mathcal{Z}_{(7)}$	*1632.24*	*-1241.58*	*390.65*	*340.09*	*296.86*
60	$\mathcal{Z}_{(8)}$	*1691.91*	–	*1691.91*	*1444.03*	*1236.26*
Total					**1704.05**	**1365.28**

We note two things:

- *The pandemic happens when the person is aged* 53 *and we see the impact in* $\mathcal{Z}_{(2)}$ *at age 54. This has to do with the convention that we assume that the deaths occur at the end on the year, hence just before the person gets 54.*
- *We see that the difference in reserves amounts to* $1704.05 - 765.56 = 938.49$ *which represents the economic loss as a consequence of the pandemic. The biggest contributor to this loss is the increased death benefit, e.g.* $926.52 - 1284.40 = 962.87$.

Matching asset portfolios *Based on the above it is now easy to calculate the cash flow matching portfolio, by just investing the different amounts of liabilities into the corresponding assets, such as buying* $24.60\mathcal{Z}_{(3)}$. *We remark that consequently we would have to sell* $-177.55\mathcal{Z}_{(1)}$. *In normal circumstances for mature businesses this will not occur, since it is a consequence that we consider a endowment insurance policy and not for example an endowment.*

Mismatch in case of a pandemic *The table below finally shows the cash flow mismatch as a consequence of the pandemic and we see that in this case the present values do not have a big impact since the main difference is at time* 1.

Age	*Unit*	*Units Normal*	*Units Stress*	*Difference Units*	*Value* $i = 2\%$	*Value* $i = 0\%$
52	$\mathcal{Z}_{(0)}$	*-1394.28*	*-1394.28*	*0.00*	*0.00*	*0.00*
53	$\mathcal{Z}_{(1)}$	*-177.55*	*-177.55*	*0.00*	*0.00*	*0.00*
54	$\mathcal{Z}_{(2)}$	*-75.24*	*926.52*	*1001.77*	*962.87*	*1001.77*
55	$\mathcal{Z}_{(3)}$	*24.60*	*24.35*	*-0.24*	*-0.23*	*-0.24*
56	$\mathcal{Z}_{(4)}$	*121.75*	*120.51*	*-1.23*	*-1.13*	*-1.23*
57	$\mathcal{Z}_{(5)}$	*215.95*	*213.76*	*-2.18*	*-1.98*	*-2.18*
58	$\mathcal{Z}_{(6)}$	*306.99*	*303.88*	*-3.11*	*-2.76*	*-3.11*
59	$\mathcal{Z}_{(7)}$	*394.65*	*390.65*	*-3.99*	*-3.48*	*-3.99*
60	$\mathcal{Z}_{(8)}$	*1709.23*	*1691.91*	*-17.31*	*-14.78*	*-17.31*
Total					**938.49**	**973.67**

Example 70 (Lapses) *In this example we want to see how lapses can influence the replicating portfolios. In order to do that we have to change the example 43 a little bit, as follows:*

- *We consider a term insurance, for a 50 year old man with a term of 10 years, and we assume that this policy is financed with a regular premium payment. Hence there are actually two different payment streams, namely the premium payment stream and the benefits payment stream. We assume that the yearly mortality is* $(1 + \frac{x-50}{10} \times 0.1)\%$. *We assume that the benefit amounts to 100000 EUR and we assume that the premium has been determined with an interest rate* $i = 2\%$.
- *In this case the premium amounts to* $P = 9562.20$ *EUR.*
- *In addition the policyholder can surrender the policy at any time and gets back 98 % of the expected future cash flows valued at the pricing interest rate of* 2%. *We remark here that this is a risk since the surrenders can happen in case the market value of the corresponding units is below the surrender value.*
- *We remark that the units have been valued with two (flat) yield curves with interest rates of* 2% *and* 4% *respectively.*

In order to calculate this example we will perform the following steps:

1. *Calculation of the cash flow matching portfolio in case of no surrenders.*
2. *Calculation of the cash flow including lapses with an average lapse rate of 7 %*
3. *Calculation of the cash flows at time 2, assuming average lapses, lapses at 25 % at time 2.*

Calculation of the cash flow matching portfolio in case of no surrenders:

Age	*Unit*	*Units for Mortality*	*Units for Premium*	*Total Units*	*Value* $i = 2\%$	*Value* $i = 4\%$
50	$\mathcal{Z}_{(0)}$	–	-9562.20	-9562.20	-9562.20	-9562.20
51	$\mathcal{Z}_{(1)}$	1000.00	-9466.57	-8466.57	-8300.56	-8140.94
52	$\mathcal{Z}_{(2)}$	1089.00	-9362.44	-8273.44	-7952.17	-7649.26
53	$\mathcal{Z}_{(3)}$	1174.93	-9250.09	-8075.16	-7609.40	-7178.79
54	$\mathcal{Z}_{(4)}$	1257.56	-9129.84	-7872.27	-7272.76	-6729.25
55	$\mathcal{Z}_{(5)}$	1336.69	-9002.02	-7665.32	-6942.72	-6300.34
56	$\mathcal{Z}_{(6)}$	1412.12	-8866.99	-7454.87	-6619.71	-5891.69
57	$\mathcal{Z}_{(7)}$	1483.67	-8725.12	-7241.45	-6304.11	-5502.90
58	$\mathcal{Z}_{(8)}$	1551.18	-8576.79	-7025.61	-5996.29	-5133.54
59	$\mathcal{Z}_{(9)}$	1614.50	-8422.41	-6807.90	-5696.55	-4783.14
60	$\mathcal{Z}_{(10)}$	–	88080.30	88080.30	72256.53	59503.90
Total					**0**	**-7368.19**

We remark that the there is considerable value in the policy if we assume no lapses, in case we earn a higher interest rate, such as 4 %.

Calculation of the cash flow matching portfolio in case of 7% surrenders:

Age	*Unit*	*Units for Mortality*	*Units for Premium*	*Total Units*	*Value* $i = 2\%$	*Value* $i = 4\%$
50	$\mathcal{Z}_{(0)}$	–	*-9562.20*	*-9562.20*	*-9562.20*	*-9562.20*
51	$\mathcal{Z}_{(1)}$	*1019.67*	*-8797.22*	*-7777.54*	*-7625.04*	*-7478.40*
52	$\mathcal{Z}_{(2)}$	*1594.01*	*-8084.65*	*-6490.63*	*-6238.59*	*-6000.95*
53	$\mathcal{Z}_{(3)}$	*2066.98*	*-7421.70*	*-5354.72*	*-5045.87*	*-4760.33*
54	$\mathcal{Z}_{(4)}$	*2449.23*	*-6805.70*	*-4356.47*	*-4024.71*	*-3723.93*
55	$\mathcal{Z}_{(5)}$	*2750.73*	*-6234.02*	*-3483.29*	*-3154.92*	*-2863.01*
56	$\mathcal{Z}_{(6)}$	*2980.77*	*-5704.13*	*-2723.35*	*-2418.26*	*-2152.31*
57	$\mathcal{Z}_{(7)}$	*3147.94*	*-5213.57*	*-2065.63*	*-1798.25*	*-1569.71*
58	$\mathcal{Z}_{(8)}$	*3260.18*	*-4759.99*	*-1499.81*	*-1280.07*	*-1095.89*
59	$\mathcal{Z}_{(9)}$	*3324.77*	*-4341.11*	*-1016.34*	*-850.42*	*-714.06*
60	$\mathcal{Z}_{(10)}$	*5486.26*	*42159.27*	*47645.53*	*39085.93*	*32187.61*
Total					**-2912.44**	**-7733.21**

We remark that at that time, the company makes still some additional gains as a consequence of the 2% surrender penalty.

Calculation of the cash flow matching portfolio in case of high surrenders: *We assume that there has been observed an exceptional lapse rate at time 2 of 25% of the portfolio.*

Age	*Unit*	*Units for Mortality*	*Units for Premium*	*Total Units*	*Value* $i = 2\%$	*Value* $i = 4\%$
50	$\mathcal{Z}_{(0)}$	–	*-9562.20*	*-9562.20*	*-9562.20*	*-9562.20*
51	$\mathcal{Z}_{(1)}$	*1019.67*	*-8797.22*	*-7777.54*	*-7625.04*	*-7478.40*
52	$\mathcal{Z}_{(2)}$	*1594.01*	*-8084.65*	*-6490.63*	*-6238.59*	*-6000.95*
53	$\mathcal{Z}_{(3)}$	*4773.16*	*-5966.47*	*-1193.30*	*-1124.48*	*-1060.84*
54	$\mathcal{Z}_{(4)}$	*1968.98*	*-5471.25*	*-3502.26*	*-3235.55*	*-2993.75*
55	$\mathcal{Z}_{(5)}$	*2211.37*	*-5011.66*	*-2800.29*	*-2536.31*	*-2301.63*
56	$\mathcal{Z}_{(6)}$	*2396.30*	*-4585.67*	*-2189.36*	*-1944.09*	*-1730.28*
57	$\mathcal{Z}_{(7)}$	*2530.70*	*-4191.30*	*-1660.60*	*-1445.65*	*-1261.92*
58	$\mathcal{Z}_{(8)}$	*2620.93*	*-3826.66*	*-1205.73*	*-1029.08*	*-881.01*
59	$\mathcal{Z}_{(9)}$	*2672.85*	*-3489.91*	*-817.05*	*-683.67*	*-574.05*
60	$\mathcal{Z}_{(10)}$	*4410.52*	*33892.74*	*38303.27*	*31422.02*	*25876.31*
Total					**-4002.67**	**-7968.76**

ALM Risk of mass lapses *Finally we want to look what happens when we have mass lapses as indicated before, but if we have invested in the cash flow matching portfolio according to average 7 % lapses. Hence we have to calculate the assets according to 7 % lapses and the liabilities according 25 % lapses.*

Age	*Unit*	*Units for Assets*	*Units for Liability*	*Total Units*	*Value* $i = 2\%$	*Value* $i = 4\%$
52	$\mathcal{Z}_{(0)}$	*-6490.63*	*6490.63*	*0*	*0*	*0*
53	$\mathcal{Z}_{(1)}$	*-5354.72*	*1193.30*	*-4161.42*	*-4079.82*	*-4001.36*
54	$\mathcal{Z}_{(2)}$	*-4356.47*	*3502.26*	*-854.21*	*-821.04*	*-789.76*
55	$\mathcal{Z}_{(3)}$	*-3483.29*	*2800.29*	*-682.99*	*-643.60*	*-607.18*
56	$\mathcal{Z}_{(4)}$	*-2723.35*	*2189.36*	*-533.99*	*-493.32*	*-456.45*
57	$\mathcal{Z}_{(5)}$	*-2065.63*	*1660.60*	*-405.02*	*-366.84*	*-332.90*
58	$\mathcal{Z}_{(6)}$	*-1499.81*	*1205.73*	*-294.08*	*-261.13*	*-232.41*
59	$\mathcal{Z}_{(7)}$	*-1016.34*	*817.05*	*-199.28*	*-173.48*	*-151.43*
60	$\mathcal{Z}_{(8)}$	*47645.53*	*-38303.27*	*9342.26*	*7973.53*	*6826.29*
Total					**1134.26**	**254.76**

Now we see that the lapses induce quite a big risk for the company since it loses in case of mass lapses almost 1 % of the face value of the policy, more concretely $1134.26 - 254.76 = 879.50$.

The above example shows very clearly how the behaviour of the policyholders can change the cash flow matching portfolio and in consequence induces a risk. As a consequence the risk minimising portfolio in the sense of $VaPo^*(x)$ for an insurance portfolio $x \in \mathcal{X}$ does also consist of additional assets offsetting the corresponding risks. In the above example the corresponding asset would be a (complex) put option, which allows to sell the bond portfolio at the predefined (book-) values. So in reality insurance companies aim to model these risks in order to determine the corresponding assets and to reduce the undesired risk.

In the example above we have assumed that at a given year 25% of the policies in force lapse. In practise one models the dynamic lapse behaviours. Eg the lapse rate is a function of the interest differential between market and book yields. Normally the corresponding lapse rates stay below 1, which is interesting. Assuming a market efficient behaviour, one would expect that there is a binary decision of the policyholders to stick to the contract or to lapse as a funtion of the beforementioned interest differential. In consequence the underlying theory how to model such policyholder behaviour is not as crisp and transparent as with the arbitrage free pricing theory, since market efficient behaviours is normally not observed. As a corollary there is a lot of model risk intrinsic to these calculations and it is important to test the results from the models with different scenarios.

Remark 71 *At the end of this section a remark on how to determine a* $VaPo^*(x)$ *for an* $x \in \mathcal{X}$ *concretely: One normally models an* $l \in \mathcal{X}$ *and simulates* $l(\omega)$ *together with some test assets* $D \subset \mathcal{Y}$ *observable prices and cash flows. We denote* $D = \{d_1, \dots d_n\}$. *Hence at the end of this process we have a vector*

$$\mathcal{W} := (l(\omega_i), d_1(\omega_i), \dots d_n(\omega_i))_{i \in I}.$$

Now the process is quite canonical:

1. *We define a distance between two* $x, y \in \mathcal{X}$, *for example by means of* $||x||$ *as defined.*
2. *We solve the numerical optimisation problem, for minimising the distance between* l *and the target* $y \in$ *span* $< \mathcal{D} >$.

We note two things:

- *The numerical procedures to determine* y *can sometimes prove to be difficult since the corresponding design matrix can be near to a singular matrix, and hence additional care is needed.*
- *In case of the* $|| \bullet ||$ *defined before, we remark that it has been deducted from the Hilbert space* $\mathcal{X}$. *Hence what we actually doing is to use the projection* $\tilde{p} : \mathcal{X} \to$ *span* $< \mathcal{D} >$, *which can be expressed by means of* $< \bullet, \bullet >$. *We remark that* $y = \tilde{p}(x)$.

Appendix D
An Introduction to Arbitrage Free Pricing

D.1 Price Systems

In this section we want to provide an introduction to modern financial market theory. The purpose of the section is not to cover each possible detail, but rather to give an overview. For an in depth study we refer to [Pli97], [HK79], [HP81] and [Duf92].

This section would be incomplete without mentioning the work of Black and Scholes [BS73] with its famous formula for pricing equity options.

D.1.1 Definitions

Firstly we need to explain the rationale for this theory. If we consider the market value of an equity price, which is modelled with a geometric Brownian motion $(S_t(\omega))$.

A European call-option is a security, which allows to purchase the underlying asset at a predefined price c (strike price at a fixed time T. At time T the value of this paper is known:

$$H = \max\left(S_T - c, 0\right).$$

A bank now wants to know the value or the price of this option at time 0. If the price is chosen in a wrong way, such as by taking the expected present value with respect to the original measure, it is possible to make a risk free gain, and we speak about arbitrage.

We consider the simplest possible economy, considering finite models. This implies that we consider discrete time. The interested reader can find analogous results in continuous time for example in [HP81]. This section should therefore illustrate the ideas and concepts of the arbitrage free pricing theory.

M. Koller, *Life Insurance Risk Management Essentials*, EAA Series,
DOI 10.1007/978-3-642-20721-1_19, © Springer-Verlag Berlin Heidelberg 2011

We consider a probability space $(\Omega, \mathcal{A}, P)$ with Ω finite. Furthermore we assume that $P(\omega) > 0 \, \forall \omega \in \Omega$. We fix a finite time horizon T, at which all trading activities stop. With $\mathcal{F}_t$ we denote the σ-algebra of the – at time t – observable outcomes. The securities are traded at times $\{0, 1, 2, \ldots, T\}$. We assume that there exist $k < \infty$ stochastic processes, which represent the development of the value of the securities $1, \ldots, k$.

$$S = \{S_t, t = 0, 1, 2, \ldots, T\} \text{ with components } S^0, S^1, \ldots S^k.$$

As usual we assume that each S^j is adapted with respect to $(\mathcal{F}_t)_t$. We interpret S_t^j as price of security j at time t. The condition of adaptedness represents the necessity to know the trajectory of S before time t, being at time t. The 0. security has a particular role. We assume that $S_t^0 = (1+r)^t$. This means that we can invest risk free at an interest rate of r. The risk free discount factor is hence given by:

$$\beta_t = \frac{1}{S_t^0}.$$

In a next step we want to understand the meaning of a trading strategy:

Definition 72 *A trading strategy is a previsible ($\phi_t \in \mathcal{F}_{t-1}$) process $\Phi = \{\phi_t, t = 1, 2, \ldots, T\}$ with components ϕ_t^k.*

We interpret ϕ_t^k as number of security k, which we hold between $[t-1, t[$. This is the reason why ϕ_t is called portfolio at time t.

Definition 1. Let X, Y be two multi–dimensional stochastic processes. In this case we denote:

$$< X_s, Y_t > = X_s \cdot Y_t = \sum_{k=0}^{n} X_s^k \times Y_t^k,$$
$$\Delta X_t = X_t - X_{t-1}.$$

In a next step we want to determine the value of a portfolio at time t:

Time	Value of the portfolio
$t-1$	$\phi_t \cdot S_{t-1}$
t^-	$\phi_t \cdot S_t$

This shows that the profit in the interval $[t-1, t[$ amounts to $\phi_t \cdot \Delta S_t$. Hence we can calculate the total profit in the interval $[0, t]$ by:

$$G_t(\phi) = \sum_{\tau=1}^{t} \phi_\tau \cdot \Delta S_\tau.$$

We set $G_0(\phi) = 0$ can call $(G_t)_{t \geq 0}$ profit process.

Proposition 73 *G is an adapted, real valued stochastic process.*

Proof. We leave the proof as an exercise.

Definition 74 *A trading strategy is called self–financing, if we have*

$$\phi_t \cdot S_t = \phi_{t+1} \cdot S_t, \qquad \forall t = 1, 2, \ldots, T-1.$$

A self-financing strategy indicates that no money is injected or withdrawn from the portfolio at any time.

Definition 75 *A trading strategy is admissible, if it is self–financing and is*

$$V_t(\phi) := \begin{cases} \phi_t \cdot S_t, \text{ falls } t = 1, 2, \ldots, T, \\ \phi_1 \cdot S_0, \text{ falls } t = 0 \end{cases}$$

positive. (With other words: we must not go bankrupt.) With Φ we denote the set of all admissible trading strategies.

Remark 76 *The idea of admissible trading strategies consists in the fact that we do not want to inject or withdraw money from the portfolio and we can only re-base the portfolio. If we could find trading strategies which have at the end always (eg $\forall \omega \in \Omega$) the same payout as an option, we would say that the value of the option is the value of the corresponding portfolio at inception.*

Definition 77 *Under a contingency claim we understand a positive random variable X, The set of all contingency claims we denote by $\mathcal{X}$.*

A random variable X is attainable, if there exists an admissible trading strategy $\phi \in \Phi$ such that

$$V_T(\phi) = X.$$

In this case we say "ϕ generates X".

Definition 78 *For an attainable contingency claim X, which is generated by ϕ we denote by*

$$\pi = V_0(\phi)$$

its price. (This price does not need to be unique, as we will see later, and it corresponds to the value of the corresponding portfolio at inception.)

D.1.2 Arbitrage

Under an arbitrage opportunity we understand

$$\phi \in \Phi \text{ with } V_0(\phi) = 0 \text{ and } V_T(\phi) \text{ positive and } P[V_T(\phi) > 0] > 0$$

(Money is created from nil). If such a strategy exists, we can create economic profit without any risk. One of the axioms of modern economy postulates the absence of such arbitrage opportunities. From this axiom we can deduct some important learning for determining the price.

In a next step we want to understand the concept of a pricing system.

Definition 79 *A map*

$$\pi : \quad \mathcal{X} \to [0, \infty[, \quad X \mapsto \pi(X)$$

is called pricing system, if the following two conditions are fulfilled:

- $\pi(X) = 0 \iff X = 0$,
- π *is linear.*

A pricing system is called consistent if the following holds:

$$\pi(V_T(\phi)) = V_0(\phi) \qquad \text{for all } \phi \in \Phi.$$

With Π we denote the set of all consistent pricing systems. With $\mathbb{P}$ we denote the set

$$\mathbb{P} = \{Q \text{ measure equivalent to } P, \text{ under which } \beta \times S \text{ is a martingale}\},$$

where we denote by β the discount factor from time t to 0. These measures are called equivalent martingale measures.

Proposition 80 *Between the sets of consistent pricing systems $\pi \in \Pi$ and the measures $Q \in \mathbb{P}$ there exists a bijection, given by*

1. $\pi(X) = \mathbb{E}^Q[\beta_T X]$.
2. $Q(A) = \pi(S_T^0 \chi_A)$ *for all* $A \in \mathcal{A}$.

Proof. For $Q \in \mathbb{P}$ we define $\pi(X) = \mathbb{E}^Q[\beta_T X]$. π is a pricing system, since P is strictly positive on Ω and Q is equivalent to P. It remains to show that π is consistent. To this end let $\phi \in \Phi$. In this case we have the following:

$$\begin{aligned} \beta_T V_T(\phi) &= \beta_T \phi_T S_T + \sum_{i=1}^{T-1} (\phi_i - \phi_{i+1}) \beta_i S_i \\ &= \beta_1 \phi_1 S_1 + \sum_{i=2}^{T} \phi_i (\beta_i S_i - \beta_{i-1} S_{i-1}), \end{aligned}$$

where we have used that ϕ is self–financing. Hence

$$\begin{aligned}
\pi\left(V_T(\phi)\right) &= \mathbb{E}^Q\left[\beta_T\, V_T(\phi)\right] \\
&= \mathbb{E}^Q\left[\beta_1\,\phi_1\,S_1\right] + \mathbb{E}^Q\left[\sum_{i=2}^{T}\phi_i\,\left(\beta_i\,S_i - \beta_{i-1}S_{i-1}\right)\right] \\
&= \mathbb{E}^Q\left[\beta_1\,\phi_1\,S_1\right] + \sum_{i=2}^{T}\mathbb{E}^Q\left[\phi_i\,\mathbb{E}^Q\left[\left(\beta_i\,S_i - \beta_{i-1}S_{i-1}\right)|\mathcal{F}_{i-1}\right]\right] \\
&= \phi_1\,\mathbb{E}^Q\left[\beta_1\,S_1\right] \\
&= \phi_1\,\beta_0\,S_0,
\end{aligned}$$

where we have used that ϕ is previsible and that $\beta\, S$ is a martingale under Q. Hence we have proofed, that π is a consistent pricing system.

Let $\pi \in \Pi$ be a consistent pricing system and define Q as above. In this case we have $Q(\omega) = \pi(S_t^0\,\chi_{\{\omega\}}) > 0$, for all $\omega \in \Omega$, since $S_t^0\,\chi_{\{\omega\}} \neq 0$. Moreover we have $\pi(X) = 0 \iff X = 0$ and hence Q is absolutely continuous with respect to P.

In a next step we need to show that Q is a probability measure. To this end we define

$$\phi^0 = 1 \qquad \text{and} \qquad \phi^k = 0 \quad \forall k \neq 0.$$

Since π is consistent we have

$$\begin{aligned}
1 &= V_0(\phi) \\
&= \pi(V_T(\phi)) \\
&= \pi(S_T^0 \cdot 1) \\
&= Q(\Omega).
\end{aligned}$$

Since the prices of positive contingency claims are positive and since Q is additive, we can show Kolmogorov's axioms, since Ω is finite. Per definition we have $Q(\omega) = \pi(S_T^0 \cdot \chi_{\{\omega\}})$ and hence also

$$\mathbb{E}[f] = \sum_\omega \pi(S_T^0 \cdot \chi_{\{\omega\}}) \cdot f(\omega) = \pi(S_T^0 \cdot \sum_\omega f(\omega)).$$

For $f = \beta_T\, X$ we hence have

$$\mathbb{E}^Q[\beta_T\, X] = \pi(S_T^0 \cdot \beta_T \cdot X) = \pi(X).$$

It remains to show that $\beta_T\, S_T^k$ is a martingale for all k. Let k be a coordinate and τ a stopping time. We define

$$\begin{aligned}
\phi_t^k &= \chi_{\{t \le \tau\}}, \\
\phi_t^0 &= \left(S_\tau^k / S_\tau^0\right)\,\chi_{\{t > \tau\}}.
\end{aligned}$$

(We hold security k until time τ and invest the proceeds into the risk free asset.) It is easy to show that this strategy ϕ is both self–financing and previsible. The following equations hold:

$$V_0(\phi) = S_0^k,$$
$$V_T(\phi) = \left(S_\tau^k / S_\tau^0\right) S_T^0$$

and moreover

$$S_0^k = \pi(S_T^0 \cdot \beta_\tau \cdot S_\tau^k)$$
$$= \mathbb{E}^Q\left[\beta_\tau \cdot S_\tau^k\right].$$

Since the above equation is valid for an arbitrary stopping time τ, we know that $\beta_T\, S_T^k$ is a martingale with respect to Q.

After this important theorem we want to list some properties without proof and we refer to [HP81] for details.

Theorem 81 *The following three statements are equivalent:*

1. *The market model does not allow for arbitrage,*
2. $\mathbb{P} \neq \emptyset$,
3. $\Pi \neq \emptyset$.

Lemma 1. *If there exists a self–financing trading strategy $\phi \in \Phi$ with*

$$V_0(\phi) = 0,\ V_T(\phi) \geq 0,\ \mathbb{E}[V_T(\phi)] > 0$$

the market model allows arbitrage.

D.1.3 Continuous Case

We assume that $\mathbb{P} \neq \emptyset$.

In a next step we need to define the different concepts:

Definition 82 • *A trading strategy ϕ is a locally bounded, previsible stochastic process.*

- *The value process with respect to a trading strategy ϕ is given by*

$$V : \Pi \to \mathbb{R}, \phi \mapsto V(\phi) = \phi_t \cdot S_t = \sum_{i=0}^{k} \phi_t^k \cdot S_t^k.$$

- *The gain process G is defined by*

$$G : \Pi \to \mathbb{R}, \phi \mapsto G(\phi) = \int_0^T \phi \, dS = \int_0^T \sum_{i=0}^{k} \phi^k dS^k.$$

- *ϕ is self–financing, if $V_t(\phi) = V_0(\phi) + G_t(\phi)$.*
- *In order to define admissible trading strategies we need the following notation:*

$$Z_t^i = \beta_t \cdot S_t^i, \qquad \textit{discounted value of security } i$$

$$G^*(\phi) = \int \sum_{i=1}^{k} \phi^i \, dZ^i, \qquad \textit{discounted profit}$$

$$V^*(\phi) = \beta V(\phi) = \phi^0 + \sum_{i=1}^{k} \phi^i \, Z^i.$$

A trading strategy is called admissible if it has the following three properties:

1. $V^*(\phi) \geq 0$,
2. $V^*(\phi) = V^*(\phi)_0 + G^*(\phi)$,
3. $V^*(\phi)$ *is a martingale under* Q.

Proposition 83 *1. The price of a contingency claim X is given by $\pi(X) = \mathbb{E}^Q[\beta_T X]$.*
2. A contingency claim is attainable $\iff$ $V^* = V_0^* + \int H dZ$ for all H.

Definition 84 *The market is complete if all integrable contingency claims are attainable.*

Since the arbitrage free pricing theory is very important we would suggest that the reader familiarised with it.

D.2 The Black-Scholes Set Up

As we have seen in the previous section, we need to pick an economic model for the calculation of option prices and we remark that there are in principle different possible choices. In this section we want to look at the Black-Scholes set up and we would like to refer to the following sources for additional information: [Dot90], [Duf88], [Duf92], [CHB89], [Per94], [Pli97].

Definition 85 (General conventions for Black-Scholes set up) *For the remainder of this chapter we use the following notation and conventions:*

- *T_x denotes the future life span of a person of age x.*
- *With $\mathcal{H}_t = \sigma(\{T > s\}, 0 \leq s \leq t)$ we denote the σ-algebras induced by T_x.*
- *For the assets we assume that their value develops according to a standardised Brownian motion W.*
- *With $\mathcal{G}_t$ we denote the σ-algebras induced by W, enlarged by all P-null sets.*

Definition 86 (Independence of financial variables) • *We assume that $\mathcal{G}_t$ and $\mathcal{H}_t$ stochastically independent. Hence we assume that the financial variables are independent of the future life span of the considered persons.*

- *With $\mathcal{F}_t = \sigma(\mathcal{G}_t, \mathcal{H}_t)$ we denote the σ-algebra generated by $\mathcal{G}_t$ and $\mathcal{H}_t$.*

Definition 87 (Black-Scholes-Model) *In this model the market consists of two investment vehicles:*

$B(t) = \exp(\delta\, t)$	*Risk free asset.*
$S(t) = S(0)\, \exp\left[\left(\eta - \frac{1}{2}\sigma^2\right) t + \sigma\, W(t)\right]$	*Units, modelled by a geometric Brownian motion.*

S solves the following stochastic differential equation (SDE):

$$dS = \eta\, S\, dt + \sigma\, S\, dW.$$

Exercise 88 *Proof the above SDE.*

In a next step we need to calculate the discounted values of B and S:

$$\begin{aligned} B^*(t) &= \tfrac{B(t)}{B(t)} = 1, \\ S^*(t) &= \tfrac{S(t)}{B(t)} = S(0)\, \exp\left[\left(\eta - \delta - \tfrac{1}{2}\sigma^2\right) t + \sigma\, W(t)\right]. \end{aligned}$$

After the definition of the investment possibilities, we need to calculate the equivalent martingale measures in order to be able to calculate the corresponding option prices. Hence we need to determine a equivalent measure Q such that S^* is a martingale with respect to Q. In order to do this we define the following Radon-Nikodym-density:

$$\xi_t = \exp\left(-\frac{1}{2}\left(\frac{\eta - \delta}{\sigma}\right)^2 t - \frac{\eta - \delta}{\sigma}\, W(t)\right) \qquad \text{for all } t \in [0, T].$$

Exercise 89 *Proof the following properties*

1. $\mathbb{E}[\xi_t] = 1$,
2. $Var[\xi_t] = \exp\left(\left(\frac{\eta-\delta}{\sigma}\right)^2 t\right) - 1$,
3. $\xi_t > 0$.

(Remark: $W(t) \sim \mathcal{N}(0,t)$.)

As a consequence of a corollary of the Girsanov-Theorem (see for example [Pro90] Theorem 3.6.21) it follows that

$$\hat{W}_t = W(t) + \frac{\eta - \delta}{\sigma} t$$

is a standardised Brownian motion under $Q = \xi \cdot P$.

After the introduction of this transformation we want to show that

$$S^*(t) = S(0) \exp\left(-\frac{1}{2}\sigma^2 t + \sigma \hat{W}(t)\right)$$

is a Q–martingale. (In this case we can calculate the prices of options as expected values with respect to Q.)

Proof. For $t, u \in \mathbb{R}, u > t$ we need to show the following equality:

$$\mathbb{E}^Q\left[S^*(u)|\mathcal{F}_t\right] = S^*(t).$$

We use the following notation: $u = t + \Delta t$, $W_u = W_t + \Delta W$ and $Z \sim \mathcal{N}(0,1)$.

$$\begin{aligned}
&\mathbb{E}^Q\left[S^*(u)|\mathcal{F}_t\right] \\
&= \mathbb{E}^Q\left[S(0) \exp\left(-\frac{1}{2}\sigma^2 t + \sigma \hat{W}(t) + (-\frac{1}{2}\sigma^2 \Delta t + \sigma \Delta\hat{W})\right) |\mathcal{F}_t\right] \\
&= S(0) \exp\left(-\frac{1}{2}\sigma^2 t + \sigma W(t)\right) \mathbb{E}^Q\left[\exp\left(-\frac{1}{2}\sigma^2 \Delta t + \sigma\sqrt{\Delta t} Z\right) |\mathcal{F}_t\right] \\
&= S^*(t).
\end{aligned}$$

We have hence shown that Q is an equivalent measure to P, with respect to which S^* is a martingale. With the words of the economist: there exists at least one equivalent pricing system.

Theorem 90 *In the above defined economy given by $(\Omega, \mathcal{A}, P)$, S and B, we can calculate the price of a mortality benefit $C(T)$ at time t by*

$$\pi_t(T) = \mathbb{E}^Q\left[\exp\left(-\delta(T-t)\right) C(T)|\mathcal{F}_t\right].$$

Remark 91 *The important difference in respect to the classical model is that we take expected values with respect to Q and not with respect to P.*

In the following we want to have a look at unit–linked insurance policies with guarantees. We use the following notation:

$C(\tau)$	Insured sum at time τ,
$N(\tau)$	Number of units at time τ,
$S(\tau)$	price of the units at time τ,
$G(\tau)$	Guaranteed amount at time τ,
$C(\tau) = \max\{N(\tau)S(\tau), G(\tau)\}$	Amount insured.

D.2.1 Endowment Policies

Proposition 92 *Given the Black-Scholes-model. In this case we can calculate the single premium for a pure endowment policy with*

$$C(T) = \max\{N(T)S(T), G(T)\}$$

by

$${}_TG_x = {}_Tp_x \left[G(T)\exp(-\delta T)\Phi(-d_2^0(T)) + S(0)N(T)\Phi(d_1^0(T))\right],$$

where

$$\Phi(y) = \frac{1}{\sqrt{2\pi}}\int_{-\infty}^{y}\exp(-\frac{x^2}{2})dx,$$

$$d_1^t(s) = \frac{\ln\left[\frac{N(s)S(t)}{G(s)}\right] + \left(\delta + \frac{1}{2}\sigma^2\right)(s-t)}{\sigma\sqrt{s-t}}, (s > t),$$

$$d_2^t(s) = \frac{\ln\left[\frac{N(s)S(t)}{G(s)}\right] + \left(\delta - \frac{1}{2}\sigma^2\right)(s-t)}{\sigma\sqrt{s-t}}, (s > t).$$

Proof. In the following we always denote by J^* the discounted value of the random variable J. The value of the endowment policy at time 0 amounts to $\mathbb{E}^Q[C^*(T)]$. If we denote by $Z = S^*(T)$, we get the following:

$${}_TG_x = {}_Tp_x\,\mathbb{E}^Q\left[\max\{N(T)Z, G^*(T)\}\right]$$

and

$$Z = S(0)\exp\left(-\frac{1}{2}\sigma^2 T + \sigma\hat{W}(T)\right) \qquad \text{with} \qquad \hat{W}(T) \sim \mathcal{N}(0, T).$$

And hence we get

$$
{}_TG_x = {}_Tp_x \int_{-\infty}^{\infty} \max \left[N(T)S(0)\exp(-\frac{1}{2}\sigma^2 T + \sigma\,\xi), G^*(T) \right] f(\xi)d\xi,
$$
$$
f(\xi) = \frac{1}{\sqrt{2\pi T}} \exp\left(-\frac{1}{2T}\xi^2\right).
$$

In a next step we set $\bar{\xi} = \frac{1}{\sigma}\left[\ln\left(\frac{G^*(T)}{N(T)S(0)}\right) + \frac{1}{2}\sigma^2 T\right]$ and remark that for all $\xi > \bar{\xi}$ we have $N(T)Z > G^*(T)$. Hence we can calculate the corresponding single premium as follows:

$$
\begin{aligned}
{}_TG_x = {}_Tp_x \Bigg(& G^*(T) \int_{-\infty}^{\bar{\xi}} f(\xi)d\xi \\
& + N(T)S(0) \int_{\bar{\xi}}^{\infty} \exp(-\frac{1}{2}\sigma^2 T + \sigma\,\xi) f(\xi)d\xi \Bigg) \\
= {}_Tp_x \Bigg(& G^*(T) \int_{-\infty}^{\bar{\xi}} f(\xi)d\xi \\
& + N(T)S(0) \int_{\bar{\xi}}^{\infty} \frac{1}{\sqrt{2\pi T}} \exp(-\frac{1}{2T}(\xi - \sigma T)^2 d\xi \Bigg).
\end{aligned}
$$

From the above equation the desired result follows.

D.2.2 Term Insurance

Proposition 93 *Given the Black-Scholes-Model, we can calculate the net single premium for a lump sum*

$$
C(t) = \max\{N(t)S(t), G(t)\}
$$

by

$$
G^1_{x:T} = \int_0^T \left(G(t)\exp(-\delta t)\Phi(-d_2^0(t)) + S(0)N(t)\Phi(d_1^0(t))_t p_x \mu_{x+t}\right) dt,
$$

where

$$
\Phi(y) = \int_{-\infty}^{y} \frac{1}{\sqrt{2\pi}} \exp(-\frac{x^2}{2})dx,
$$
$$
d_1^t(s) = \frac{\ln\left[\frac{N(s)S(t)}{G(s)}\right] + \left(\delta + \frac{1}{2}\sigma^2\right)(s-t)}{\sigma\sqrt{s-t}},
$$

$$d_2^t(s) = \frac{\ln\left[\frac{N(s)S(t)}{G(s)}\right] + \left(\delta - \frac{1}{2}\sigma^2\right)(s-t)}{\sigma\sqrt{s-t}},$$

for $s > t$.

Exercise 94 *Proof the above proposition using the methods learnt for the endowment cover.*

D.3 Thiele's Differential Equation

In order to deduct Thiele's differential equation we need to introduce premium payments in a first step. By $\bar{p}(t)$ we denote the premium density at time t. As a consequence of the equivalence principle we have the following two equations:

$$_TG_x = \int_0^T \bar{p}(t)\exp(-\delta\, t)_t p_x dt,$$

respectively

$$G^1_{x:T} = \int_0^T \bar{p}(t)\exp(-\delta\, t)_t p_x dt.$$

In this section we want to treat the pure endowment cover and the term insurance separately. For these two types of insurance cover the mathematical reserves are given as follows:

$$\begin{aligned}\text{Endowment: } V(t) &= {}_{T-t}p_{x+t}\pi_t(T)\\ &\quad - \int_t^T \bar{p}(\xi)\exp(-\delta(\xi - t))_{\xi-t}p_{x+t}\, d\xi.\end{aligned}$$

$$\begin{aligned}\text{Term insurance: } V(t) &= \int_t^T \left(\pi_t(\xi)\mu_{x+\xi} - \bar{p}(\xi)\exp(-\delta(\xi - t))\right)\\ &\quad \times_{\xi-t}p_{x+t}\, d\xi,\end{aligned}$$

where

$$\begin{aligned}\pi_t(s) &= G(s)\exp(-\delta(s-t))\Phi(-d_2^t(s))\\ &\quad + N(s)S(t)\Phi(d_1^t(s)),\\ d_1^t(s) &= \frac{\ln\left[\frac{N(s)S(t)}{G(s)}\right] + \left(\delta + \frac{1}{2}\sigma^2\right)(s-t)}{\sigma\sqrt{s-t}},\end{aligned}$$

$$d_2^t(s) \quad = \quad \frac{\ln\left[\frac{N(s)S(t)}{G(s)}\right] + \left(\delta - \frac{1}{2}\sigma^2\right)(s-t)}{\sigma\sqrt{s-t}},$$

for $s > t$.

Remark 95 • *In contrast to the classical case the reserves are not anymore deterministic, but crucially depend on the value of the underlying asset S.*

- *We remark that we need now to apply Itô-calculus, where we have for the pure continuous for a Brownian motion the following:*

$$df(W) = f'\,dW + \frac{1}{2}f''\,ds.$$

For the two insurance types we have the following theorem:

Theorem 96 *1. The differential equation for the market value of a pure endowment policy is given by:*

$$\frac{\partial V}{\partial t} = \bar{p}(t) + (\mu_{x+t} + \delta)\,V(t) - \frac{1}{2}\sigma^2 S(t)^2 \frac{\partial^2 V}{\partial S^2} - \delta\,S(t)\frac{\partial V}{\partial S}.$$

2. The differential equation for the market value of a term insurance policy is given by:

$$\frac{\partial V}{\partial t} = \bar{p}(t) + (\mu_{x+t} + \delta)\,V(t) - C(t)\mu_{x+t} - \frac{1}{2}\sigma^2 S(t)^2 \frac{\partial^2 V}{\partial S^2} - \delta\,S(t)\frac{\partial V}{\partial S}.$$

Before proofing the theorem we would like to add some comments:

Remark 97 *1. For $\mu_{x+t} = \bar{p}(t) = 0\,\forall t$ we get the Black-Scholes-formula.*

2. The first term of the above differential equation correspond to the classical set–up. This relates to the dependence on premiums, mortality and interest. As a consequence of the fluctuation of the value of the units S we get an additional term: $-\frac{1}{2}\sigma^2 S(t)^2 \frac{\partial^2 V}{\partial S^2} - \delta\,S(t)\frac{\partial V}{\partial S}$.

Proof. Since

$$\pi_t^*(T) = \exp(-\delta t)\pi_t(T),$$

we get the following equation as a consequence of the definition of V:

$$V(t) = {}_{T-t}p_{x+t}\pi_t^*(T)\exp(\delta t) - \int_t^T \bar{p}(\xi)\exp(-\delta(\xi - t))_{\xi-t}p_{x+t}d\xi$$

and hence

$$\pi_t^*(T) = \Psi(t)\left[V(t) + \int_t^T \bar{p}(\xi)\exp(-\delta(\xi - t))_{\xi-t}p_{x+t}d\xi\right],$$

where

$$\Psi(t) = \frac{\exp(-\delta t)}{{}_{T-t}p_{x+t}}.$$

Since π_t^* is a function of S and t, we can apply the Itô-formula on the function $\pi_t^*(t, S)$:

$$\begin{aligned} dY_t &= U(t + dt, X_t + dX_t) - U(t, X_t) \\ &= \left(U_t dt + \frac{1}{2} U_{xx} b^2 dt \right) + U_x dX_t \\ &= \left(U_t + \frac{1}{2} U_{xx} b^2 \right) dt + U_x \, b \, dB_t \end{aligned}$$

and we get:

$$d\pi^* = \left(\frac{\partial \pi^*}{\partial t} + \frac{\partial \pi^*}{\partial S} \, a + \frac{1}{2} \frac{\partial^2 \pi^*}{\partial S^2} \, b^2 \right) dt + \frac{\partial \pi^*}{\partial S} b \, d\hat{W},$$

knowing that

$$dS = \delta S(t) dt + \sigma S(t) d\hat{W}.$$

Hence we have $a = \delta S(t)$ and $b = \sigma S(t)$. In a next step we want to determine the different terms for the above formula:

$$\begin{aligned} \frac{\partial \pi_t^*}{\partial S} &= \Psi(t) \, \frac{\partial V}{\partial S}, \\ \frac{\partial^2 \pi_t^*}{\partial S^2} &= \Psi(t) \, \frac{\partial^2 V}{\partial S^2}. \end{aligned}$$

In order to calculate $\frac{\partial \pi^*}{\partial t}$, we first calculate:

$$\begin{aligned} \frac{\partial}{\partial t} {}_{\xi-t}p_{x+t} &= \mu_{x+t} \, {}_{\xi-t}p_{x+t}, \\ \frac{\partial}{\partial t} \Psi(t) &= \left(\frac{A}{B} \right)' = \frac{A'}{B} - \frac{A}{B^2} B' \\ &= -\left(\mu_{x+t} + \delta \right) \Psi(t). \end{aligned}$$

If we now compile the different parts we get:

$$\begin{aligned} \frac{\partial \pi^*}{\partial t} &= \frac{\partial \Psi}{\partial t} \left(V(t) + \int_t^T \bar{p}(\xi) \exp(-\delta(\xi - t))_{\xi - t}p_{x+t} dt \right) \\ &+ \Psi(t) \left(\frac{\partial V}{\partial t} + \frac{\partial}{\partial t} \int_t^T \bar{p}(\xi) \exp(-\delta(\xi - t))_{\xi - t}p_{x+t} dt \right) \\ &= \Psi(t) \left(\frac{\partial V}{\partial t} - \left(\mu_{x+t} + \delta \right) V(t) - \bar{p}(t) \right), \end{aligned}$$

by using the chain rule

$$\frac{\partial}{\partial t}\int_t^T \bar{p}(\xi)\exp(-\delta(\xi - t))_{\xi-t}p_{x+t}dt.$$

Finally we get:

$$\begin{aligned}\pi_s^*(T) = \pi_t^*(T) &+ \int_t^s \Psi(\xi)\frac{\partial V}{\partial S}\sigma S d\hat{W}(\xi) \\ &+ \int_t^s \Psi(\xi)\left[\frac{\partial V}{\partial S}\delta S + \frac{1}{2}\sigma^2 S^2 \frac{\partial^2 V}{\partial S^2} - (\mu_{x+\xi} + \delta)V(\xi) \right.\\ &\left. + \frac{\partial V}{\partial t}(\xi)\bar{p}(\xi)\right]d\xi.\end{aligned}$$

Sine $\pi_\cdot^*(T)$ is a martingale, the drift term is zero, and get the required result:

$$\frac{\partial V}{\partial t} = \bar{p}(t) + (\mu_{x+t} + \delta)\, V(t) - \frac{1}{2}\sigma^2 S(t)^2 \frac{\partial^2 V}{\partial S^2} - \delta\, S(t)\frac{\partial V}{\partial S}.$$

Exercise 98 *Proof the second part of the theorem.*

Appendix E
An Introduction to Stochastic Integration

The aim of this appendix is to provide all necessary definition and results with respect to Martingales and stochastic integration. Since some of the underlying properties and theorems require a lot of advanced mathematics, we do not aim to proof the different theorems. For valuable literature we refer to [Pro90] and [IW81].

E.1 Stochastic Processes and Martingales

Definition 99 *A probability space $(\Omega, \mathcal{A}, P)$ is called filtered, if $\mathbb{F} = (\mathcal{F}_t)_{t \geq 0}$ there exists a family of $\sigma-$algebras with*

1. $\mathcal{F}_0 \supset \{A \in \mathcal{A} | P(A) = 0\}$,
2. $\mathcal{F}_s \subset \mathcal{F}_t$ *for* $s \leq t$.

The filtration is called continuous from the right side, if $\mathcal{F}_t = \bigcap_{t' > t} \mathcal{F}_{t'}, \forall t \geq 0$.

Definition 100 *A random variable $T : \Omega \to [0, \infty]$ is called* stopping time, *if $\{T \leq t\} \in \mathcal{F}_t$ for all $t \in \mathbb{R}_+$.*

Proposition 101 *T is a stopping time if and only if $\{T < t\} \in \mathcal{F}_t$ for all $t \in \mathbb{R}_+$. ([Pro90] Thm. 1.1.1.)*

Definition 102 *Let X, Y be two stochastic processes. X and Y are called* modifications *if*

$$X_t = Y_t \qquad P\text{–almost everywhere } \forall t.$$

X and Y are called identical,

$$X_t = Y_t, \forall t \qquad P\text{–almost everywhere}.$$

M. Koller, *Life Insurance Risk Management Essentials*, EAA Series,
DOI 10.1007/978-3-642-20721-1_20, © Springer-Verlag Berlin Heidelberg 2011

Definition 103 *1. A stochastic process is called càdlàg (continue à droite, limites à gauche), if its trajectories are right-continuous, with limits from the left.*

2. *A stochastic process is called càglàd its trajectories are left-continuous, with limits from the right.*
3. *A stochastic process is called* adapted, *if $X_t \in \mathcal{F}_t$ (X_t is $\mathcal{F}_t$-measurable).*

Proposition 104 *1. Let Λ be an open set and X an adapted càdlàg-process. In this case is $T := \inf\{t \in \mathbb{R}_+ : X_t \in \Lambda\}$ a stopping time.*

2. *Let S, T be two stopping times and $\alpha > 1$. In this case the following random variables are also stopping times $\min(S, T)$, $\max(S, T)$, $S + T$, $\alpha \cdot T$.*

Proof. [Pro90] Thm. 1.1.3 and Thm. 1.1.5.

Definition 105 *Let $(\Omega, \mathcal{A}, (\mathcal{F}_t)_{t\geq 0}, P)$ be a filtrated Probability space. A stochastic process is called Martingale, if*

- *$X_t \in L^1(\Omega, \mathcal{A}, P)$, d.h. $\mathbb{E}[|X_t|] < \infty$,*
- *For $s < t$ follows $\mathbb{E}[X_t|\mathcal{F}_s] = X_s$.*

Remark 106 *I we replace "$=$" in the above equation by "$\leq$" (resp. "$\geq$"), X is called Super-martingale (resp. Sub-martingale).*

Theorem 107 *Let X be a super-martingale. In this case the following conditions are equivalent*

1. *The map $T \to \mathbb{R}, t \mapsto \mathbb{E}[X_t]$ is right continuous.*
2. *There exists a unique modification Y of X, which is càdlàg.*

Proof. [Pro90] Thm. 1.2.9.

Proposition 108 *Let X be a martingale. In this case there exists a unique modification Y of X, which is càdlàg.*

Theorem 109 (Doob's stopping theorem) *Let X be a right-continuous martingale, which is closed by X_∞, i.e. $X_t = \mathbb{E}[X_\infty|\mathcal{F}_t]$. Moreover let S and T be two stopping times with $S \leq T$ P–a.e. Then we have the following:*

1. *X_S, $X_T \in L^1(\Omega, \mathcal{A}, P)$,*
2. *$X_S = \mathbb{E}[X_T|\mathcal{F}_S]$.*

Proof. [Pro90] Thm. 1.2.16.

Definition 110 *Let X be a stochastic process and T a stopping time. With $(X_t^T)_{t\geq 0}$ we denote the stopped stochastic process X_t^T, defined by $X_t^T = X_{\min(t,T)}$ for $t \geq 0$.*

Theorem 111 (Jensen-inequality) *Let $\phi : \mathbb{R} \to \mathbb{R}$ a convex function and $X \in L^1(\Omega, \mathcal{A}, P)$ with $\phi(X) \in L^1(\Omega, \mathcal{A}, P)$. Moreover let $\mathcal{G}$ be a σ-algebra. In this case we have the following inequality*

$$\phi \circ \mathbb{E}[X|\mathcal{G}] \leq \mathbb{E}[\phi(X)|\mathcal{G}].$$

Proof. [Pro90] Thm. 1.2.19.

E.2 Stochastic Integral

The aim of this section is to provide a short introduction into the theory of stochastic integrals. We closely follow [Pro90].

In principle we can consider the stochastic integration of semi-martingales as trajectory wise Stieltjes-integration, as we know them from typical lectures in analysis. The idea is to form the integral as a limit of sums of the form

$$\sum f(T_k)\ (T_{k+1} - T_k)$$

for finer re-participations. In the following let $(\Omega, \mathcal{A}, (\mathcal{F}_t)_{t\geq 0}, P)$ be a filtrated probability space satisfying the usual regularity conditions.

Definition 112 1. *A stochastic process H is called simple predictable, if it is of the form*

$$H_t = H_0 \cdot \chi_{\{0\}}(t) + \sum_{i=1}^{n} H_i \cdot \chi_{]T_i,T_{i+1}]}(t)$$

with

$$0 = T_1 \leq \ldots \leq T_{n+1} < \infty$$

a finite family of stopping times and $H_i \in \mathcal{F}_t$, $(H_i)_{i=0,\ldots,n}$ P-a.e finite.

With $\mathbb{S}$ we denote the set of all simple predictable stochastic processes and with $\mathbb{S}_u$ the set $\mathbb{S}$, equipped with the topology of uniform convergence in (t,ω) on $\mathbb{R} \times L^\infty(\Omega, \mathcal{A}, P)$.

2. *With $\mathbb{L}^0$ we denote the vector space of all finite, real-valued random variables, equipped with the topology induced by the convergence in probability.*

In a next step we define a sense for the expression $\int H\,dX$ for certain processes $(X_t)_{t\in\mathbb{R}}$ and $(H_t)_{t\in\mathbb{R}}$. In order that such an operator I_X devotes the name integral, we would expect that it is linear and fulfils a sort of the Lebesgue theorem.

We require for the convergence theorem the following continuity: If H^n converges uniformly to H, we require that $I_X(H^n)$ converges in probability to $I_X(H)$.

For a stochastic process X we then define $I_X : \mathbb{S} \to \mathbb{L}^0$ as follows:

$$I_X(H) = H_0\, X_0 + \sum_{i=1}^{n} H_i \left(X^{T_i} - X^{T_{i+1}}\right),$$

where

$$H_t = H_0 \cdot \chi_{\{0\}}(t) + \sum_{i=1}^{n} H_i \cdot \chi_{]T_i,T_{i+1}]}(t).$$

The above definition of $I_X(H)$ is independent from its representation of H.

Definition 113 (Total Semimartingale) *A stochastic process $(X_t)_{t\geq 0}$ is called total Semi-martingale, if we have the following:*

1. *X càdlàg and*
2. *I_X is a continuous map from $\mathbb{S}^u$ to $\mathbb{L}^0$.*

Definition 114 (Semimartingale) *A stochastic process $(X_t)_{t\geq 0}$ is called Semi-martingale, if X^t (cf. definition 110) is a total semi-martingale for all $t \in [0,\infty[$.*

Remark 115 *Semi-martingales are hence defined as "good" integrators.*

The following proposition summarises the most important properties of the operator I_X:

Proposition 116 *1. The set of all semi-martingales is a vector space.*

2. *Let Q be a measure which is absolutely continuous with respect to P. In this case each P-semi-martingale is also a Q-semi-martingale.*
3. *For a sequence $(P_n)_{n\in\mathbb{N}}$ a probability measures, for which $(X_t)_{t\geq 0}$ is a P_n-semi-martingale, we define $R = \sum_{n\in\mathbb{N}} \lambda_n\, P_n$, with $\sum_{n\in\mathbb{N}} \lambda_n = 1$. In this case $(X_t)_{t\geq 0}$ is also R-semi-martingale.*
4. *(Stricker's Theorem) Let X be semi-martingale with respect to the filtration $(\mathcal{F}_t)_{t\geq 0}$ and let $(\mathcal{G}_t)_{t\geq 0}$ a sub-filtration of $(\mathcal{F}_t)_{t\geq 0}$ such that X is adapted with respect to $(\mathcal{G}_t)_{t\geq 0}$. In this case X is a $\mathcal{G}$-semi-martingale.*

Proof. The above properties follow from the definition of semi-martingales. The proofs can be found in [Pro90] chapter II.2.

In a next step we want to characterise the class of semi-martingales.

Theorem 117 *Each adapted process with càdlàg-paths and finite variation on compact sets is a semi-martingale.*

Proof. This proposition follows from the fact that

$$|I_X(H)| \leq ||H||_u \int_0^\infty |dX_s|,$$

where we denote with $\int_0^\infty |dX_s|$ the total variation.

Theorem 118 *Each quadratic integrable martingale with càdlàg-paths is a semi-martingale.*

Proof. Let X be a quadratic integrable martingale with $X_0 = 0$, $H \in \mathbb{S}$. In order to show the continuity of the operator I_X, it is sufficient to proof the following inequality:

$$\begin{aligned}
E\left[(I_X(H))^2\right] &= E\left[\left(\sum_{i=0}^{n} H_i \left(X^{T_i} - X^{T_{i+1}}\right)\right)^2\right] \\
&= E\left[\sum_{i=0}^{n} H_i^2 \left(X^{T_i} - X^{T_{i+1}}\right)^2\right] \\
&\leq ||H||_u^2\, E\left[\sum_{i=0}^{n} \left(X^{T_i} - X^{T_{i+1}}\right)^2\right] \\
&= ||H||_u^2\, E\left[\sum_{i=0}^{n} \left({X^{T_i}}^2 - {X^{T_{i+1}}}^2\right)\right] \\
&= ||H||_u^2\, E\left[{X_{T^{n+1}}}^2\right] \\
&\leq ||H||_u^2\, E\left[{X_{T^{\infty}}}^2\right].
\end{aligned}$$

Example 119 *The Brownian motion is a semi-martingale.*

After characterising semi-martingales we want to enlarge in a next step the class of integrands. A very suitable class are càglàd-processes. We choose them in order that the proofs remain relatively simple.

Definition 120 *With $\mathbb{D}$ (resp. $\mathbb{L}$) we denote the set of all adapted càdlàg (resp. càglàd)-processes. With $b\mathbb{L}$ we denote all $X \in \mathbb{L}$, with bounded paths.*

Until now we have seen the topology of uniform convergence (on $\mathbb{S}_u$) and the topology of convergence in probability on $\mathbb{L}^0$. we introduce another topology:

Definition 121 *For $t \geq 0$ and a stochastic process H we define*

$$H_t^* = \sup_{0 \leq s \leq t} |H_s|.$$

A sequence $(H^n)_{n \in \mathbb{N}}$ converges uniformly on compact subsets in probability (we refer this topology as ucp-topology) to H, if

$$(H^n - H)_t^* \to 0$$

in probability for $n \to \infty$ and all $t \geq 0$.

With $\mathbb{D}_{ucp}$, $\mathbb{L}_{ucp}$ and $\mathbb{S}_{ucp}$ we denote the respective sets, equipped with the above defined topology.

Remark 122 *1. The ucp-topology can be defined by a metric. An equivalent metric is for example:*

$$d(X, Y) = \sum_{i=1}^{\infty} \frac{1}{2^n} E\left[\min(1, (X - Y)_n^*\right].$$

2. $\mathbb{D}_{ucp}$ is a complete metric space.

In order to extend I_X, we need the following theorem:

Theorem 123 *The vector space $\mathbb{S}$ is dense with respect to the ucp-topology in $\mathbb{L}$.*

Proof. [Pro90] Thm. 2.4.10.

If we now can show that I_X is continuous, we can extend I_X. In order to do this, we define:

Definition 124 *For $H \in \mathbb{S}$ and X a semi-martingale we define $J_X : \mathbb{S} \to \mathbb{D}$ by*

$$J_X(H) = H_0 X_0 + \sum_{i=0}^{n} H_i \left(X^{T_i} - X^{T_{i+1}}\right),$$

where

$$H_t = H_0 \cdot \chi_{\{0\}}(t) + \sum_{i=1}^{n} H_i \cdot \chi_{]T_i, T_{i+1}]}(t).$$

With $H_i \in \mathcal{F}_{T_i}$, $0 = T_1 \leq \ldots \leq T_{n+1} < \infty$ stopping times.

Definition 125 (Stochastic Integral) *For $H \in \mathbb{S}$ and X a càdlàg-process we call $J_X(H)$ stochastic integral of H with respect to X and denote*

$$H \cdot X := \int H_s \, dX_s := J_X(H).$$

After the definition of the stochastic integral $\mathbb{S}$, we want to extend it to $\mathbb{L}$. In order to do that we need the following theorem:

Theorem 126 *For a semi-martingale X the map $J_X : \mathbb{S}_{ucp} \to \mathbb{D}_{ucp}$ is continuous. The extension of J_X on $\mathbb{S}_{ucp}$ is also called a stochastic integral and we use the notation introduced in definition 125.*

Proof. [Pro90] Thm. 2.4.11.

Remark 127 *In order to extend J_X to $\mathbb{D}$, we use the fact that the space $\mathbb{D}_{ucp}$ is a complete metric space.*

With

$$H \cdot X_t := \int_0^t H_s \, dX_s := \int_{[0,t]} H_s \, dX_s$$

denote the stochastic process $J_X(H) = \int H_s \, dX_s$, at $t \geq 0$.

E.3 Properties of the Stochastic Integral

After having defined the stochastic integral we want to have a look at its properties.

Proposition 128 *1. Let T be a stopping time. In this case we have $(H \cdot X)^T = H \cdot \chi_{[0,T]} \cdot X = H \cdot X^T$.*

2. Let $G, H \in \mathbb{L}$ and X a semi-martingale. in this case $Y := H \cdot X$ is also a semi-martingale. Moreover we have:

$$G \cdot Y = G \cdot (H \cdot X) = (G \cdot H) \cdot X.$$

Proof. [Pro90] Thm. 2.5.12 and 2.5.19.

Definition 129 *For a càdlàg-process X we denote*

$$X_-(t) = \lim_{s \uparrow t} X(s),$$
$$\Delta X(t) = X(t) - X_-(t).$$

Definition 130 *A random partition σ of $\mathbb{R}$ is a finite sequence of stopping times with*

$$0 = T_0 \leq T_1 \leq \ldots \leq T_n < \infty.$$

A sequence $(\sigma_n)_{n \in \mathbb{N}}$ of random partitions of $\mathbb{R}$ converges to the identity, if the following conditions are fulfilled:

1. $\lim_{n \to \infty} (\sup_k T_k^n) = \infty$ *P-a.e.,*
2. $||\sigma_n|| := \sup_k |T_{k+1}^n - T_k^n|$ *converges P-a.e. to* 0.

For a process Y and a random partition σ we define

$$Y^\sigma := Y_0 \cdot \chi_{\{0\}} + \sum_k Y_{T_k} \cdot \chi_{]T_k, T_{k+1}]}.$$

Remark 131 *It is easy to show that*

$$\int Y_s^\sigma \, dX_s = Y_0 \, X_0 + \sum_k Y_{T_k} \left(X^{T_{k+1}} - X^{T_k} \right)$$

for all semi-martingales X and for all Y in $\mathbb{S}$, $\mathbb{D}$ and $\mathbb{L}$.

With the help of random partition we can calculate the stochastic integral as follows

Theorem 132 *Let X be a semi-martingale, $Y \in \mathbb{D}$ and $(\sigma_n)_{n \in \mathbb{N}}$ a sequence of random partitions, which converges to the identity. In this case*

$$\int_{0+} Y_s^{\sigma_n} \, dX_s = \sum_k Y_{T_k^n} \left(X^{T_{k+1}^n} - X^{T_k^n} \right)$$

converges with respect to the ucp-topology to the stochastic integral $\int (Y_-) \, dX$.

Proof. [Pro90] Thm. 2.5.21.

Definition 133 *Let X and Y be two semi-martingales. In this case we denote*

$$[X, X] = ([X, X]_t)_{t \geq 0} \text{ the quadratic variation process,}$$

$$[X, X] := X^2 - 2 \int X_- dX,$$

resp

$$[X, Y] := XY - \int X_- \, dY - \int Y_- \, dX$$

the covariance process.

Proposition 134 *Let X be a semi-martingale. Then we have the following:*

1. $[X, X]$ *is càdlàg, monotonously increasing and adapted.*
2. $[X, X]_0 = X_0^2$ *and* $\Delta[X, X] = (\Delta X)^2$.
3. *For a sequence* $(\sigma_n)_{n\in\mathbb{N}}$ *of random partitions converging to* 1*, we have the following*

$$X_0^2 + \sum_i (X^{T_{i+1}^n} - X^{T_i^n})^2 \longrightarrow [X, X] \text{ with respect to ucp for } n \to \infty.$$

4. *Let T be a stopping time. In this case we have* $[X^T, X] = [X, X^T] = [X^T, X^T] = [X, X]^T$.

Proof. [Pro90] Thm. 2.6.22.

Remark 135 • *The map* $(X, Y) \mapsto [X, Y]$ *is bilinear and symmetric.*

• *We have the following polarisation identity:*

$$[X, Y] = \frac{1}{2}\left([X + Y, X + Y] - [X, X] - [Y, Y]\right).$$

Proposition 136 *The bracket process* $[X, Y]$ *of two semi-martingales X and Y has paths of bounded variation on compact sets and is a semi-martingale.*

Proof. [Pro90] Cor. 2.6.1.

Proposition 137 (Partial Integration)

$$d(XY) = X_- \, dY + Y_- \, dX + d[X, Y].$$

Proof. [Pro90] Cor. 2.6.2.

Proposition 138 *Let M be a local martingale. In this case 1 and 2 are equivalent and 3 follows from 1 and 2.*

1. *M is a martingale with* $\mathbb{E}[M_t^2] \leq \infty \; \forall t \geq 0$,
2. $E\left[[M, M]_t\right] < \infty \; \forall t \geq 0$,
3. $\mathbb{E}[M_t^2] = E\left[[M, M]_t\right] \forall t \geq 0$.

Proof. [Pro90] Cor. 2.6.4.

Theorem 139 *Let X, Y be two semi-martingales and $H, K \in \mathbb{L}$. Then we have the following:*

1. $[H\cdot X, K\cdot Y]_t = \int_0^t H_s K_s d[X,Y]_s \forall t \geq 0$,
2. $[H\cdot X, H\cdot X]_t = \int_0^t H_s^2 d[X,X]_s \forall t \geq 0$.

Proof. [Pro90] Thm. 2.6.29.

Theorem 140 (Itô-ormula) *Let X be a semi-martingale, $f \in C^2(\mathbb{R})$. In this case the Itô-formula holds:*

$$f(X_t) - f(X_0) = \int_{0+}^t f'(X_s^-)dX_s + \frac{1}{2}\int_{0+}^t f''(X_s^-)d[X,X]_s^{cont} + \sum_{0<s\leq t}\{f(X_s) - f(X_s^-) - f'(X_s^-)\Delta X_s\}.$$

Proof. [Pro90] Thm. 2.7.32.

Remark 141 *For a function $f \in C^2(\mathbb{R})$, we have*

$$f(t) - f(0) = \int_0^t f'(s)ds.$$

For stochastic integration there are two additional terms. The term

$$\frac{1}{2}\int_{0+}^t f''(X_s^-)d[X,X]_s^{cont}$$

is a consequence of the quadratic variation of the process and

$$\sum_{0<s\leq t}\{f(X_s) - f(X_s^-) - f'(X_s^-)\Delta X_s\}$$

is induced by the jumps.

Proposition 142 (Variable transformation) *Let V be a stochastic process with bounded variation and right continuous paths. For $f \in C^1(\mathbb{R})$ the process $(f(V_t))_{t\geq 0}$ has bounded variation and the we have the following:*

$$f(V_t) - f(V_0) = \int_{0+}^t f'(V_{s_-})\,dV_s + \sum_{0<s\leq t}\left(f(V_s) - f(V_{s_-}) - f'(V_{s_-})\Delta V_s\right).$$

Proposition 143 (Itô-Formula) *Let X be a continuous martingale and $f \in C^2(\mathbb{R})$. In this case $f(X)$ is a semi martingale and we have:*

$$f(X_t) - f(X_0) = \int_{0+}^t f'(X_s)dX_s + \frac{1}{2}\int_{0+}^t f''(X_s)d[X,X]_s.$$

Appendix F
CERA Comparison

This section will provide a comparison between the topics covered in this book and the respective requirements of the International Actuarial Association IAA, in order to meet the requirements of the Global Enterprise Risk Management Designation Recognition Treaty.

F.1 Enterprise Risk Management Concept and Framework

	Requirements	Reference
(a)	Describe the concept of ERM, the drivers behind it and the resulting value to organisations. (2-3)	1
(b)	Explain the principal terms in ERM. (2-3)	1
(c)	Analyse an appropriate framework for an organisation's enterprise risk management and an acceptable governance structure. (4-5)	15
(d)	Evaluate an organisation's risk management culture including: risk consciousness, accountabilities, discipline, collaboration, incentive compensation, and communication. (4-5)	1 & 15
(e)	Demonstrate an understanding of governance issues including market conduct, audit, and legal risk. (3-4)	15
(f)	Demonstrate an understanding of risk frameworks in regulatory and other environments (e.g. Basel II, Solvency II, Sarbanes-Oxley, COSO, Aus/NZ 4360, ISO 31000) and their underlying principles. (3-4)	14
(g)	Demonstrate an understanding of the perspectives of regulators, rating agencies, stock analysts, and company stakeholders and how they evaluate the risks and the risk management of an organisation. (3-4)	14
(h)	Propose how an ERM process can create value for an organisation through better assessment of the organisation's risk profile, possible reduction in economic capital, improvement in rating, etc. (5)	1
(i)	Relate the risk and return trade-offs that result from changes in the organisation's risk profile. (3-4)	5 – 13

DOI 10.1007/978-3-642-20721-1_21,

F.2 ERM Process (Structure of the ERM Function and Best Practices)

	Requirements	Reference
(j)	Demonstrate how to articulate an organisation's risk appetite, quantified risk tolerances, risk philosophy and risk objectives. (3-4)	4
(k)	Demonstrate how to articulate a desired risk profile and appropriate risk filters. (3-4)	4
(l)	Assess the overall corporate risk exposure arising from financial and non-financial risks. (6)	5 – 13
(m)	Compare the relevance of risk measurement and management to various stakeholders including customers, regulators, government, company directors, professional advisors, shareholders and the general public. (4)	5 – 13
(n)	Demonstrate an understanding of contagion and how it affects different stakeholders. (3-4)	5 – 13
(o)	Evaluate the elements of a successful risk management function and a structure for an organisation's risk management function. (4-5)	1 & 15
(p)	Determine how financial and other risks and opportunities influence the selection of strategy and how ERM can be appropriately embedded in an entity's strategic planning. (4-5)	5 – 13
(q)	Demonstrate the application of a risk control process such as the Risk Management Control Cycle or other similar approach. (3)	1
(r)	Propose ERM solutions or strategies to address real (case study) and hypothetical situations. (5-6)	12

F.3 Risk Categories and Identification

	Requirements	Reference
(s)	Explain what is meant by risk and uncertainty. (2)	1
(t)	Describe different definitions and concepts of risk. (2)	1
(u)	Discuss risk taxonomy. (2-3)	1
(v)	Investigate and interpret financial and non-financial risks faced by an entity, including but not limited to: currency risk, credit risk, spread risk, liquidity risk, interest rate risk, equity risk, hazard/insurance risk, pricing risk, reserving risk, other product risk, operational risk, project risk and strategic risk. (3-4)	5 – 13

F.4 Risk Modelling and Aggregation of Risks

	Requirements	Reference
(w)	Demonstrate how each of the financial and non-financial risks faced by an entity can be amenable to quantitative analysis. (3-4)	5 – 13
(x)	Demonstrate enterprise-wide risk aggregation techniques incorporating the use of correlation. (3-4)	5 – 13
(y)	Evaluate and select appropriate copulas as part of the process of modelling multivariate risks. (4-5)	9
(z)	Demonstrate the use of scenario analysis and stress testing in the risk measurement process. (3-4)	5 – 13
(aa)	Examine the use of extreme value theory to help model risks. (4)	
(bb)	Demonstrate the importance of the tails of distributions, tail correlations, and low frequency / high severity events. (3-4)	5 – 13
(cc)	Demonstrate an understanding of model and parameter risk. (3-4)	5 – 13
(dd)	Evaluate and select appropriate models to handle diverse risks, including the stochastic approach. (4-5)	5 – 13

F.5 Risk Measures

	Requirements	Reference
(ee)	Apply risk metrics to quantify major types of risk exposure and tolerances in the context of an integrated risk management process. (3-4)	5 – 13
(ff)	Demonstrate the properties of risk measures (e.g. VaR and TVaR) and their limitations. (3-4)	4
(gg)	Analyse quantitative financial and insurance data using modern statistical methods (including asset prices, credit spreads and defaults, interest rates, incidents, causes and losses). (4-5)	5 – 13
(hh)	Evaluate best practices in risk measurement, modelling, and management of various financial and non-financial risks faced by an entity. (4-5)	5 – 13
(ii)	Analyse credit risk as related to fixed income securities. (4-5)	6

F.6 Risk Management Tools and Techniques

	Requirements	Reference
(jj)	Relate the rationale for managing risk and the selection of the appropriate degree of hedging of risk. (3-4)	5 – 13
(kk)	Demonstrate risk optimisation and the impact on an organisation's value of an ERM strategy. (3-4)	5 – 13
(ll)	Demonstrate means for transferring risk to a third party, and estimate the costs and benefits of doing so. (3)	5 – 13
(mm)	Demonstrate means for reducing risk without transferring it. (3-4)	5 – 13
(nn)	Demonstrate how derivatives, synthetic securities, and financial contracting may be used to reduce risk or to assign it to the party most able to bear it. (3-4)	5 – 13
(oo)	Determine an appropriate choice of hedging strategy for a given situation (e.g., reinsurance, derivatives, financial contracting), which balances benefits with inherent costs, including exposure to credit risk, basis risk, moral hazard, and other risks. (4-5)	5 – 13
(pp)	Demonstrate an understanding of the practicalities of market risk hedging, including dynamic hedging. (3-4)	5 – 13
(qq)	Define credit risk as related to derivatives; define credit risk as related to reinsurance ceded; define counter-party risk and demonstrate the use of comprehensive due diligence and aggregate counter-party exposure limits. (3-4)	5 – 13
(rr)	Apply funding and portfolio management strategies to control equity and interest rate risk, including key rate risks. Explain the concepts of immunisation including modern refinements and practical limitations. (3-4)	5 – 13
(ss)	Analyse application of ALM principles to the establishment of investment policy and strategy including asset allocation. (4-5)	5 – 13
(tt)	Identify and interpret other key risks (e.g. operational, strategic, legal, and insurance risks) and uncertainty and demonstrate possible mitigation strategies. (3-4)	5 – 13

F.7 Economic Capital

	Requirements	Reference
(uu)	Interpret the concept of economic measures of value (e.g., EVA, embedded value, economic capital) and demonstrate their uses in corporate decision-making processes. (3-4)	5 – 13
(vv)	Apply risk measures and demonstrate how to use them in economic capital assessment. (3-4)	5 – 13
(ww)	Propose techniques of allocating/appropriating the "cost" of risk/capital/hedge strategy to business units in order to gauge performance (e.g. returns on marginal capital). (5-6)	5 – 13
(xx)	Develop an economic capital model for a representative financial firm. (5-6)	5 – 13

References

Bau91. H. Bauer. *Wahrscheinlichkeitstheorie*. De Gruyter, 1991.

Bau92. H. Bauer. *Mass- und Integrationstheorie*. De Gruyter, 1992.

BGH+86. N. L. Bowers, H. U. Gerber, J. C. Hickman, D. C. Jones, and C. J. Nesbitt. *Actuarial Mathematics*. Society of Actuaries, 1986.

BM01. D. Brigo and F. Mercurio. *Interest Rate Models – Theory and Practise*, Springer, 2001.

BP80. D. R. Bellhouse and H. H. Panjer. Stochastic modelling of interest rates with applications to life contingencies. *Journal of Risk and Insurance*, 47:91–110, 1980.

BS73. F. Black and M. Scholes. The pricing of options and corporate liabilities. *Journal of Political Economy*, 81:637–654, 1973.

Büh92. H. Bühlmann. Stochastic discounting. *Insurance Mathematics and Economics*, 11/2:113–127, 1992.

Büh95. H. Bühlmann. Life insurance with stochastic interest rates. In G. Ottaviani, editor, *Financial Risk in Insurance*, pages 1–24. Springer, 1995.

CHB89. J. Cox, C. Huang, and S. Bhattachary. Option pricing theory and its applications. In G. Constantinides, editor, *Theory of Valuation*, pages 272–288. Rowman and Littlefield Publishers, 1989.

CFO08. CFO Forum. *Market Consistent Embedded Value*. http://www.cfoforum.nl/embedded_value.html.

CIR85. J. Cox, J. Ingersoll, and S. Ross. A theory of term structure of interest rates. *Econometrica*, 53:385–408, 1985.

Con91. J. Conway. *A Course in Functional Analysis*, Springer, 2 edition, 1991.

CW90. K. L. Chung and R. J. Williams. *Introduction to Stochastic Integration*. Birkhäuser, 2 edition, 1990.

DAV09. DAV Unterarbeitsgruppen "Rechnungsgrundlagen der Pflegeversicherung" und "Todesfallrisiko". *Herleitung der Rechnungsgrundlagen DAV 2008 P für die Pflegerenten(zusatz)versicherung*, und *Raucher- und Nichtrauchersterbetafeln für Lebensversicherungen mit Todesfallcharakter*, und *Herleitung der Sterbetafel DAV 2008 T für Lebensversicherungen mit Todesfallcharakter*. *Blätter der DGVFM*, 30/1:31–140, 141–187, 189–224, 2009.

Doo53. J. L. Doob. *Stochastic Processes*. Wiley, 1953.

Dot90. M. Dothan. *Prices in Financial Markets*. Oxford University Press, 1990.

DS57. N. Dunford and J. T. Schwartz. *Linear Operators Parts 1, 2 and 3*. Wiley – Interscience, 1957.

Duf88. D. Duffie. *Security Markets: Stochastic Models*. Academic Press, 1988.

Duf92. D. Duffie. *Dynamic Asset Pricing Theory*. Prinston University Press, 1992.

DVJ88. D. J. Daley and D. Vere-Jones. *An Introduction to the Theory of Point Processes*. Springer, 1988.

M. Koller, *Life Insurance Risk Management Essentials*, EAA Series,
DOI 10.1007/978-3-642-20721-1, © Springer-Verlag Berlin Heidelberg 2011

Fel50. W. Feller. *An Introduction to Probability Theory and Its Applications*. Wiley, 1950.

FHR00. M. Frenkel, U. Hommel and M. Rudolf. *Risk Management – Challange and Opportunity*, Springer, 2000.

Fis78. M. Fisz. *Wahrscheinlichkeitsrechnung und mathematische Statistik*. Deutscher Verlag der Wissenschaften, 1978.

Ger95. H. U. Gerber. *Life Insurance Mathematics*. Springer, 2 edition, 1995.

GK06. J. Graham and D. Kaye. *A Risk Management Approach to Business Continuity*, Rothstein Associates, 2006.

HK79. J. M. Harrison and D. Kreps. Martingales and multiperiod security markets. *Journal of Economic Theory*, 20:381–401, 1979.

HN96. O. Hesselager and R. Norberg. On probability distributions of present values in life insurance. *J. Insurance Math. Econom.*, 18/1:135–142, 1996.

Hoe69. J. M. Hoem. Markov chain models in life insurance. *Blätter der Deutschen Gesellschaft für Versicherungsmathematik*, 9:91–107, 1969.

HP81. J. M. Harrison and S. R. Pliska. Martingales, stochastic integrals and continuous trading. *Stochastic Processes and their Applications*, 11:215–260, 1981.

Hua91. C. Huang. Lecture notes on advanced financial econometrics. Technical report, Sloan School of Management, MIT, Massachusetts, 1991.

Hul97. J. C. Hull. *Options, Futures and Other Derivatives*. Prentice Hall, 1997.

IW81. N. Ikeda and S. Watanabe. *Stochastic Differential Equations and Diffusion Processes*. North-Holland, 1981.

JP04. J. Jacot and P. Protter. *Probability Essentials*, Springer, 2004.

KP92. P. Kloeden and E. Platen. *Numerical Solution of Stochastic Differential Equations*, volume 23 of *Applications of Mathematics*. Springer, 1992.

Kol10. M. Koller. *Stochastische Modelle in der Lebensversicherung*, Springer, 2 edition, 2010.

KS88. I. Karatzas and S. E. Shreve. *Brownian Motion and Stochastic Calculus*. Springer, 1988.

KM03. P. Koch and P. Merino. *A Discrete Introduction – Mathematical Finance and Probability*, Birkhäuser, 2003.

LC92. R. Lee and L. Carter. Modelling and forecasting U.S. mortality. *Journal of the American Statistical Association*, 87:659–671, 1992.

Mol92. C. M. Moller. Numerical evaluation of Markov transition probabilities based on the discretized product integral. *Scand. Actuarial J.*, pages 76–87, 1992.

Mol95. C. M. Moller. A counting process approach to stochastic interest. *Insurance Mathematics and Economics*, 17:181–192, 1995.

NM96. R. Norberg and C. M. Moller. Thiele's differential equation by stochastic interest of diffusion type. *Scand. Actuarial J.*, 1996/1:37–49, 1996.

Nor90. R. Norberg. Payment measures, interest and discounting. *Scand. Actuarial J.*, pages 14–33, 1990.

Nor91. R. Norberg. Reserves in life and pension insurance. *Scand. Actuarial J.*, pages 3–24, 1991.

Nor92. R. Norberg. Hattendorff's theorem and Thiele's differential equation generalized. *Scand. Actuarial J.*, pages 2–14, 1992.

Nor94. R. Norberg. Differential equations for higher order moments of present values in life insurance. *Insurance: Mathematics and Economics*, pages 171–180, 1994.

Nor95a. R. Norberg. Stochastic calculus in actuarial science. Working paper, Laboratory of Actuarial Mathematics University of Copenhagen, 1995.

Nor95b. R. Norberg. A time-continuous Markov chain interest model with applications to insurance. *J. Appl. Stoch. Models and Data Anal.*, 11:245–256, 1995.

Nor96a. R. Norberg. Addendum to Hattendorff's theorem and Thiele's differential equation generalized. *Scand. Actuarial J.*, pages 2–14, 1996.

Nor96b. R. Norberg. Bonus in life insurance: Principles and prognoses in a stochastic environment. Working paper, Laboratory of Actuarial Mathematics University of Copenhagen, 1996.

Nor98. R. Norberg. Vasicek beyond the normal. Working paper, Laboratory of Actuarial Mathematics University of Copenhagen, 1998.

Par94a. G. Parker. Limiting distributions of the present value of a portfolio. *ASTIN Bulletin*, 24/1:47–60, 1994.

Par94b. G. Parker. Stochastic analysis of a portfolio of endowment insurance policies. *Scand. Actuarial J.*, 2:119–130, 1994.

Par94c. G. Parker. Two stochastic approaches for discounting actuarial functions. *ASTIN Bulletin*, 24/2:167–181, 1994.

PDHO09. E. Pittaco, M. Denuit, S. Haberman, A. Olivieri. *Modelling longevity for pensions and annuity business*, Oxford University Press, 2009.

Ped89. G. Pedersen. *Analysis Now*, Springer, 1989.

Per94. S. A. Persson. *Pricing Life Insurance Contracts under Financial Uncertainty*. PhD thesis, Norwegian School of Economics and Business Administration, Bergen, 1994.

Pli97. S. R. Pliska. *Introduction to Mathematical Finance. Discrete Time Models*. Blackwell Publishers, 1997.

Pro90. P. Protter. *Stochastic Integration and Differential Equations*, volume 21 of *Applications of Mathematics*. Springer, 1990.

PT93. H. Peter and J. R. Trippel. Auswertungen und Vergleich der Sterblichkeit bei den Einzelkapitalversicherungen der Schweizerischen Lebensversicherungs- und Rentenanstalt in den Jahren 1981–1990. *Mitteilungen der Schweiz. Vereinigung der Versicherungsmathematiker*, pages 23–44, 1993.

PTVF. W. Press, S. Teukolsky, W. Vetterling and B. Flannery. *Numerical Recipes in C – The Art of Scientific Computing*, Cambridge University Press, 2 edition, 1994.

RH90. H. Ramlau-Hansen. Thiele's differential equation as a tool in product development in life insurance. *Scand. Actuarial J.*, pages 97–104, 1990.

Rog97. L. C. G. Rogers. The potential approach to the term structure of interest rates and foreign exchange rates. *Mathematical Finance*, pages 157–176, 1997.

ECSO2. Website of the European Commission. Various material in relation to Solvency II. http://ec.europa.eu/internal_market/insurance/solvency/index_en.htm.

Vas77. O. Vasicek. An equilibrium characterisation of the term structure. *Journal of Financial Economics*, 5:177–188, 1977.

WBF10. M. Wüthrich and H. Bühlmann and H. Furrer. *Market-Consistent Actuarial Valuation*. Springer, 2 edition, 2010.

WH86. H. Wolthuis and I. Van Hoek. Stochastic models for life contingencies. *Insurance Mathematics and Economics*, 5:217–254, 1986.

Wil86. A. D. Wilkie. Some applications of stochastic interest models. *Journal of the Institute of Actuaries Student Soc.*, 29:25–52, 1986.

Wil95. A. D. Wilkie. More on a stocastic asset model for actuarial use. *British Actuarial Journal*, 1, 1995.

Wol85. H. Wolthuis. Hattendorf's theorem for a continuous-time Markov model. *Scand. Actuarial J.*, 70, 1985.

Wol88. H. Wolthuis. *Savings and Risk Processes in Life Contingencies*. PhD thesis, University of Amsterdam, 1988.

Index

M. Koller, *Life Insurance Risk Management Essentials*, EAA Series,
DOI 10.1007/978-3-642-20721-1, © Springer-Verlag Berlin Heidelberg 2011

MIX
Papier aus verantwortungsvollen Quellen
Paper from responsible sources
FSC® C105338

If you have any concerns about our products,
you can contact us on
ProductSafety@springernature.com

In case Publisher is established outside the EU,
the EU authorized representative is:
Springer Nature Customer Service Center GmbH
Europaplatz 3, 69115 Heidelberg, Germany

Printed by Libri Plureos GmbH
in Hamburg, Germany